The Essence of a Genius

A Tribute to Yoichiro Nambu

World Scientific Series in 20th Century Physics

Published

Vol. 46 *The Essence of a Genius: A Tribute to Yoichiro Nambu*
by L. Brink and P. Ramond

Vol. 45 *The Finite Quantum Many-Body Problem: Selected Papers of Aage Bohr*
edited by Ricardo A. Broglia

Vol. 44 *Symmetries in Nature — The Scientific Heritage of Louis Michel*
edited by T. Damour, I. Todorov and B. Zhilinksii

Vol. 43 *Nambu: A Foreteller of Modern Physics*
edited by T. Eguchi and M. Y. Han

Vol. 42 *Selected Papers of K. C. Chou*
edited by Y.-L. Wu

Vol. 41 *Many-Body Theory of Molecules, Clusters, and Condensed Phases*
edited by N. H. March and G. G. N. Angilella

Vol. 40 *Murray Gell-Mann — Selected Papers*
edited by H. Fritzsch

Vol. 39 *Searching for the Superworld — A Volume in Honour of Antonino Zichichi on the Occasion of the Sixth Centenary Celebrations of the University of Turin, Italy*
edited by S. Ferrara and R. M. Mössbauer

Vol. 38 *Matter Particled — Patterns, Structure and Dynamics — Selected Research Papers of Yuval Ne'eman*
edited by R. Ruffini and Y. Verbin

Vol. 37 *Adventures in Theoretical Physics — Selected Papers with Commentaries*
by Stephen L. Adler

Vol. 36 *Selected Papers (1945–1980) with Commentary*
by Chen Ning Yang

Vol. 35 *A Career in Theoretical Physics (Second Edition)*
by P. W. Anderson

Vol. 34 *Quantum Mechanics in Phase Space: An Overview with Selected Papers*
edited by C. K. Zachos, D. B. Fairlie and T. L. Curtright

Vol. 33 *Formation and Evolution of Black Holes in the Galaxy — Selected Papers with Commentary*
edited by H. A. Bethe, G. E. Brown and C.-H. Lee

Vol. 32 *In Conclusion — A Collection of Summary Talks in High Energy Physics*
edited by J. D. Bjorken

Vol. 31 *From the Preshower to the New Technologies for Supercolliders — In Honour of Antonino Zichichi*
edited by B. H. Wiik, A. Wagner and H. Wenninger

Vol. 30 *Selected Papers of J. Robert Schrieffer — In Celebration of His 70th Birthday*
edited by N. E. Bonesteel and L. P. Gor'kov

Vol. 29 *Selected Works of Emil Wolf (with Commentary)*
edited by E. Wolf

Vol. 28 *The Legacy of Léon Van Hove*
edited by A. Giovannini

For information on Vols. 1–24, please visit http://www.worldscientific.com/series/wsscp

World Scientific Series in 20th Century Physics Vol. 46

The Essence of a Genius

A Tribute to Yoichiro Nambu

Lars Brink

Department of Physics, Chalmers Institute of Technology
S-41216 Göteborg, Sweden

Pierre Ramond

Department of Physics, Institute for Fundamental Theory,
University of Florida, Gainesville, Florida 32611, USA

World Scientific

NEW JERSEY • LONDON • SINGAPORE • BEIJING • SHANGHAI • HONG KONG • TAIPEI • CHENNAI • TOKYO

Published by

World Scientific Publishing Co. Pte. Ltd.
5 Toh Tuck Link, Singapore 596224
USA office: 27 Warren Street, Suite 401-402, Hackensack, NJ 07601
UK office: 57 Shelton Street, Covent Garden, London WC2H 9HE

British Library Cataloguing-in-Publication Data
A catalogue record for this book is available from the British Library.

World Scientific Series in 20th Century Physics — Vol. 46
THE ESSENCE OF A GENIUS
A Tribute to Yoichiro Nambu

ISBN 978-981-127-719-1 (hardcover)
ISBN 978-981-127-720-7 (ebook for institutions)
ISBN 978-981-127-721-4 (ebook for individuals)

For any available supplementary material, please visit
https://www.worldscienti ic.com/worldscibooks/10.1142/13435#t=suppl

Foreword

Yoichiro Nambu was one of the giants in physics of the last century. Bruno Zumino is famous for his statement on Nambu: *"he is always ten years ahead of us so I tried to understand his work in order to contribute to a new area which will flourish ten years from now. Contrary to my expectation, however, it took ten years to understand them."* This statement is still true in an even broader sense. His profound ideas about fundamental physics are still playing an important role and are rediscovered over and over again.

Some can be found in the 1995 book *Broken Symmetry*, edited by Tohru Eguchi and Kazuhiko Nishijima, a subset of his collected works which they selected together with Nambu. It is still in demand and contains unique material. Instead of producing a second edition, we have selected in this book some of his most famous papers. Included are his first two papers, on the Lamb shift and on the solution to the two-dimensional Lenz–Ising model. These were written in the late 1940s when he was a graduate student living in his office and surviving mostly on potatoes. This is a time when he was always hungry but shared his living with other graduate students, making the physics department of Tokyo University into a round-the-clock academy. The papers are remarkable both from an historic and a physics perspective.

Each paper begins with a background section that describes the physics that influenced Nambu before describing his paper(s) and then explains the paper, and finally its impact on the development of modern physics. References are by no means complete, they are given only to the extent that they explain the physics climate before Nambu's contribution.

One of Nambu's most important contribution is his explanation of superconductivity. We describe the history of superconductivity and give a fairly extensive description of the Bardeen–Cooper–Schrieffer theory. Nambu realized immediately when the theory was announced that gauge invariance seemed to be broken by the Cooper pairs, and it took him more than two years to understand how gauge invariance was still conserved. The key result was that the symmetry was nonlinearly realized and that the symmetry of the ground state was spontaneously broken. For the future development of superconductivity, its significance was that BCS theory was consistent, which condensed matter physicists had already assumed, but was not yet proven.

Nambu, however, realized that spontaneous symmetry breaking could be a key ingredient in particle physics. For this work, he got a late but very well-deserved Nobel Prize in 2008. Maybe it was because he was so much ahead of his contemporaries and so prolific that he had to wait so long. His success was quick though. He provided a deep understanding of chiral symmetry breakdown and the small mass of the pion, key elements in the strong interactions of elementary particles. One might say that most of nuclear physics rests on this approach. The section on the conference report *A Superconductor Model of Elementary Particles and its Consequences* from 1960, which had long disappeared gives the key ideas, Nambu once explained that in conference proceedings he could write more extensively and also present ideas that were not fully worked out. This led to the famous papers with Giovanni Jona-Lasinio which explain the basis for the strong interactions among hadrons.

We also have a section on the Han–Nambu paper, where they introduce three triplets of quarks and eight gluons to bind the quarks, the precursor to QCD.

We end by discussing Nambu's three conference reports which played a seminal role in the development of string theory. In the first, he suggests the Veneziano model as a theory of relativistic bosonic strings and proves that its scattering amplitudes factorize. In the second, he introduces quarks at the end of open strings and proposes the interactions as a string splitting into two. In the third, duality and hadrodynamics, he proposes the Nambu–Goto action for a relativistic string, which was the starting point for many of the further developments. He could not attend but fortunately had sent a manuscript that was handed out to the participants. (LB was there!) He had typed at least most of it himself so it was not meant to be the final version.

As mentioned in the preface, Nambu's journal papers are not easy to understand. Like the Japanese gentleman he was, he did not want to humiliate the reader with obvious statements. This is our interpretation but it fits very well with his character. He was a very polite, considerate human being with a self-reliance inside that he did not show on the outside. He was probably the only one in the highest stratum of physicists who was respected by everybody. Franois Englert said at the press conference when he got his Nobel Prize that Nambu was the only colleague that he did not call by his/her first name, but always addressed him as Professor Nambu, as did Murray Gell-Mann.

It is interesting that the conference report from 1960 that was mentioned above disappeared from circulation and was thought to be lost until Tohru Eguchi managed to find an old copy and could transcribe it in a readable form. In the report, Nambu spells out what became pion physics with chiral symmetry breaking and a pion as a fermion–antifermion pair with masses of a few MeV — four years before the introduction of quarks. This book and the Eguchi–Nishijima book on some of Nambu's collected works are essentially the only place to find this report.

When Yoichiro Nambu died in 2015 after a long life, one of the pillars of modern physics was gone. His influence has not diminished over the years. His achievements are still in focus in modern physics. The fact that he was so far ahead of most other physicists of his days makes us want to discover more and more of the legend. He will be remembered and admired for many years to come. We hope that this book will help newer generations to learn from one of the old masters.

Lars Brink and *Pierre Ramond*

Acknowledgments

In writing this book, we have benefited from the help of A. P. Balachandran, Korkut Bardakçi, Jeremy Bernstein, Thomas Curtright, Herbi Dreiner, François Englert, Paul Frampton, David Gross, David Horn, Jihn-eui Kim, Axel Kleinschmidt, Makoto Kobayashi, Kimyeong Lee, Viatcheslav Mukhanov, Raphaël Nepomechie, Hermann Nicolai, Hiroshi Ooguri, Sacha Polyakov, Karin Reich, Hirotaka Sugawara, and Frank Wilczek.

Contents

Chapter 1: Nambu and the Lamb Shift

On October 25, 1948, *Prog. Theo. Phys. Vol. IV, No. 1, Jan. Mar. (1949)* received

The Level Shift and the Anomalous Magnetic Moment of the Electron,

the paper where Nambu computes $(g-2)$ and the *Lamb shift* using Tomonaga's formulation of QED.

Background

Yôichirō Nambu came back to Tokyo University after the war. Nambu was always a very modest person never boasting about his achievements, but it is clear that he must have been an outstanding undergraduate, graduating in 2.5 years since it was cut short by half a year. He came back as a research associate which corresponded to a graduate studentship. He had been given this position at graduation. He came back to a devastated Tokyo and for three years he lived in his office sleeping on his desk mainly living on potatoes. It is difficult for us to understand how hard life was, the years after the war.

He shared the office with Zirō Koba (1915–1973), a student of Shin'ichirō Tomonaga (1906–1979) who was a professor at the nearby Tokyo Bunrika Daigaku. In this way, Nambu learned about Tomonaga's formulation of Quantum Electrodynamics. There were a number of students in the same conditions as Nambu, and since they all lived in their offices, there were lively physics discussions during the long evenings. It was a remarkable group of young students and many of them later had very distinguished careers.

The older theoretical physicists at Tokyo University were mostly engaged in research in statistical physics and condensed matter theory, and it was natural for Nambu to take up a study of the two-dimensional Ising model which had been solved by Lars Onsager (1903–1976) in 1944. He was well into a new formulation and solution of the model in the fall of 1947 when news reached Japan about the experimental discovery of the Lamb shift. He was then asked by Tomonaga to study this problem and put aside the work on the Ising model. Two years later, he took it up again and published the result. This is the story discussed in the next chapter.

The experiment of Lamb and Retherford and the Shelter Island conference

The war efforts had attracted new young people to research in fundamental physics in USA. When the war was over, the universities were being filled up with young excellent people together with their seniors. It did take time though to get new research going. Some of the young stars had difficulties to focus on new ideas and were not publishing anything in the first year or two after their return. "Theoreticians were in disgrace" — as the young precocious student Murray Gell-Mann later remembered.

It all changed in the early summer of 1947 when a new kind of conference was organized at Shelter Island, June 2–4. Only 24 people were invited and it included all the rising stars as well as the leaders of the previous generation. The highlight of the conference was the talk by Willis Lamb (1913–2008) who together with Robert Retherford (1912–1981) had measured the energy difference between the $2^2S_{\frac{1}{2}}$ and the $2^2P_{\frac{1}{2}}$ levels of hydrogen.[1] According to the Dirac theory there should be no difference, but the experiment that used modern radar technique developed during the war showed a difference of about 1000 MHz. The Dirac theory was not complete. We needed full relativistic Quantum Electrodynamics, where physical results could be computed in a perturbation expansion.

Such calculations had been tried as early as in 1930 by Ivar Waller (1898–1991) and J. Robert Oppenheimer (1904–1967), but the problem was that they generated infinities. Both Werner Heisenberg (1901–1976) and Hendrik Kramers (1894–1952) attributed these infinities to effects from other interactions such as the strong interactions. It was now clear that the problem had to be solved within QED itself. Already on the train back from Shelter Island, Hans Bethe (1906–2005) made a calculation[2] where he used the physical values for the electron self energy, i.e. its mass, charge, and the massless photon. He used some non-relativistic approximations but got a value quite close to the experimental value. The race was now on and the history of the events that followed are well known. Bruce French (1921–2002) and Victor Weisskopf (1908–2002) were the first to calculate[3] the Lamb shift correctly. They did not publish right away, since their values did not agree with Julian Schwinger's (1918–1994) nor with Richard Feynman's (1918–1988). Both had made some mistakes. Schwinger had published a letter to the editor of *Physical Review* submitted on December 30, 1947 where he outlined his program and then gave the details in four papers the following year. His value of the energy shift was 1051 MHz. The measured value today is 1058 MHz, Schwinger's success was formidable. He had also worked out the anomalous magnetic moment. In January, the American Physical Society met in New York and Schwinger reported his results and was the obvious star.

[1] Willis E. Lamb and Robert C. Retherford, *Phys. Rev.* **72**, 241 (1947).
[2] H. A. Bethe, *Phys. Rev.* **72**, 339 (1947).
[3] J. B. French and V. F. Weisskopf, *Phys. Rev.* **75**, 1240 (1949).

It was now clear that QED had to be renormalized and all infinities had to be packed into the three parameters which were fixed by their experimental values. In fact, gauge invariance was not so much discussed so QED was treated like a normal quantum field theory.

When Schwinger was working in a canonical formulation, Feynman was setting up his own formalism[4] starting from his formulation of quantum mechanics in terms of path integrals. He devised a space time formulation representing the perturbative quantum field theory with diagrams. During the next two years, he showed how this powerful technique allowed him to compute scattering amplitudes, the anomalous magnetic moment and the Lamb shift very efficiently. One key point was his introduction of the Feynman propagator with antiparticles "going backwards in time". The multitude of his results showed that his formalism was correct and after some time it became the standard procedure for performing calculations in QED, overtaking Schwinger's canonical formalism. The two formulations were so different, but Freeman Dyson (1923–2020) managed to show[5] that they were in fact equivalent.

Interestingly enough, the Swiss physicist E. C. G. Stückelberg (1905–1984) and his student Dominique Rivier (1918–1998) had submitted a paper[6] about renormalization of QED using the same propagator as Feynman's. The paper was rejected with a report saying that this was not a research paper but merely a program. As far as we know, the referee's name was never revealed. The final proofs of QED being renormalizable and a consistent quantum field theory were given by Abdus Salam[7] (1926–1996) and by John Ward[8] (1924–2000) in 1951.

QED and the Lamb shift in Japan

The Japanese physicists lived in their own bubble in the after-war Japan. They had essentially no access to foreign scientific journals and very few contacts with the world outside. The seniors like Yukawa (1907–1981), Nishina (1890–1951), Sakata (1911–1970) and Tomonaga had some private contacts with the research outside, but they were very much home bound.

Tomonaga had set up a program to renormalize QED and had essentially solved it by 1947, developing his own formalism. News about the Lamb shift and the new activities in USA, however, did not come through scientific exchange but through a short article in *Time Magazine* dated Sept 29. It described the breakthrough with few details but ended with *"But Nobelman I.I. Rabi (1898–1988) compared Lamb & Retherford's criticism of the Dirac theory with Einstein's modification of*

[4]R. P. Feynman, *Phys. Rev.* **76**, 769 (1949); *Phys. Rev.* **80**, 440 (1950).
[5]F. J. Dyson, *Phys. Rev.* **75**, 486 (1949).
[6]D. Rivier and E. C. G. Stückelberg, *Phys. Rev.* **74**, 218 (1948).
[7]Abdus Salam, *Phys. Rev.* **84**, 426 (1951).
[8]J. C. Ward, *Phys. Rev.* **84**, 897 (1951).

Newton's laws of motion". It is not clear when the Japanese could read the article. How quickly could a journal reach Japan before air mail? Our estimate is that the Japanese read the article in mid-October, and then Tomonaga assembled his troups. It must have been clear to Tomonaga that he had the formalism and the knowledge to compute $(g-2)$ and the *Lamb shift*.

As he started his calculation with his group, he asked Nambu to do the computation on his own. He must have been aware of Nambu's brilliance already by now even though he had not published any papers. He also sent his papers on the renormalization of QED to Oppenheimer, whom he had met during his stay in Europe in the late 1930's. This was very important for the future.

Nambu must have jumpstarted as did the other Japanese physicists. On November 24–25 the Physical Society of Japan organized a symposium in Kyoto on "The theory of elementary particles". Some 26 papers were presented and Nambu presented one together with Kubo on *"On the Level Shift of Hydrogen Atom"*, where they essentially sketched a calculation that mimicked Bethe's own. They did not publish it but a similar paper appeared in the *Physical Review* by T. A. Welton (1918–2010).[9]

Nambu's Calculation of the Lamb shift

It took Nambu about a year to finish his calculation and submit it for publication.[10] He started by pointing out that Tomonaga's method can be applied to the conventional perturbation formalism. He saw that both the electron self-energy as well as the coupling have to be subtracted, not discussing gauge invariance. He could now follow Koba and Tomonaga's calculation of radiative corrections for the elastic scattering of the electron.[11] He used the graphical representation introduced by Koba for an electron interacting with its own electromagnetic field.

His perturbative calculation which treated electrons and positrons separately, was relativistic except for the external states. He could then read off two kinds of matrix elements. He represented all terms with some momentum space diagrams to read off seven quite complicated terms, two of which are infinite and correspond to expected divergencies. He renormalized them by introducing counterterms. To solve the very complicated integrals he had to restrict himself to small initial and final momenta compared with the rest mass of the electron, which does not affect the result much. After some quite arduous calculations he computed the anomalous magnetic moment and got Schwinger's result. Finally he computed the Lamb shift value of 1087 MHz. It was much higher than experiment, but Nambu had neglected the vacuum polarization of the external field. In a "note added in proof", he corrects

[9]T. A. Welton, *Phys. Rev.* **74**, 1157 (1948).
[10]Y. Nambu, *Prog. Theor. Phys.* Vol. IV, No. 1, 82 (1949).
[11]Z. Koba and S. Tomonaga, *Prog. Theor. Phys.* **3**, 290 (1948).

his error, which shifts the value downwards to 1060mc, much closer to experiment and slightly higher than Schwinger's.

In the same volume as Nambu's, Tomonaga *et al.* published[12] a similar result. It differed from Nambu's who had used perturbation theory *ab initio*, while the Tomonaga group had first used a"relativistic canonical transformation" to cancel the mass singularity and only then applied perturbation theory. Tomonaga's paper was received by PTP one month earlier than Nambu's, yet it is published a month later!

In the midst of the post Shelter Island this paper-frenzy, the papers by Tomonaga and his collaborators were mentioned, but not the actual calculation of the Lamb shift by Tomonaga and collaborators as well as Nambu's. It will not be the last time that his work was ignored.

The paper shows a mastery with heavy calculations which would be a trademark of Nambu in his career. It is a remarkable achievement of a young graduate student living in his office with no previous publications, most often hungry, with very little contacts with the outside world. It is clear that Tomonaga's method was cumbersome. The great breakthrough for calculations was Feynman's introduction of his diagrams and his choice of propagator. In some reminiscences, Feynman remembered how he could solve the photon–photon scattering with one term while Tomonaga had needed over 700 terms. (He might have overstated the case.) This makes Nambu's achievement even more impressive.

The Japanese role in the development

On the American side, the heroes were Feynman and Schwinger and few mentioned Tomonaga's role. The one who did acknowledge Tomonaga was Oppenheimer. At the Solvay conference in 1948, gathering a mature and very respected collection of the old leaders of modern physics, he gave a talk on the new developments. In it, he wrote "It is probable that, at least to order e^2, more than one covariant formalism can be developed. Thus Stueckelberg's four-dimensional perturbation theory would seem to offer a suitable starting point, as also do the related algorithms of Feynman. But a method originally suggested by Tomonaga, and independently developed and applied by Schwinger, would seem, apart from its practicality, to have the advantage of considerable generality and a complete conceptual consistency. It has been shown by Dyson how Feynman's algorithms can be derived from Tomonaga's equations." (He had Tomonaga's name misspelt, which might have been the fault of the person who typed it.)

In 1965, the Nobel Prize for physics was shared by Feynman, Schwinger and Tomonaga. It is interesting to study the nomination pattern for the three. Feynman had altogether 48 nominations over the years with 13 in 1965. Schwinger had

[12]H. Fukuda, Y. Myamoto, and S. Tomonaga, *Prog. Theor. Phys.* Vol. IV, No. 2, Apr. Jun., 121 (1949).

correspondingly 30 over the years and four the final year while Tomonaga had 15 over the years and one in the final year. It is to be noted though that in the award ceremony speech, Prof. Ivar Waller makes it clear that Tomonaga had started his program already in 1943 and had fully understood the method of renormalization, when he finally heard about the Lamb shift and could use it to compute the correct value. He did not say it, but an interpretation is that he could have got the Prize alone.

82

Progress of Theoretical Physics Vol. IV, No. 1, Jan.~Mar., 1949.

The Level Shift and the Anomalous Magnetic Moment of the Electron.

Yôichiro NAMBU.
Department of Physics, Tokyo University.

(Received Oct. 25, 1948)

§ 1. Introduction.

The recent achievements both in experiment and theory have confirmed that the reaction of the radiation field is a really observable phenomenon and that one can calculate this effect in our present formalism of quantum theory if one employs some subtraction prescription. Thus Bethe(1) computed successfully the level shift of the hydrogen atom revealed by the experiment of Lamb and Retherford.(2) Following Bethe's idea, Tomonaga(3) has developed, independently of American authors, a so-called "self-consistent" subtraction method which aims at disposing of the infinities in a self-consistent manner and obtaining finite results for various processes. In fact he and collaborators could show that the radiative corrections for the elastic scattering of the electron and the Compton scattering were made finite by this method.

Although Tomonaga's theory has recourse to a relativistic canonical transformation which may be regarded as a generalization of the transformation used by Bloch and Nordsieck and by Pauli and Fierz, yet his method can equally be applied to the conventional perturbation formalism. We have here calculated the radiative correction of an electron moving in an external (electromagnetic) field along the line of the ordinary perturbation theory. To get definite numerical results we had to content ourselves with non-relativistic approximation for the initial and final states.

According to Tomonaga there are three kinds of infinities, that is, the electron's self energy, or the mass type infinity, the photon's self energy, or the vacuum polarization type infinity,(4) and an infinity depending on the external field. We shall see that the subtraction of the first and the third type infinity is necessary in our case.

A similar calculation making use of the canonical transformation has been carried out by Fukuda, Miyamoto and Tomonaga.[5] Comparison of the obtained results is not without interest in view of the provisional character of the present theory.

§ 2. The perturbation scheme.

The outline of the calculation developed here is more or less identical with that which Dancoff[6] followed in his estimation of the radiative correction for the elastic scattering of the electron and which has also been discussed by Koba and Tomonaga,[7] and Endo, Kinoshita and Koba in the light of the new subtraction theory. The difference lies, however, in that they investigated the scattering cross section and therefore considered only transitions between positive energy states while we are now going to ask the transition matrix itself, thus including the transitions for the positrons as well as the creation and annihilation of pairs. Moreover we want to extend the calculation to the case of a magnetic external potential, or in general, an external static electromagnetic field

$$V-(\alpha A)\equiv H.$$

The state of an electron (or more precisely, of an electron field) in interaction with its own electromagnetic field may conveniently be expressed by the following transition scheme[8]:

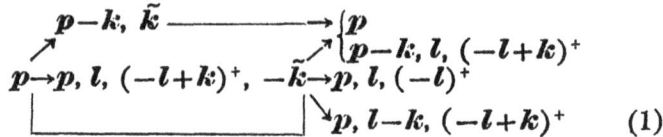

$$\begin{array}{l} \quad \boldsymbol{p}-\boldsymbol{k},\ \tilde{\boldsymbol{k}} \longrightarrow \left\{\begin{array}{l}\boldsymbol{p} \\ \boldsymbol{p}-\boldsymbol{k},\ \boldsymbol{l},\ (-\boldsymbol{l}+\boldsymbol{k})^{+}\end{array}\right. \\ \nearrow \qquad\qquad\qquad\qquad \nearrow \\ \boldsymbol{p}\rightarrow\boldsymbol{p},\ \boldsymbol{l},\ (-\boldsymbol{l}+\boldsymbol{k})^{+},\ -\tilde{\boldsymbol{k}}\rightarrow\boldsymbol{p},\ \boldsymbol{l},\ (-\boldsymbol{l})^{+} \\ \qquad\qquad\qquad\qquad\quad \searrow \boldsymbol{p},\ \boldsymbol{l}-\boldsymbol{k},\ (-\boldsymbol{l}+\boldsymbol{k})^{+} \end{array} \tag{1}$$

where $\boldsymbol{p}$ is the momentum of the electron in the unperturbed state, $\tilde{\boldsymbol{k}}$ is that of the photon emitted, and $\boldsymbol{l}$ and $(-\boldsymbol{l}+\boldsymbol{k})^{+}$ are momenta of the virtual pair. A corresponding scheme may be drawn for the positron. What is to be considered next is the matrix element of the external field H with respect to the transition of the electron from momentum $\boldsymbol{p}$ to momentum $\boldsymbol{q}$ when the interaction with the electromagnetic field is taken into account according to the above scheme.

In the e^2-approximation here concerned, there are two kinds of matrix elements, namely

(i) first order in e: $(\boldsymbol{p}, s;\ 0\,|H|\,\boldsymbol{q}, t;\ \tilde{\boldsymbol{k}}, \lambda_k)$, (2)

(ii) second order in e: $(p, s;\ 0\,|H|\,q, t;\ 0)$, (3)

where s and t denote the spin variable of the electron, and λ_k the polarization state of the emitted photon. The first kind corresponds to the Bremsstrahlung of the electron in the field H. The required correction to the zero order matrix H_{pq} comes out of the combined effect of the first order correction due to (3) and the second order correction due to (2).

Let the interaction Hamiltonian be given by

$$H=-e\int\psi^*\alpha\psi A\,dv+\frac{e^2}{2}\int\psi^*\psi\cdot\psi^{*\prime}\psi'/|r-r'|\cdot dv\,dv'+\int\psi^*(V-\alpha A)\psi\,dv, \quad (4)$$

where the first two terms are the interaction of the electron field with its own electromagnetic field, while the third is the interaction with the external potential, the coefficient of which we need not assume to be small.

By decomposing the field variables into Fourier components, the matrix element (2) is immediately obtained:

$$(p;0|H|q;\tilde{k},\lambda_k)=-\frac{e}{2\pi}\sum_{\pm}\pm\left\{\frac{\alpha_\lambda\cdot\lambda^\pm(p-q)H_{p-k,q}}{E_p\mp E_{p-k}-k}+\frac{H_{p,q-k}\lambda^\pm(q-k)\alpha_\lambda}{E_q\mp E_{q-k}-k}\right\}. \quad (5)$$

For the calculation of the matrix elements (3) it is convenient to classify various transition types into several groups as was done by Dancoff. Here, however, the classification is not identical with his, a fuller account of which will be given in the next section. We shall further introduce diagrams illustrating various possible paths connecting the initial and final state. In these diagrams the abscissae mean the momentum of the electron, while the ordinates are used to discriminate positive and negative energy states. Dotted lines imply transitions due to emission or absorption of a photon (Coulomb interaction included), and full lines those due to the external field.

(A) and (B)

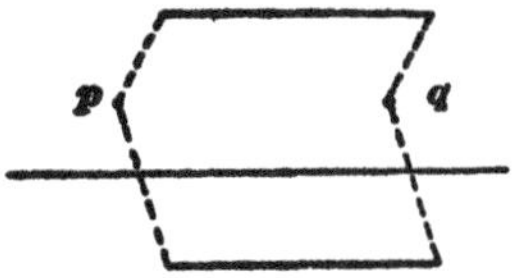

(A) $p\to p-k,\ \tilde{k}\overset{H}{—}q-k,\ \tilde{k}\leftarrow q$

$+\to+$

(B) $p\to p,\ q,\ (-q-k)^+,\ \tilde{k}—q,\ p,\ (-p-k)^+,\ \tilde{k}\leftarrow q$

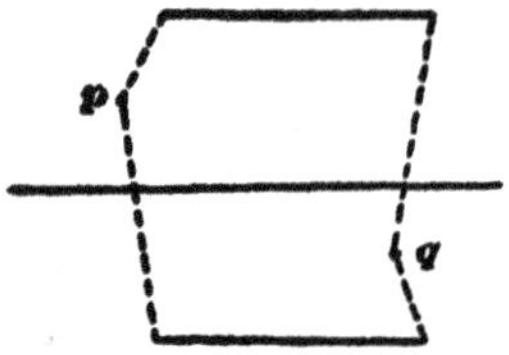

$$+\to-\quad \begin{array}{l} p,\ (-q)^{+}\to p-k,\ (-q)^{+},\ \tilde{k}\text{——}q-k,\ (-q)^{+},\ \tilde{k}\leftarrow\cdot \\ p,\ (-q)^{+}\to p,\ (-q-k)^{+},\ \tilde{k}\text{——}p,\ (-p-k)^{+},\ \tilde{k}\leftarrow\cdot \end{array}$$

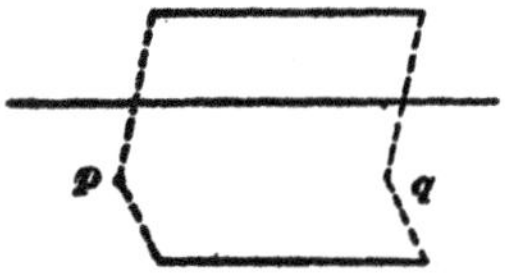

$$-\to-\quad \begin{array}{l} (-q)^{+}\to(-q)^{+},\ (-p)^{+},\ p-k,\ \tilde{k}\text{——}(-q)^{+},\ (-p)^{+},\ q-k,\ \tilde{k}\leftarrow(-p)^{+} \\ (-q)^{+}\to(-q-k)^{+},\ \tilde{k}\text{——}(-p-k)^{+},\ \tilde{k}\leftarrow(-p)^{+} \end{array}$$

(C)

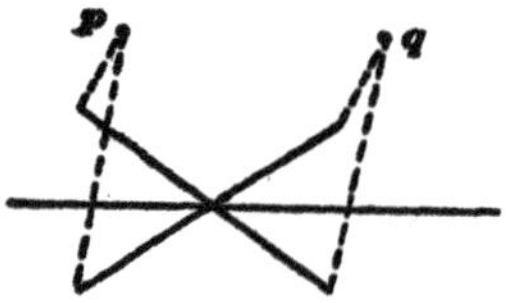

$$+\to+\ p\begin{array}{l}\nearrow p-k,\ \tilde{k} \\ \searrow p,\ q,\ (-q+k)^{+},\ -\tilde{k}\end{array}\rightarrow p-k,\ q,\ (-q+k)^{+}\overset{H}{\text{——}}q$$

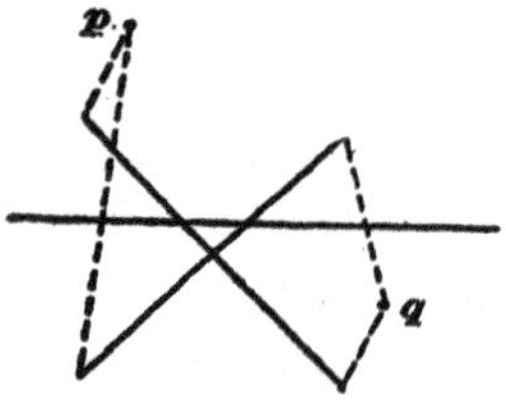

$$+\to-p,(-q)^{+}\begin{array}{l}\nearrow p-k,\ (-q)^{+},\ \tilde{k} \\ \searrow p,(-q+k)^{+},-\tilde{k}\end{array}\rightarrow p-k,\ (-q+k)^{+}\text{——}\cdot$$

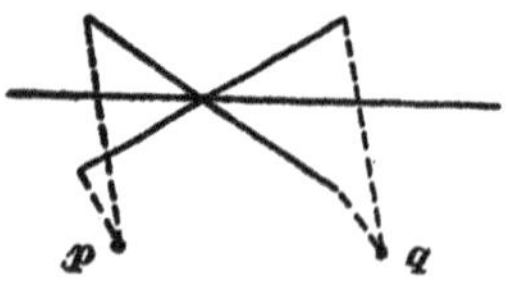

$$-\to-(-q)^{+}\begin{array}{l}\nearrow(-p)^{+},(-q)^{+},p-k,\tilde{k} \\ \searrow(-q+k)^{+},\ -\tilde{k}\end{array}\rightarrow(-p)^{+},\ (-q+k)^{+},\ p-k,\ \text{—}(-p)^{+}$$

(D)

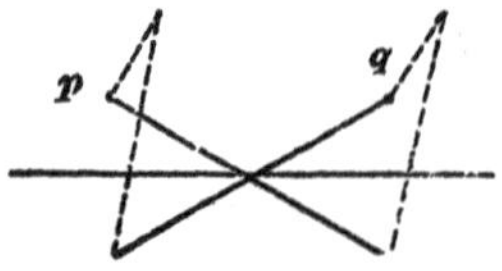

$$+\to+\quad p\to p,\ q-k,\ (-q)^{+},\ \tilde{k}\overset{H}{\text{——}}q-k,\ \tilde{k}\to q$$

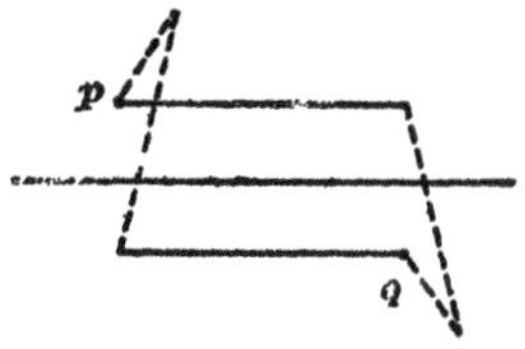

$+\to-\quad p,\ (-q)^+\to p,\ (-q+k)^+,\ -\tilde{k}\text{——}q,$
$(-q+k)^+,\ -\tilde{k}\leftarrow.$

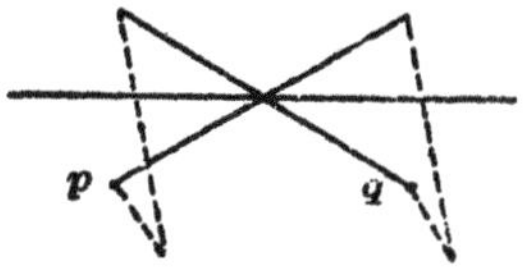

$-\to-\quad (-q)^+\to(-q+k)^+,\ -\tilde{k}\text{——}q,$
$(-q+k)^+,\ (-p)^+,\ -\tilde{k}\leftarrow(-p)^+$

(E)

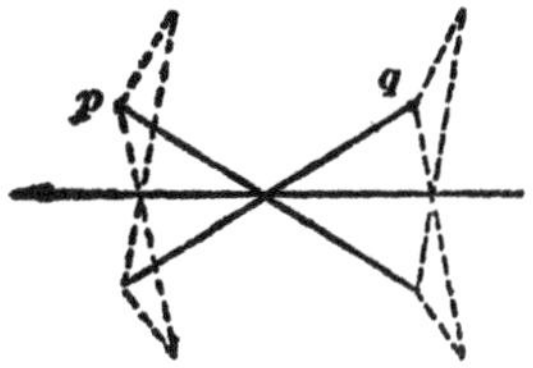

$+\to+\quad p \nearrow p,\ q-k,\ (-q)^+,\ \tilde{k} \searrow$
$p \searrow p,\ q,\ (-q+k)^+,\ -\tilde{k} \nearrow \to p,\ q,\ (-q)^+\overset{H}{\text{——}}q$

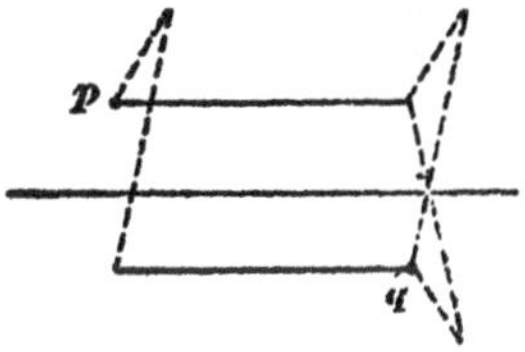

$+\to-\quad p,\ (-q)^+\overset{H}{\text{——}}q,$
$(-q)^+\leftarrow \nearrow q-k,\ (-q)^+,\ \tilde{k}\ \nwarrow$
$\searrow (-q+k)^+,\ q,\ -\tilde{k} \swarrow$

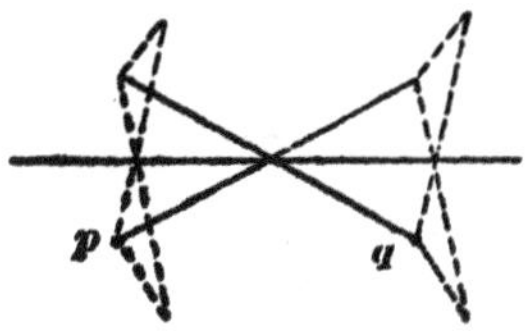

$-\to-\quad (-q)^+\text{——}q,\ (-p),$
$(-q)^+\leftarrow \nearrow q-k,\ (-q)^+,\ (-p)^+,\ \tilde{k} \nwarrow (-p)^+$
$\searrow q,(-q+k)^+,(-p)^+,-\tilde{k} \swarrow$

(F)

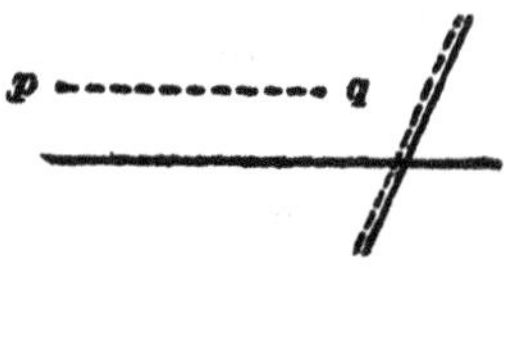

$+\leftarrow+\quad p\to q,\ \widetilde{p-q}\overset{H}{\text{——}}q,\ l-p+q,\ (-l)^+,\ \widetilde{p-q}\leftarrow q$
$p \nearrow q,\ \widetilde{p-q} \searrow$
$p \searrow p,\ l+p-q,\ (-l)^+,\ -\widetilde{p+q} \nearrow \to q,\ l+p-q,$
$(-l)^+\overset{H}{\text{——}}q$

$+\rightarrow-$

$p,\ (-q)^+\rightarrow\widetilde{p-q}$ —— $l-p+q,\ (-l)^+,\ \widetilde{p-q}\leftarrow\cdot$

$\overrightarrow{}$

$p,\ (-q)^+ \nearrow p-q$

$\searrow p,(-q)^+,\ l+p-q,(-l)^+,\ -\widetilde{l+p}$ $\rangle\rightarrow$

$l+p-q,\ (-l)^+$—.

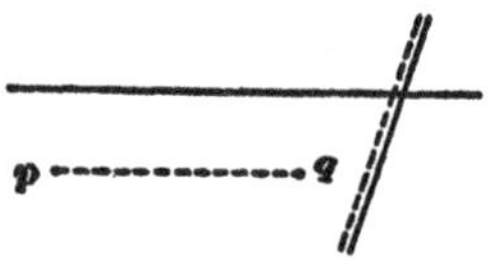

$--\rightarrow-$

$(-q)^+\rightarrow(-p)^+,\ \widetilde{p-q}\overset{H}{—}(-p)^+,\ l-p+q,$

$(-l)^+,\ \widetilde{p-q}\leftarrow(-p)^+$

$(-q)^+ \nearrow (-p)^+,\ \widetilde{p-q}$

$\searrow (-q)^+,\ l+p-q,(-l)^+,-\widetilde{p+q}$ $\rangle\rightarrow$

$(-p)^+,\ l+p-q,\ (-l)^+$——$(-p)^+$

$$\text{(A)}\quad \frac{1}{4\pi^2}\frac{e^2}{\hbar c}\sum_{\pm,\pm'}\int\frac{d\mathbf{k}}{k^3}\frac{\lambda^\pm(p)(\boldsymbol{\alpha}\times\mathbf{k})\lambda^+(p-k)H_{pq}\lambda^+(q-k)(\boldsymbol{\alpha}\times\mathbf{k})\lambda^{\pm'}(q)}{(\pm E_p-E_{p-k}-k)(\pm' E_q-E_{q-k}-k)}$$

$$-\frac{1}{2}\frac{1}{4\pi^2}\cdot\frac{e^2}{\hbar c}\sum_{\pm}\left\{\int\frac{d\mathbf{k}}{k^3}\frac{\lambda^\pm(p)(\boldsymbol{\alpha}\times\mathbf{k})\lambda^+(p-k)(\boldsymbol{\alpha}\times\mathbf{k})\lambda^\pm(p)}{(\pm E_p-E_{p-k}-k)(\pm E_p-E_{p-k}-k)}H_{pq}+\text{conj.}\right\} \tag{6}$$

$$\text{(B)}\quad \frac{1}{4\pi^2}\frac{e^2}{\hbar c}\sum_{\pm,\pm'}\int\frac{d\mathbf{k}}{k^3}\frac{\lambda^\pm(p)(\boldsymbol{\alpha}\times\mathbf{k})\lambda^-(p-k)H_{pq}\lambda^-(q-k)(\boldsymbol{\alpha}\times\mathbf{k})\lambda^{\pm'}(q)}{(\mp E_p-E_{p-k}-k)(\mp' E_q-E_{q-k}-k)}$$

$$-\frac{1}{2}\frac{1}{4\pi^2}\cdot\frac{e^2}{\hbar c}\sum_{\pm}\left\{\int\frac{d\mathbf{k}}{k^3}\frac{\lambda^\pm(p)(\boldsymbol{\alpha}\times\mathbf{k})\lambda^-(p-k)(\boldsymbol{\alpha}\times\mathbf{k})\lambda^\pm(p)}{(\mp E_p-E_{p-k}-k)(\mp E_p-E_{p-k}-k)}H_{pq}+\text{conj.}\right\} \tag{7}$$

$$\text{(C}_1\text{)}\quad -\frac{1}{4\pi^2}\cdot\frac{e^2}{\hbar c}\sum_{\pm,\pm'}\left\{\int\frac{d\mathbf{k}}{k^3}\frac{\lambda^\pm(p)(\boldsymbol{\alpha}\times\mathbf{k})\lambda^-(p-k)H_{pq}\lambda^+(q-k)(\boldsymbol{\alpha}\times\mathbf{k})\lambda^{\pm'}(q)}{(\mp' E_q-E_{q-k}-k)(\pm' E_q-E_{q-k}\mp E_p-E_{p-k})}+\text{conj.}\right\}$$

$$-\frac{1}{4\pi^2}\cdot\frac{e^2}{\hbar c}\sum_{\pm,\pm'}\left\{\int\frac{d\mathbf{k}}{k^3}\frac{\lambda^\pm(p)(\boldsymbol{\alpha}\times\mathbf{k})\lambda^+(p-k)H_{pq}\lambda^-(q-k)(\boldsymbol{\alpha}\times\mathbf{k})\lambda^{\pm'}(q)}{(\mp' E_q-E_{q-k}-k)(\pm E_p-E_{p-k}\mp' E_q-E_{q-k})}+\text{conj.}\right\} \tag{8}$$

$$\text{(C}_2\text{)}\quad -\frac{1}{2}\cdot\frac{1}{4\pi^2}\cdot\frac{e^2}{\hbar c}\sum_{\pm,\pm'}\left\{\int\frac{d\mathbf{k}}{k^2}\frac{\lambda^\pm(p)\lambda^-(p-k)H_{pq}\lambda^+(q-k)\lambda^{\pm'}(q)}{\pm' E_q-E_{q-k}\mp E_p-E_{p-k}}+\text{conj.}\right\}$$

$$-\frac{1}{2}\cdot\frac{1}{4\pi^2}\cdot\frac{e^2}{\hbar c}\sum_{\pm,\pm'}\left\{\int\frac{d\mathbf{k}}{k^2}\frac{\lambda^\pm(p)\lambda^+(p-k)H_{pq}\lambda^-(q-k)\lambda^{\pm'}(q)}{\pm E_p-E_{p-k}\mp' E_q-E_{q-k}}+\text{conj.}\right\}$$

$$\text{(D)}\quad -\frac{1}{4\pi^2}\cdot\frac{e^2}{\hbar c}\sum_{\pm}\left\{\int\frac{d\boldsymbol{k}}{k^3}\frac{\lambda^\pm(p)(\boldsymbol{\alpha}\times\boldsymbol{k})\lambda^\pm(p-k)(\boldsymbol{\alpha}\times\boldsymbol{k})\lambda^\mp(p)}{(E_p-E_{p-k}-k)(-E_p-E_{p-k}-k)}H_{pq}+\text{conj.}\right\}\quad (9)$$

$$\text{(E)}\quad -\frac{1}{4\pi^2}\cdot\frac{e^2}{\hbar c}\sum_{\pm}\left\{\int\frac{d\boldsymbol{k}}{k^3}\frac{\lambda^\pm(p)(\boldsymbol{\alpha}\times\boldsymbol{k})(\lambda^\pm(p-k)-\lambda^\mp(p-k))(\boldsymbol{\alpha}\times\boldsymbol{k})\lambda^\mp(p)}{(-2E_p)(-E_p-E_{p-k}-k)}H_{pq}+\text{conj.}\right\}$$

$$-\frac{1}{4\pi^2}\cdot\frac{e^2}{\hbar c}\sum_{\pm}\left\{\int\frac{d\boldsymbol{k}}{k^2}\frac{\lambda^\pm(p)(\lambda^\pm(p-k)-\lambda^\mp(p-k))\lambda^\mp(p)}{-2E_p}H_{pq}+\text{conj.}\right\}\quad (10)$$

$$\text{(F)}\quad +\frac{1}{4\pi^2}\cdot\frac{e^2}{\hbar c}\sum_{\pm,\pm'}\left\{\int\frac{d\boldsymbol{l}}{|\boldsymbol{p}-\boldsymbol{q}|^3}\frac{\lambda^\pm(p)(\boldsymbol{\alpha}\times\overline{\boldsymbol{p}-\boldsymbol{q}})\lambda^{\pm'}(q)}{\pm E_p\mp' E_q-|\boldsymbol{p}-\boldsymbol{q}|}\cdot\frac{Tr[\lambda^+(\boldsymbol{l}-\boldsymbol{p}+\boldsymbol{q})(\alpha\times\overline{\boldsymbol{p}-\boldsymbol{q}})\lambda^-(l)H_{pq}]}{-E_l-E_{l-p+q}-|\boldsymbol{p}-\boldsymbol{q}|}+\text{conj.}\right\}$$

$$+\frac{1}{4\pi^2}\cdot\frac{e^2}{\hbar c}\sum_{\pm,\pm'}\left\{\int\frac{d\boldsymbol{l}}{|\boldsymbol{p}-\boldsymbol{q}|^3}\frac{\lambda^\pm(p)(\boldsymbol{\alpha}\times\overline{\boldsymbol{p}-\boldsymbol{q}})\lambda^{\pm'}(q)}{\pm E_p\mp' E_q-|\boldsymbol{p}-\boldsymbol{q}|}\cdot\frac{Tr[\lambda^-(l)(\boldsymbol{\alpha}\times\overline{\boldsymbol{p}-\boldsymbol{q}})\lambda^+(l+p-q)H_{pq}]}{\pm E_p-E_l\mp' E_q-E_{l+p-q}}+\text{conj.}\right\}$$

$$+\frac{1}{4\pi^2}\cdot\frac{e^2}{\hbar c}\sum_{\pm,\pm'}\frac{d\boldsymbol{l}}{|\boldsymbol{p}-\boldsymbol{q}|^3}\frac{\lambda^\pm(p)(\boldsymbol{\alpha}\times\overline{\boldsymbol{p}-\boldsymbol{q}})\lambda^{\pm'}(q)}{-E_l-E_{l+p-q}-|\boldsymbol{p}-\boldsymbol{q}|}\cdot\frac{Tr[\lambda^-(l)(\boldsymbol{\alpha}\times\overline{\boldsymbol{p}-\boldsymbol{q}})\lambda^+(l+p-q)H_{pq}}{\mp E_p-E_l\mp' E_q-E_{l+p-q}}+\text{conj.}\right\}$$

$$+\frac{1}{4\pi^2}\cdot\frac{e^2}{\hbar c}\sum_{\pm,\pm'}\int\frac{d\boldsymbol{l}}{|\boldsymbol{p}-\boldsymbol{q}|^2}\frac{\lambda^\pm(p)\lambda^{\pm'}(q)Tr[\lambda^-(l)\lambda^+(l+p-q)H_{pq}]}{\pm E_p-E_l\mp' E_q\mp' E_{l+p-q}}.\quad (11)$$

§ 3. Subtraction of infinite terms.

The terms (A) and (B) in the last section correspond to (A) and (B) in Dancoff's paper; that is, (A) includes only even transitions by H, where intermediate states are all positive, and (B) is the same but with all negative intermediate states. (C) corresponds to odd transitions, or Dancoff's (b), (b′), (c), (c′) of (C), where H plays a rôle either at the first or at the last stage of the transition chain, while (D) has, being also odd, H in the middle (Dancoff's (a) and (a′)).

(E) is the infinite mass terms. In fact, we can show that these terms are of the same form as caused by a mass type perturbation $\delta m\beta$. To subtract this part we may add a counter term $-\delta m\beta$, where δm is the diverging self energy of the electron:

$$\delta m = \lim_{K\to\infty} \frac{e^2}{\pi\hbar c}\left(\frac{3}{2}\ln 2K - \frac{1}{4}\right). \tag{12}$$

But we shall here neglect entirely this part (E), assuming that it is exactly equivalent to a perturbation due to the self energy of the electron.

The last part (F) is responsible for the infinite polarization of vacuum caused by the external field. This should also be subtracted as a whole by a similar reasoning that it simply amounts to a renormalization of the coupling constant or reinterpretation of the external field.[10]

Thus we are left with the terms (A), (B), (C) and (D), which we now proceed to evaluate.

§ 4. Calculation for a non-relativistic electron.

The integrations involved in the terms (A) to (D) are in general so complicated that it seems very difficult to obtain a rigorous expression of the correction which is valid throughout the whole range of the electron's energy. We have therefore restricted ourselves to non-relativistic cases, that is, small initial and final momenta compared with the rest mass of the electron, although intermediate states may have any high energy and hence must be treated relativistically.

We expand all quantities appearing in these integrals in powers of $\boldsymbol{p}$ and $\boldsymbol{q}$ and retain only terms up to qudratic. Thus for instance,

$$E_{p-k} = \sqrt{m^2 + p^2 + k^2 - 2(\boldsymbol{pk})} = E_k\{1 - (\boldsymbol{pk})/E_k^2 + p^2/2E_k^2 - (\boldsymbol{pk})^2/2E_k^4\},$$

$$\frac{1}{E_p - E_{p-k} - k} \equiv \frac{1}{F_p} = \frac{1}{F_0}\left\{1 - \frac{(\boldsymbol{pk})}{EF_0} + \frac{(\boldsymbol{pk})^2}{E^2F_0^2} + \frac{P^2E^2 - (\boldsymbol{pk})^2}{2E^3F_0} - \frac{p^2}{2mF_0}\right\} \tag{13}$$

Then we can carry out the integration in the photon momentum $\boldsymbol{k}$ in terms of elementary functions. Each individual term may be eventually divergent as $\int^\infty dk/k$, but after combining all such contributions we see that these divergencies just cancel. In the case of the Coulomb type external potential V this occurs already for the individual parts (A) and (B), no other terms being divergent, while for the magnetic type potential $\boldsymbol{\alpha A}$ convergence is obtained only after the combination of (A), (B) and (C).

We add all the matrix elements obtained for transitions $+\to+$, $+\to$

$-$, $-\rightarrow+$ and $-\rightarrow-$ together, and collect similar terms from (A) to (D). We give here the results for the Coulomb type and the magnetic type separately since they have different forms.

(i) Coulomb type, V. The correction consists of three terms:

$$\frac{e^2}{\pi\hbar c}V_{p,q}\times\begin{cases}\left(x-\frac{1}{2}\right)\beta(\boldsymbol{a},\boldsymbol{p}-\boldsymbol{q}), & (14)\\ \left(\frac{1}{3}\ln 2\varepsilon-\frac{7}{24}\right)(\boldsymbol{p}-\boldsymbol{q})^2, & (14')\\ \left(-\frac{1}{12}-\frac{1}{3}\ln 2\right)i\boldsymbol{\sigma}\cdot(\boldsymbol{p}\times\boldsymbol{q}), & (14'')\end{cases}$$

where ε is the cut-off momentum of the photon on the low energy side. This term arises from (A) and (B), and corresponds to the well-known infra-red catastrophe. This will be compensated if we add a correction due to (2). The term denoted by x is also divergent on the low energy side, but has another origin. It is that term which arises from the transition $+\rightarrow-$ or $-\rightarrow+$ in (C_2), and is related with the fact that the Born approximation becomes invalid for low energies in the case of Coulomb interaction.

(ii) Magnetic type, $\boldsymbol{aA}$. There arise more terms than in the former case, namely $\boldsymbol{a}_n$, $\beta(\boldsymbol{p}+\boldsymbol{q},\boldsymbol{n})$, $i\beta\boldsymbol{\sigma}\cdot(\overline{\boldsymbol{p}-\boldsymbol{q}}\times\boldsymbol{n})$, $(\boldsymbol{pq})\boldsymbol{a}_n$, $(\boldsymbol{p}^2+\boldsymbol{q}^2)\boldsymbol{a}_n$, $(\boldsymbol{pn})(\boldsymbol{qa})+(\boldsymbol{qn})(\boldsymbol{pa})$, $(\boldsymbol{pn})(\boldsymbol{pa})+(\boldsymbol{qn})(\boldsymbol{qa})$ and $i\rho_1(\boldsymbol{p}\times\boldsymbol{q})\cdot\boldsymbol{n}$, where $\boldsymbol{n}$ is the unit vector in the direction of $\boldsymbol{A}$. We give here the coefficients only for terms up to the first order in $\boldsymbol{p}$ or $\boldsymbol{q}$.

$$\frac{e^2}{\pi\hbar c}A_{p,q}\times\begin{cases}\left(\frac{1}{6}-\frac{2}{3}\ln 2K+x\right)\boldsymbol{a}_n, & (15)\\ \left(\frac{4}{9}+\frac{1}{3}\ln 2K-\frac{x}{2}\right)\beta(\boldsymbol{p}+\boldsymbol{q},\boldsymbol{n}), & (15')\\ \left(\frac{25}{36}+\frac{1}{3}\ln 2K-\frac{x}{2}\right)i\beta\boldsymbol{\sigma}\cdot(\overline{\boldsymbol{p}-\boldsymbol{q}}\times\boldsymbol{n}). & (15'')\end{cases}$$

x is of the same nature as before, while the logarithmic terms are divergent on the ultra-violet side. This divergence, however, need not be taken up seriously. For these various kinds of terms are not rigorously relativistic in spite of their apparent Dirac matrix expression, and if we reduce them to Pauli approximation, the divergencies then cancel each other.

§ 5. Derivation of the correction formulae.

Now we transform the above results to a representation in the ordinary space :

$$\begin{aligned} &\beta(\alpha, p-q)\, V_{p,q} \rightarrow -\rho_2(\sigma E), \\ &i\sigma\cdot(p\times q)\, V_{p,q} \rightarrow -E\cdot(\sigma\times p), \\ &(p-q)^2 V_{p,q} \rightarrow -\Delta V, \\ &i\beta\sigma\cdot(\overline{p-q}\times n)A \rightarrow \beta(\sigma H). \end{aligned} \tag{16}$$

Looking at this, we notice that the electron possesses an additional interaction with the magnetic field which may be interpreted as a correction to the magnetic moment of one Bohr magneton. The above results, however, have not a relativistically invariant form. Indeed there are some terms in the correction to αA whidh have no partners for V. The coefficients of $\rho_2(\sigma E)$ and $\rho_3(\sigma H)$ are also different. This may be attributed to our non-relativistic treatment in the process of calculation as well as in the entire formalism.

The above mentioned unpaired terms (15) and (15') have non-vanishing values for $p=q$, and the first of them has the same form as the original interaction αA. Therefore we can omit this term by renormalizing the external field or its coupling constant. The same will be done for the other term if we divide it into αA and $\beta(\sigma H)$ in the order of the Pauli approximation and retain only the second term :

$$\beta(p+q,\, n) = 2u_n - i\beta\sigma\cdot(\overline{p-q}\times n). \tag{17}$$

Then the additional magnetic moment becomes

$$\frac{1}{4}\,\frac{e^2}{\pi\hbar c}\,\frac{\hbar e}{mc} = \frac{1}{2\pi}\,\frac{e^2}{\hbar c}\mu\,. \tag{18}$$

This is in agreement with Schwinger's result.[9]

Before we apply the above results to the level shift of the hydrogen atom we have to take into account the contribution from (2), or the emission of real photons. The calculation is similar but more complicated because it is a fourth order perturbation. But as we know that this correction is large only in the neighbourhood of the infra-red lim t, we use Bethe's non-relativistic formula[1] for the contribution of low energy photons. Then the lower limit of integration in (14′) and the upper limit in Bethe's formula cancel each other. Further, if we make Pauli approximation,

$$\lambda^+(p)\beta(\alpha, p-q)\lambda^+(q) = -\frac{1}{2}(p-q)^2 + i\sigma(p\times q). \quad (19)$$

The first term is added to the main term (14′) which does not depend on the spin variable. The second term is combined with (14″), and represents a spin-orbit interaction. In the case of the hydrogen-like atom,

$$\boldsymbol{E} = -e^2 Z\boldsymbol{r}/r^3, \quad (20)$$

$$\begin{aligned}\boldsymbol{E}\cdot(\alpha\times\boldsymbol{p}) &= \sigma\cdot(\boldsymbol{p}\times\boldsymbol{E}) = -e^2 Z\sigma\cdot(\boldsymbol{p}\times\boldsymbol{r})/r^3 = -e^2 Z(\sigma\boldsymbol{l})/r^3 \\ &= -e^2 Z\{j(j+1)-l(l+1)-3/4\}/2r^3, \end{aligned} \quad (21)$$

where l is the orbital, and j the total angular momentum. Thus we get for the level shift the following formula:

$$\begin{aligned}\delta E_{nlj} = \frac{8}{3\pi}\left(\frac{e^2}{\hbar c}\right)^3 Z^4 R_y\bigg\{&\left(\ln\frac{mc^2}{\langle E_0-E\rangle_{AV}} - \ln 2 + \frac{1}{8}\right)\frac{1}{n^3}\delta_{lc} \\ &+ \left(\frac{1}{2}\ln 2 + \frac{7}{8}\right)\frac{j(j+1)-l(l+1)-3/4}{n^3 l(l+1)(2l+1)}(1-\delta_{lo})\bigg\}. \end{aligned} \quad (22)$$

Especially the level difference of $2S_{1/2}$ and $2P_{1/2}$ becomes for $Z=1$

$$\begin{aligned}\delta E_{2S_{1/2}} - \delta E_{2P_{1/2}} = \frac{1}{3\pi}\left(\frac{e^2}{\hbar c}\right)^3 Z^4 R_y\bigg(&\ln\frac{mc^2}{\langle E_0-E\rangle_{AV}} \\ &-\frac{5}{6}\ln 2 + \frac{5}{12}\bigg) = 1019 \text{ mc}. \end{aligned} \quad (23)$$

§ 6. Discussion.

Our results do not completely agree with the calculation of Fukuda, Miyamoto and Tomonaga, who used the cananical transformation method. One might doubt that the origin of this discrepancy lies in the ambiguity of the subtraction prescription and the non-relativistic character of the calculation. In fact, if we actually calculate the part (E), we see that it is not completely compensated by the self energy term (E′):

$$(\text{E}) = -\frac{3}{4}\ln 2K + \frac{7}{8}, \qquad (\text{E}') = \frac{3}{4}\ln 2K - \frac{1}{8} \quad (24)$$

Moreover, tracing back the calculation, we find a difference already existing in the starting expressions (A) to (D) The corresponding expression derived by the canonical transformation has the part (D) smaller by a factor 1/2, and contains unfamiliar terms which have no equivalent in our case :

$$-\frac{1}{2}\int\frac{1}{2k^3}\frac{\lambda^-(p)(\alpha\times k)\lambda^+(p-k)(\alpha\times k)\lambda^+(p)}{(E_p-E_{p-k}-k)(-E_p-E_{p-k}-k)}dk\cdot H_{p,q}+\text{conj.}$$

$$-\frac{1}{2}\int\frac{1}{2k^3}\frac{\lambda^-(p)(\alpha\times k)\lambda^-(p-k)(\alpha\times k)\lambda^-(p)}{(E_p-E_{p-q}-k)(-E_p-E_{p-k}-k)}dk\cdot H_{p,q}+\text{conj.}\quad (25)$$

This situation may be explained as follows. Hitherto we have been doing with the static perturbation. This is sufficient for the calculation of the diagonal part of an observable, e.g. the self energy in the ordinary sence But in such cases as ours in which the transition due to the self energy and its elimination is concerned, the time dependent perturbation is required for a more careful definition of the self energy. The perturbed wave function is then given by

$$a_n^{(2)}(t)=a_n^{(0)}+\frac{H_{nn_0}}{E_{n_0}-E_n}a_{n_0}^{(0)}(\exp i(E_{n_0}-E_n)t/\hbar-1)$$
$$+\sum_{n'}\frac{H_{nn'}H_{n'n_0}}{E_{n_0}-E_{n'}}a_{n_0}^{(0)}\left[\frac{\exp i(E_{n_0}-E_n)t/\hbar-1}{E_{n_0}-E_n}-\frac{\exp i(E_{n'}-E_n)t/\hbar-1}{E_{n'}-E_n}\right].\quad (26)$$

We divide it in two parts :

$$a_n^{(2)}(t)=\Big[a_n^{(0)}+H_{nn_0}a_{n_0}^{(0)}(\exp i(E_{n_0}-E_n)t/\hbar-1)/(E_{n_0}-E_n)$$
$$+\frac{1}{2}\sum_{n'}\frac{H_{nn'}H_{n'n_0}a_{n_0}^{(0)}}{(E_{n_0}-E_{n'})(E_{n'}-E_n)}(\exp i(E_{n'}-E_n)t/\hbar-1)$$
$$\cdot(\exp i(E_{n_0}-E_{n'})/\hbar-1)\Big]$$
$$+\frac{1}{2}\sum_{n'}\frac{H_{nn'}H_{n'n}a_{n_0}^{(0)}}{(E_{n_0}-E_{n'})(E_{n'}-E_n)}\Big[\frac{2E_{n'}-E_{n_0}-E_n}{E_{n_0}-E_n}(\exp i(E_{n_0}-E_n)t/\hbar-1)$$
$$-(\exp i(E_{n'}-E_n)t/\hbar-\exp i(E_{n_0}-E_{n'})t/\hbar)\Big].\quad (27)$$

The first is the renormalized wave function while the second is *defined* to be the perturbation due to the self energy term. This procedure, however, is different from that of the static perturbation in which the renormalized wave function would be

$$a_n^{(2)}=\left[a_n^{(0)}+\frac{H_{nn_0}}{E_{n_0}-E_n}a_n^{(0)}+\sum_{\substack{n'\\ n'\neq n_0}}\frac{H_{nn'}H_{n'n_0}}{(E_{n_0}-E_{n'})(E_{n_0}-E_n)}\right]$$
$$\times\left[\sum_{n'}\left(a_{n'}^{(0)}+\frac{H_{n'n_0}}{E_{n_0}-E_{n'}}a_{n_0}^{(0)},\ a_{n'}^{(0)}+\frac{H_{n'n_0}}{E_{n_0}-E_{n'}}a_{n_0}^{(0)}\right)\right]^{-1/2}. \quad (28)$$

This means a non-linear hence non-unitary, correspondence between the perturbed and unperturbed wave function.

If we make necessary modifications according to the above circumstance, the level shift formulae become

$$\delta E_{n'j}=\frac{8}{3\pi}\left(\frac{e^2}{\hbar c}\right)^3 Z^4 R_y\left\{\left(\ln\frac{mc^2}{\langle E_0-E\rangle_{AV}}-\ln 2+\frac{7}{8}\right)\frac{1}{n^3}\delta_{l0}\right.$$
$$\left.+\left(\frac{1}{2}\ln 2+\frac{1}{8}\right)\frac{j(j+1)-l(l+1)-3/4}{n^3 l(l+1)(2l+1)}(1-\delta_{l0})\right\}, \quad (29)$$

$$\delta E_{2S1/2}-\delta E_{2P1/2}=1087\ \text{mc.}^* \quad (30)$$

The auther expresses cordial thanks to Professor Tomonaga who suggested this work and to Messrs Fukuda, Miyamoto and Tani for kind helps and valuable discussions.

References.

(1) Bethe, Phys. Rev. **72** (1947), 339.

(2) Lamb and Retherford, Phys. Rev. **72** (1947), 241.

(3) Tati and Tomonaga, Prog. Theor. Phys. **3** (1948), 211, 391; Tomonaga, Phys. Rev. **74** (1948) 224.

(4) On this point, see a note by J. R. Oppenheimer supplemented to Tomonaga's letter, and Tati and Tomonaga's remark at the end of their paper, (3).

(5) Fukuda, Miyamoto and Tomonaga, to be published in this journal.

(6) Dancoff, Phys. Rev. **55** (1939), 959.

(7) Koba and Tomonaga, Prog. Theor. Phys. **3** (1948), 216ff, 290.

(8) Endo, Kinoshita and Koba, Prog. Theor. Phys. in press.

(9) Schwinger, Phys. Rev. **73** (1948). 416.

* **Added in Proof**

Complete neglect of the vacuum polarization due to the external field is open to question. A deviation from mere charge-renormalizing correction would modify the level shift, Eq. (30), to 1060mc.

Chapter 2: Nambu and the Ising Model

On September 1, 1949, *Prog. Theo. Phys.* **5**, *1, (1950)* received

A Note on the Eigenvalue Problem in Crystal Statistics,

by Yoichiro Nambu, with the mention that

"The main part of the present work had been completed nearly two years ago. It is through the kindness of Professor Husimi and Mr. Syôzi of Osaka University that the author enjoys the opportunity of publishing this note."

Nambu had learned in 1948, from *"Lectures at the Annual Meeting of the Physical Society"* by Kodi Husimi who had made progress along the same lines in his hitherto unpublished work. Eleven years his senior, Professor Husimi was a well-known physicist. With Itiro Syôzi, he was about to publish *"The Statistics of Honeycomb and Triangular Lattice I"*.[13] This may explain Nambu's gracious acknowledgment.

Why did Nambu wait two years to publish? After the war, Japan was not allowed to receive scientific publications from abroad, and in fact, Japanese physicists had learned of the Lamb–Retherford experiment and Schwinger's calculation from the September 29, 1947 issue of *Time Magazine.*

Professor Tomonaga asked Nambu to compute the Lamb shift on his own. Nambu put aside his work on the Ising model to concentrate on the new project, and the result was submitted on October 26, 1948, *"The Level Shift and the Anomalous Magnetic Moment of the Electron"* has been discussed in the previous chapter as a virtuoso performance of young Nambu.

He returned to his work on the square Ising Model as a quantum system of N Qubits (before taking the thermodynamics limit). It reduced Lars Onsager's (1903–1976) formidable calculation[14] of the eigenvalues of the transfer matrix to a few pages.

In the paper, Nambu considered three different physical situations, a linear array of one spin, the isotropic $X\ Y$ model, and the square Ising Model. Given Nambu's deep insights, one should pay attention to the last sentence of the introduction:

[13]K. Husimi, I. Syozi, *Prog. Theo. Phys.* **5**, 177 (1950).
[14]L. Onsager, *Phys. Rev.* **65**, 117 (1944).

Though as yet no substantial application has been attempted, nor anything physically new has been derived, it may be hoped that it will do some benefit for those who are interested in such problems.

Background

It is not possible to describe the scientific path of the Ising model without mentioning Wilhelm Lenz. Born on February 8, 1888 in Frankfurt am Main, he completed his habilitation at the University of Munich in 1914 under Arnold Sommerfeld (1868–1951). He stayed as a Lecturer until 1920 when he was appointed Ausserordentlitcher (a.o.) Professor for Theoretical Physics in Rostock.

The same year, a professorship in theoretical physics opened up at the University of Hamburg, and he applied for it. With high recommendations from Sommerfeld,[15] he was appointed in 1921, and spent the rest of his career at the University of Hamburg. He died in 1957.

A century later, Lenz's name is mostly associated with the Laplace–Runge–Lenz vector. Not everyone knows of his seminal contribution to the Ising model, taught in every graduate course on Statistical Mechanics. At Rostock in 1920, Lenz submitted a short paper[16]: *"Beitrag zum Verständnis der magnetischen Erscheinungen in festen Körpern"*, where he suggested that ferromagnetism could be modeled by an array of elementary magnets pointing in two opposite directions, capable of flipping between them under the influence of interactions with neighboring sites.

Einstein thought Lenz's paper was important,[17] but it did not elicit much interest, at that time when so many fundamental breakthroughs were happening.[18]

Building on Lenz's paper, his doctoral student Ernst Ising (1900–1998) published his thesis[19] *"Beitrag zur Theorie des Ferromagnetismus"*, where he provided an analytic solution of his advisor's model for a one-dimensional lattice only, and found no transition to ferromagnetism. He suggested (erroneously) the same lack of transition for a two-dimensional lattice; his paper was not cited until 1927 by Werner Heisenberg.

In the 1930's, Lawrence Bragg (1890–1971) and Evan Williams (1903–1945), Hans Bethe (1906–2005), and Rudolph Peierls[20] (1907–1995) found by various approximations a ferromagnetic phase transition, but none could provide analytic evidence.

[15]Witness his influence: Lenz's successor at Rostock, was Otto Stern who later moved to Hamburg.

[16]W. Lenz, *Physik. Z.* **XXI**, 613–615 (1920).

[17]As quoted by K. Reich, A. Einstein, *Collected Papers* 9, pp. 18–20.

[18]The paper following Lenz's in *Physik. Z.* was by Wolfgang Pauli on Quantum Mechanics and the Magneton.

[19]E. Ising, *Zeits. f. Physik* **31**, 253 (1925).

[20]W. L. Bragg and E. J. Williams, *Proc. Soc. London Ser. A* **145**, 699 (1934); H. Bethe, *Proc. Royal Soc. A* **150** 552 (1935); R. Peierls, *Proc. Camb. Phil. Soc.* **32**, 477 (1936).

Further progress by Hendrik Kramers (1894–1952) and Gregory Wannier[21] (1911–1983), and independently by Ryogo Kubo[22] (1920–1995), expressed the partition function in terms of the largest eigenvalues of a matrix, known today as the "transfer matrix".

Kramers and Wannier[23] solved, as Ising did earlier, the one-dimensional nearest-neighbor model using their transfer matrix method, and formulated the model on a square and on a "screw" (helical) lattice, but could not solve either analytically. They discovered a duality between low and high temperatures, suggesting a critical temperature.

Enter Lars Onsager. Born in 1903 near Oslo, he graduated in 1925 from the Norwegian Institute of Technology as a chemical engineer. He was Peter Debye's (1884–1966) assistant at Zurich's ETH until 1928 when he was hired by Johns Hopkins University. Dismissed a semester later for his inability to lecture freshmen, he migrated to Brown University as instructor, this time lecturing to graduate students. He lasted four years at Brown, dismissed for lack of funds, a consequence of the Great Depression. In the meantime, his reputations as an extraordinary scientist was growing: Yale offered him in 1933 the prestigious postdoctoral Sterling fellowship, and an Assistant Professorship a year later, all without a Ph.D.! In 1935, his dissertation on "Deviations of Ohm's law in Weak Electrolytes", was accepted by the Yale Chemistry Department only through the intervention of the chair of mathematics! His career then followed the usual academic ladder, culminating in 1945, with Yale's J. W. Gibbs Professorship.

In a phenomenal *tour de force*, Onsager,[24] provided a complete solution of the Ising model on a square lattice, and verified the Kramers–Wannier suggestion for the transition temperature. As had happened earlier in his career, nobody understood his proof, but everyone believed the results. It was a seminal achievement: when asked after the war if anything new had happened in fundamental physics, Pauli replied "not much, except for Onsager's solution of the Ising Model". Five years later, his student Bruria Kaufman (1926–2010), published[25] a readable account, with an impressive use of group-theory to diagonalize the transfer matrix.

Onsager's paper, entitled *"Crystal Statistics. I. A Two-Dimensional Model with an Order-Disorder Transition,"* focused on second-order phase transitions, where both sides of the transition have different symmetries and some physical parameters vanish. He envisaged the lattice as a linear collection of sites, together with a similar collection of columns. For each row, he proceeded along the path pioneered by Kramers and Wannier for the linear lattice, and then connected them vertically:

[21]H. A. Kramers and G. H. Wannier, *Phys. Rev.* **60** (1941) 252.

[22]R. Kubo, *Bousserion-Kenkiyu* **1**, 1 1943.

[23]Theirs was a long-distance collaboration separated by both the Atlantic Ocean and the war.

[24]op. cit.

[25]B. Kaufman, *Phys. Rev.* **76**, 1232 (1949).

A linear array (row) of N lattice points is described by a $(2^N \times 2N)$ K-W transfer matrix,

$$V_j = \sqrt{2\sinh(2H)}e^{H^* C_j}, \quad H = J/kT, \quad \tanh H^* \equiv e^{-2H},$$

where $j = 1, 2, \ldots, N$ and J is the interaction strength between neighboring sites. The operators C_j satisfy $C_j^2 = 1$, and commute at different sites, $[C_j, C_k] = 0$, $j \neq k$. The full row transfer matrix follows

$$V_1 = \prod V_j = (2\sinh(2H)^{N/2} e^{H^* B}, \qquad B = \sum_{j=1}^{N} = C_j.$$

To complete the square lattice, Onsager added N rows stacked on top of one another, producing N vertical columns of sites, with a different interaction strength J' between vertical neighbors. Each site of the column is assigned a charge s_j, with the relevant transfer matrix representing each column being

$$V_2 = e^{H' A}, \qquad A = \sum_{j=1}^{N} s_j s_{j+1}.$$

A site has therefore two fermionic charges s_j and C_j with

$$s_i^2 = C_j^2 = 1, \qquad \{s_j, C_j\} = 0,$$

and operators at different sites that commute with one another.

To construct the full transfer matrix, Onsager started with one column and used V_1 to produce a new column. The vertical interactions are then generated by V_2. These combined operations result in two completed columns with $V = V_2 V_1$. The addition of two more columns requires $V_2 V_1 V_2 V_1$, and so on. At the end, the $(N \times N)$ array Ising model is represented by the full $(2^N \times 2^N)$ transfer matrix,

$$V = (2\sinh(2H)^{N/2} e^{H' A} e^{H^* B}.$$

Its largest eigenvalue determines the partition function in the thermodynamic limit. Their calculation is Onsager's formidable achievement.

This formula is the starting point of Nambu's paper which we present in some detail.

Nambu's Crystal Statistics

Single Spin Linear Array

In this first section, Nambu discussed the conventional description of the one-dimensional array of N identical particles with a different two-valued spin at each site, $n = 1, 2, \ldots, N$, and nearest-neighbor interactions,

$$P = \sum_{n=1}^{N} P_{n,n+1} = \sum_{n=1}^{N} \frac{1 + \sigma_n \sigma_{n+1}}{2}.$$

Inspired by quantum field theory, he introduced a different Fermi oscillator at each site,

$$\{a_n, a_n^\dagger\} = 1, \qquad [a_n, a_m^\dagger] = 0, \quad n \neq m,$$

commuting with one another at different sites, with periodic boundary conditions. He obtained,

$$P = \sum_{n=1}^{N} [a_n^\dagger a_{n+1} + a_{n+1}^\dagger a_n + 2a_n^\dagger a_n a_{n+1}^\dagger a_{n+1} - 2a_n^\dagger a_n + 1].$$

Now comes Nambu's fundamental observation: P, is a function of quadratic combinations, and is the same whether the operators at different sites commute or anticommute. In an audacious leap, Nambu suggested an alternate description of P in terms of new ladder operators which **anticommute at different sites**,

$$\{a_n, a_m\} = \{a_n^\dagger, a_m^\dagger\} = 0; \qquad \{a_n, a_m^\dagger\} = \delta_{m,n},$$

for all n, m. P now lives in a much smaller $2N$-dimensional Hilbert space, rather than in a 2^N-dimensional space of the conventional approach. This should lead to the same physics.

The rest of the paper is the exploitation of his assumption, first for P, then for the isotropic X–Y model, and culminating in a much simpler solution of the Ising model for both square and "screw" (helical) arrays.

Nambu expressed P in terms of the Fourier transforms

$$\tilde{a}_k = \frac{1}{\sqrt{N}} \sum_n a_n \eta^k, \quad \tilde{a}_k^\dagger = \frac{1}{\sqrt{N}} \sum_n a_n^\dagger \eta^{-k},$$

with $\eta = e^{\frac{2\pi i}{N}}$, and found,

$$P = \sum_{k=1}^{N} \left[\tilde{a}_k^\dagger \tilde{a}_k \eta^{-k} - \tilde{a}_k^\dagger \tilde{a}_k + c.c.\right] + \sum_{k,l=1}^{N} \tilde{a}_k^\dagger \tilde{a}_k \tilde{a}_l^\dagger \tilde{a}_l \delta(k - l \pm 1).$$

P is clearly the Hamiltonian sum of a quadratic "kinetic" term and a quartic expression describing a "short-range" potential. For large N, the potential can be neglected and P describes "free" N Bloch spin waves with energies $\epsilon_k = 2(\cos \frac{2\pi}{N} k - 1)$, $k = 1, 2, \ldots, N$. Nambu concluded that it was
"... a good approximation when the magnetization is nearly complete (low temperature)".

The Isotropic X–Y Model

To simplify the notation, Nambu replaced the ladder operators by $2N$ "real coordinates" $(a_n^\dagger + a_n)$, $i(a_n^\dagger - a_n)$, that is Grassmann coordinate and momentum at each site. They span an orthogonal basis, the "Nambu basis", in a $2N$-dimensional vector space,

$$\{x_n, x_m\} = 2\delta_{rs}, \quad n, m = 1, 2, \ldots 2N.$$

The permutation operator of the "Isotropic X–Y" model includes two Pauli spin matrices, σ_x and σ_y,

$$P_{X-Y} = \sum_{n=1}^{N} (\sigma_{n,x}\sigma_{n+1,x} + \sigma_{n,y}\sigma_{n+1,y}) \equiv \sum_{n=1}^{N} (A_n + B_n).$$

The new operators A_n and B_n commute, except at adjacent sites where they anti-commute,

$$\{A_n, B_{n\pm1}\} = 0,$$

and obey the constraints $A_n^2 = B_n^2 = 1$. These algebraic requirements are solved by expressing A_n and B_n as quadratic combinations of "Nambu's basis" coordinates $\{x_n\}$,

$$A_n = ix_{2n}x_{2n+1}, \quad B_n = ix_{2n-1}x_{2n+2}, \qquad n = 1, 2, \ldots, N,$$

so that A_n links adjacent sites and B_n hops over three sites. For N even, the constraints collapse into one, $1 = \prod A_n = \prod B_n \equiv x$.

Nambu now introduced two different Fourier transforms for even and odd sites,

$$\tilde{x}_k \equiv \frac{1}{\sqrt{2N}} \sum_{n=1}^{N} x_{2n}\eta^{nk}, \qquad \tilde{y}_k \equiv \frac{1}{\sqrt{2N}} \sum_{n=1}^{N} x_{2n+1}\eta^{nk}.$$

They describe for each k two fermion oscillators,

$$\{\tilde{x}_k, \tilde{x}_{-l}\} = \{\tilde{y}_k, \tilde{y}_{-l}\} = \delta_{kl}, \quad \{\tilde{x}_k, \tilde{y}_{-l}\} = 0,$$

where k, l run from $-N$ to N in integer steps. Then

$$P_{X-Y} = -2\sum_{k=1}^{N} \left(\tilde{x}_k\tilde{y}_{-k} + \tilde{y}_k\tilde{x}_{-k}\eta^{2k}\right) = -4\sum_{k=1}^{N/2} z_k \sin\frac{2\pi k}{N}$$

where $z_k \equiv \tilde{x}_k\tilde{y}_{-k}\eta^{-k} - \tilde{x}_{-k}\tilde{y}_k\eta^k$ is a sum of quadratic forms in $\tilde{x}_k$ and $\tilde{y}_l$, which can be readily diagonalized.

For each k, Nambu found that z_k has four eigenvalues, $\epsilon_k = 0, 0, 1, -1$, so that

$$P_{X-Y} = 4\sum_{k=1}^{N/2} \epsilon_k \sin\frac{2\pi k}{N},$$

restricted by the one boundary condition $x = 1$, (N even). With a display of legerdemain, Nambu restricted the number of non-zero eigenvalues,

$$x = 1 \quad \longrightarrow \quad \prod_{n=1}^{N/2}(1 - 2\epsilon_k^2) = (-1)^{N/2},$$

which completes the solution of the isotropic X–Y model.

The two-dimensional Ising Model

Nambu's started from Onsager's expression (neglecting the prefactor) for the square Ising model,

$$\mathcal{H} = \exp\left[H' \sum_{n=1}^{N} s_n s_{n+1}\right] \exp\left[H^* \sum_{n}^{N} c_n\right]$$

where the spins satisfy, $s_n^2 = c_n^2 = 1$, $\{ s_n, c_n \} = 0$, and commute with those at different sites.

Nambu introduced new variables,

$$S_n \equiv s_n s_{n+1}, \qquad C_n \equiv c_n,$$

which commute with one another except at adjacent sites where they anticommute

$$\{ S_n, C_{n\pm 1} \} = 0,$$

with boundary conditions, $S \equiv S_1 S_2 \cdots S_N = 1$, $C \equiv C_1 C_2 \cdots C_N = \pm 1$.

S_n and C_n are now expressed in the "Nambu basis" $\{x\}$,

$$S_n = i x_{2n} x_{2n+1}, \quad C_n = i x_{2n-1} x_{2n},$$

for even N with boundary conditions,

$$C = i^N x_1 x_2 x_3 x_4 \cdots x_{2N-1} x_{2N} \equiv X, \quad S = i^N x_2 x_4 x_2 x_5 \cdots x_{2N} x_1 = -X.$$

Nambu's computation starts with the transfer matrix,

$$\mathcal{H} = \exp\left[iH' \sum x_{2n} x_{2n+1}\right] \exp\left[iH^* \sum x_{2n-1} x_{2n}\right] \equiv \mathcal{H}_2 \mathcal{H}_1,$$

in the $\{x\}$ basis. It is a product of operators,

$$\mathcal{U} = e^{\theta x_n x_m}, \quad n \neq m, \quad e^{\theta/2\, x_n x_m} = \cos\theta + \sin\theta\, x_n x_m,$$

which describe a rotation by θ in the $(x_n - x_m)$ plane.

Using periodicity, $x_{2n+1} = x_1$,

$$\mathcal{H}_1 = \exp\left[iH^*(x_1 x_2 + x_3 x_4 + \cdots x_{2N-1} x_{2N})\right],$$

$$\mathcal{H}_2 = \exp\left[iH'(x_2 x_3 + x_4 x_5 + \cdots x_{2N} x_1)\right],$$

so that $\mathcal{H}_1$ rotates "odd–even" pairs (x_{2n-1}, x_{2n}),

$$\mathcal{H}_1 : \quad \begin{pmatrix} x_{2n-1} \\ x_{2n} \end{pmatrix} \quad \longrightarrow \quad \begin{pmatrix} y_{2n-1} \\ y_{2n} \end{pmatrix} = R(2iH^*) \begin{pmatrix} x_{2n-1} \\ x_{2n} \end{pmatrix},$$

while $\mathcal{H}_2$ rotates "even–odd" pairs (y_{2n}, y_{2n+1}),

$$\mathcal{H}_2 : \quad \begin{pmatrix} y_{2n} \\ y_{2n+1} \end{pmatrix} \quad \longrightarrow \quad \begin{pmatrix} z_{2n} \\ z_{2n+1} \end{pmatrix} = R(2iH') \begin{pmatrix} y_{2n} \\ y_{2n+1} \end{pmatrix}$$

where $(u = 2H', 2H^*)$,

$$R(iu) = \begin{pmatrix} \cosh(u) & i\sinh(u) \\ -i\sinh(u) & \cosh(u) \end{pmatrix}$$

Nambu saw that $\mathcal{H} = \mathcal{H}_2\mathcal{H}_1$ can be separated into products of these two carefully chosen operations, amounting to a linear transformation on the original pair,

$$\begin{pmatrix} x_{2n-1} \\ x_{2n} \end{pmatrix} \longrightarrow \begin{pmatrix} z_{2n-1} \\ z_{2n} \end{pmatrix} \equiv \lambda \begin{pmatrix} x_{2n-1} \\ x_{2n} \end{pmatrix},$$

where λ is the eigenvalue. He rewrote it as two equations in terms of the coefficients

$$a = i\cosh(2H^*)\sinh(2H'), \quad b = i\sinh(2H^*)\cosh(2H')$$

$$c = -\sinh(2H^*)\sinh(2H'), \quad d = \cosh(2H^*)\cosh(2H')$$

which satisfy $ab - cd = 0$, $a^2 + b^2 + c^2 + d^2 = 1$,

$$-bx_{2n-1} + (d-\lambda)x_{2n} + ax_{2n+1} + cx_{2n+2} = 0,$$
$$-ax_{2n} + (d-\lambda)x_{2n+1} + bx_{2n+2} + cx_{2n-1} = 0.$$

Nambu's elegant solution is to introduce two matrices and an eigenfunction,

$$A = \begin{pmatrix} a & c \\ \lambda - d & -b \end{pmatrix}, \quad B = \begin{pmatrix} b & \lambda - d \\ c & -a \end{pmatrix}, \quad \psi_n = \begin{pmatrix} x_{2n-1} \\ x_{2n} \end{pmatrix},$$

so that the eigenvalue equations become one matrix equation,

$$A\psi_{n+1} = B\psi_n,$$

resulting in a recursion relation (A is not singular),

$$\psi_{n+1} = A^{-1}B\psi_n \equiv D\psi_n.$$

It is readily solved,

$$\psi_{n+1} = D^n\psi_1,$$

and the periodicity constraint $\psi_{N+1} = \psi_1$ leads to the characteristic equation,

$$\det(1 - D^N) = 0.$$

It is solved by means of the "well-known" identity,

$$1 - D^N = \prod_{k=1}^{N}(\eta^k - D), \qquad \eta = e^{\frac{2\pi i}{N}},$$

which reduces to N equations,

$$\det(\eta^k - D) = 0 \quad \longrightarrow \quad |A\eta^k - B| = 0. \quad k = 1, 2 \ldots, N.$$

Explicitly,

$$\begin{vmatrix} \eta^k a - b & \eta^k c - (\lambda - d) \\ \eta^k(\lambda - d) - c & -\eta^k b + a \end{vmatrix} = 0.$$

This simple quadratic equation,

$$\lambda^2 - 2\lambda\big[d + c\cos\varphi_k\big] + 1 = 0, \qquad \varphi_k = \frac{2\pi k}{N},$$

has two solutions for each k,

$$\lambda_{k\pm} = \cosh(2\gamma_k) \pm \sinh(2\gamma_k),$$

with

$$\cosh 2\gamma_k = d + c\cos\varphi_k = \cosh(2H^*)\cosh(2H') - \sinh(2H^*)\sinh(2H')\cos\varphi_k,$$

the same formula as Onsager's Eq. (95) of his paper:

$$\frac{1}{2}\sum_{r=1}^{n}\gamma_{2r-1}$$
$$= \frac{1}{2}\sum_{r=1}^{n}\cosh^{-1}\left[\cosh 2H'\cosh 2H^* - \sinh 2H'\sinh 2H^*\cos((2r-1)\pi/2n)\right],$$

with largest eigenvalue,

$$H_{\max} = \exp\left[\sum_k |\gamma_k|\right].$$

By a simple series of steps, Nambu had duplicated Onsager's result! It is a conceptual result, the Ising model realized from N Qbits.

Nambu also pointed out that his method applies *mutatis mutandis* (when necessary changes are made) to certain variants of Onsager model such as the honeycomb lattice of Husimi and Syôzi.[26]

When he tried to apply his method to the three-dimensional Ising lattice, Nambu found that not all operators were exponentials of quadratics (i.e. rotations), some are exponentials of quartics, such as $e^{ax_1x_2x_3x_4}$. In view of Nambu's many prescient comments, it might be interesting to follow his path, although no analytic solution has ever been found.

Kramers-Wannier "Screw" Lattice

In their attempt to find an analytic solution for Ising's model, Kramers and Wannier had argued in 1941 that it was simpler to describe the lattice in terms of one string of spins, lying on the wires of an infinite solenoid, which they called the *"screw lattice"*. Nambu noted that

"This model seems more convenient for general purposes than that used by Onsager."

To transform Onsager's expression into the Kramers–Wannier helical string model, Nambu rearranged the interaction as

$$\mathcal{H} = e^{iH^*x_1x_2}\prod_{n=1}^{N}\mathcal{H}_n e^{iH'x_{2N}x_1}, \qquad \mathcal{H}_n = e^{iH^*x_{2n+1}x_{2n+2}}\,e^{iH^*x_{2n}x_{2n+1}}.$$

[26]K. Husimi and I. Syôzi, op. cit.

Neglecting the two boundary terms, he started from,

$$\mathcal{H} = \prod_{n=1}^{N} \mathcal{H}_n.$$

First step was to express a displacement operator P as a product of rotations,

$$x_n \quad \longrightarrow \quad e^{-x_n x_{n+1}\pi/4}\, x_n\, e^{x_n x_{n+1}\pi/4} = x_{n+1},$$

from which

$$\mathcal{H}_{n+1} = P\,\mathcal{H}_n\,P^{-1} = P^{n+1}\mathcal{H}_0\,P^{-n-1}, \quad \mathcal{H}_0 = e^{iH' x_1 x_2}\,e^{iH^* x_{2N} x_1} = \mathcal{H}_N,$$

by periodicity. The wavefunctions

$$\Psi_n = \mathcal{H}_n \mathcal{H}_{n-1} \cdots \mathcal{H}_1 \Psi_0,$$

obey the recursion relation, $\Psi_{n+1} = \mathcal{H}_{n+1}\,\Psi_n$, "*a Schrödinger equation for a discrete time variable!*"

The modified eigenfunction $\Psi'_n = P^{-n}\Psi_n$, also satisfies a recursion relation,

$$\Psi'_{n+1} = \mathcal{H}_0 P^{-1}\,\Psi'_n \equiv \mathcal{A}\Psi'_n,$$

but the shift operator, $\mathcal{H}_0 P^{-1}$ does not depend on n. Then,

$$\Psi'_N = \mathcal{A}^N \Psi'_0, \quad \longrightarrow \quad \mathcal{A}^N = 1,$$

and the eigenvalues are roots of unity, $\lambda^N = 1$.

The eigenvalues are determined from the "eigenoperator" equation (see the next section),

$$\mathcal{A}\,X\,\mathcal{A}^{-1} = \lambda X.$$

Its solution is expanded as a linear combination in the Nambu basis,

$$X = \sum_{n=1}^{2N} \alpha_n\, x_n.$$

After inserting this expansion in the eigenoperator equation, it is written in terms of three coefficients, a, b, c,

$$X = \sum_{n=1}^{N-1} [a x_{2n-1} + b x_{2n}]\lambda^{n-1} + c x_{2N} + a x_{2N-1}\lambda^{N-1},$$

reducing the eigenvalue operator equation to three coupled algebraic equations,

$$\begin{aligned}
\lambda^N x &= \cosh 2H^* \cosh 2H' x - i \sinh 2H^* \cosh 2H' y + i \sinh 2H' z,\\
\lambda z &= i \sinh 2H^* x + \cosh 2H' y,\\
\lambda^{N-1} y &= -i \sinh 2H' \cosh 2H^* x - \sinh 2H' \sinh 2H^* y + \cosh 2H' z.
\end{aligned}$$

By eliminating the real variables $\mathbf{x}, \mathbf{y}, \mathbf{z}$ Nambu arrived at the consistency equation

$$\lambda^{2N} + \sinh 2H^* \sinh 2H'(\lambda + \lambda^{-1}) - 2\cosh 2H^* \cosh 2H' + \lambda^{-N} = 0,$$

whose solution yields the eigenvalues. Setting $\lambda = e^{2\gamma}$, it reduces to

$$\cosh 2N\gamma = \sinh 2H^* \sinh 2H'(\lambda + \lambda^{-1}) - 2\cosh 2H^* \cosh 2H' \cosh\gamma,$$

to be solved for γ. Nambu assumed $2\gamma = 2\gamma_0 + i\omega$, with $\omega = k\pi/N$, $k = 1, 2 \ldots, 2N$. Comparing the real and imaginary parts yields

$$\pm\cosh 2\Gamma = \cosh 2H^* \cosh 2H' - \sinh 2H^* \sinh 2H' \cos\omega.$$

Since the rhs is positive, it follows that

$$\cosh 2\Gamma = \cosh 2H^* \cosh 2H' - \sinh 2H^* \sinh 2H' \cos\omega, \quad \omega = \frac{k\pi}{N}, \quad k = 1, 2, \ldots, N,$$

which is Onsager's formula, for large N, $N\gamma_0 \to \Gamma$.

Additional Remarks

In the solution for the screw lattice, Nambu emphasized a new mathematical method to solve eigenvalue problems which he had earlier used in his papers on "*Third Quantization*". He defined an "eigenoperator" X whose commutator with the operator of interest is proportional to itself,

$$[H, X] = \lambda X.$$

Stated without proof are its properties:

- λ is the difference of two eigenvalues, $\lambda = E_n - E_m$.
- X transforms an eigenvector Ψ_m of H into another Ψ_n with eigenvalue $\lambda_n = E_m + \lambda$.
- The product of the two eigenoperators $X_2 X_1$ is again an eigenoperator with eigenvalue $\lambda = \lambda_1 + \lambda_2$, transforming an eigenvector to another one.
- When H has a simple structure, a general eigenoperator X will be factorized into a product of eigenoperators

$$X = X_1 X_2 \cdots X_k, \quad \text{with eigenvalues } \; e^{\lambda} = e^{\lambda_1 + \lambda_2 + \ldots \lambda_k}.$$

The Ising Model Today

Today,[27] more than 800 researchers reference the Ising model every year; it is to Statistical Mechanics what the Standard Model is to Particle Physics. It has opened fundamental insight into the physics of phase transitions, and provided a bridge to conformal symmetry at critical points and critical exponents, and even to String Theory, where it is a Conformal Theory associated with the Virasoro algebra with critical exponent of one-half.

Lars Onsager and the Nobel Prize

In 1931, Onsager found the connection that led him to formulate equations that came to be known as the reciprocal relations, later known as the "Onsager relations". This allowed a complete description of irreversible processes. About the time that he became famous for his solution of the Lenz–Ising Model, his "Onsager relations" were fully understood, and it became clear that they were of the highest calibre. In 1952, George Rushbrooke (1915–1995) was the first to nominate Onsager for his work on irreversible thermodynamics.

---

[27] S. G. Brush, *Rev. Mod. Phys.* **39**, 883–893 (1962); M. Niss, *Arch. Hist. Exact Sci.* **59**, 267–318 (2005); ibid **63**, 243–287 (2009); "The Fate of Ernst Ising and the Fate of His Model", T. Ising, R. Folk, R. Kenna, B. Berche, and Y. Holovatch, *Journ. Phys. Stud.* **21**(3) 3002 (19 p.) (2017) ArXiv:1706.01764.

It was soon[28] understood that his solution of the Ising model described the physics on both sides of a phase transition in terms of one mathematical scheme, another fundamental advance worthy of a Nobel Prize.

After the first nomination in physics, nominations started to drop in also in chemistry. In the 1960's the number of nominations started to grow a lot. By the year 1968, when he finally got the chemistry prize, altogether 93 nominations had been submitted, 48 in physics and 45 in chemistry. We do not know how many if any were for the solution of the Ising Model, but we believe that almost all were for irreversible thermodynamics. Even without this discovery, the nominations for his solution of the Ising Model had begun to accumulate.

Why did Onsager get the prize so late? We can only speculate that he being consistently nominated both in physics and chemistry delayed the responsibility of which committee to take the prize. We should also remember that in the 1960s, it was harvest time for all the discoveries in particle physics of the post-war era, and physicists were happy to leave the prize to the chemistry committee.

Onsager was a man of great insights and also of few words; witness his shortest ever banquet speech at the Nobel ceremony:

Your Majesty, Your Royal Highness, Ladies and Gentlemen,

I have received the greatest honor of my life — and the greatest surprise. Never did I dream that the Nobel Prize could be awarded for the reciprocal relations. Not that I felt bashful about the work, but how does one equate a principle with a discovery? It is not the naked principle but rather the philosophy that goes with it; by and by, that is fruitful.

But a joy it is, so much greater because all those who ever shared anything with me, whether in youthful play or in earnest endeavor, they are so glad, all of them!
Min hjertlige takk!

When Nambu had published his work, he had already set up his own group in Osaka with a group of aspiring physicists, including Kazuhiko Nishijima, and their focus was on the field of particle physics. He left the Ising Model feeling that he had only provided his own solution which did not take the field further. In a sense, his solution came too early, which is very typical for Nambu. The coming years, he was so busy especially after his move to the US, he did not really have time to further study the Ising Model and its generalizations. We are convinced he could have turned his insight into more gold.

[28]When asked if there had been progress in fundamental physics during the war, Pauli responded "nothing much, except for Onsager's solution."

1

Progress of Theoretical Physics Vol. V, No. 1, Jan.~Feb., 1950.

A Note on the Eigenvalue Problem in Crystal Statistics.

Yôichirô NAMBU

Department of Physics, Tokyo University.

(Received September 1, 1949)

§ 1. Introduction.

In recent years there have been some remarkable developments in the mathematical theory of order-disorder transition. The problem of determining the partition function for a lattice with simple interaction model was found, by several authors,[1] to be reduced to an eigenvalue problem which is easier to handle with, and which, with suitable approximations, would reveal more refined details of the phase transition than the former theories. This method of attack culminated in the work of Onsager,[2] who succeeded in giving the rigorous solution of the above mentioned eigenvalue problem for two-dimensional simple square Ising lattice. The mathematical tools involved, however, are so hopelessly complicated that one would quite simply lose sight in the jungle of hypercomplex numbers. A considerable improvement on this mathematical point has been attained by Husimi and Syôzi[3] who used it to solve the eigenvalue problem for the honeycomb type lattice. The present author, independently of them, has reached more or less similar ideas and considerations which it is the purpose of the present paper to expose in brief detail. Though as yet no substantial applications has been attempted, nor anything physically new has been derived, it may be hoped that it will do some profit for those who are interested in such problems.

§ 2. Preliminary Considerations.

Let us take N identical particles 1, 2,$\cdots N$, arranged in order, and consider the interchange $P_{n,n+1}$ of two adjacent particles n and $n+1$ ($n=1, 2,\cdots N$). Such operators are met with in the theory of ferromagnetism in which the eigenvalues of the operator (for one-dimensional case)

$$P=\sum_n P_{n,n+1}=\sum \frac{1+\sigma_n\cdot\sigma_{n+1}}{2} \tag{1}$$

are asked. Here σ_n means the spin matrix $(\sigma_x, \sigma_y, \sigma_z)$ at the nth site. As P commutes with $\sum_n \sigma_{nz}=2S-N$, where S is the total number of spins oriented to $+z$ direction, we can work within a subspace in which S is kept fixed. Then

P is equivalent to an operator affecting the arrangement of S particles over N sites. Each site is either occupied or not occupied by one of these particles. If we introduce such operators, that change the unoccupied nth site into occupied, and *vice versa*, by

$$a_n^+=\begin{pmatrix} 0 & 0 \\ 1 & 0 \end{pmatrix}, \quad a_n=\begin{pmatrix} 0 & 1 \\ 0 & 0 \end{pmatrix}, \tag{2}$$

then P can be written in terms of them as follows:

$$\begin{aligned} P&=\sum (a_n^+a_{n+1}+a_{n+1}^+a_n+a_n^+a_na_{n+1}^+a_{n+1}+a_na_n^+a_{n+1}a_{n+1}^+) \\ &=\sum (a_n^+a_{n+1}+a_{n+1}^+a_n-2a_n^+a_n+2a_n^+a_na_{n+1}^+a_{n+1})+N. \end{aligned} \tag{3}$$

The operators (2) remind one of the operators familiar in the second quantization, the only difference being the lack of anticommutativity between different sites. But we can easily see that the introduction of the sign functions necessary to make the a_n's and a_n^+'s anticommute does not impair the relation (3), so that we are permitted to put

$$[a_n,\ a_m^+]_+=\delta_{nm}, \quad [a_n,\ a_m]_+=[a_n^+,\ a_m^+]_+=0. \tag{4}$$

Now we make use of the Fourier transformation to rewrite (3) as

$$\begin{aligned} P&=\sum_{n,m} (a_n^+a_m\delta(n-m+1)-a_n^+a_m\delta(n-m)+a_n^+a_na_m^+a_m\delta(n-m+1)+\text{conj.} \\ &=\frac{1}{N}\sum_{n,m}\sum_k \left(a_n^+e^{-\frac{2\pi i}{N}nk}a_me^{+\frac{2\pi i}{N}mk}e^{-\frac{2\pi i}{N}k}-a_n^+e^{-\frac{2\pi i}{N}nk}a_me^{+\frac{2\pi i}{N}mk}+\text{conj.}\right) \\ &\qquad +\sum_{n\ m} a_n^+a_na_m^+a_m\delta(n-m\pm 1) \\ &=\sum_k\left(a_k^+a_ke^{-\frac{2\pi i}{N}k}-a_k^+a_k+\text{conj.}\right)+\sum a_n^+a_na_m^+a_m\delta(n-m\pm 1), \end{aligned} \tag{5}$$

where

$$a_k=\frac{1}{\sqrt{N}}\sum_n a_ne^{\frac{2\pi i}{N}nk}, \quad a_k^+=\frac{1}{\sqrt{N}}\sum_n a_ne^{-\frac{2\pi i}{N}nk}. \tag{5'}$$

There is a striking analogy to the ordinary quantum mechanics. P is just the Hamiltonian of interacting identical systems for which the "kinetic energy" and the "interaction potential" are given respectively by

$$2\left(\cos\frac{2\pi}{N}k-1\right) \quad \text{and} \quad \delta(x_i-x_j\pm 1), \tag{6}$$

where k and x take discrete values. The kinetic terms alone lead to the well known Bloch spin waves whereas the interaction terms cause scattering of them. The latter can be neglected when the number of particles are small because the interaction force is of short range, which corresponds to the fact that the notion of spin wave is a good approximation when the magnetization is nearly complete (low temperature).

As the above example shows, the introduction of a set of anticommuting operators is particularly convenient because of its orthogonal property[4]. Instead of the creation and annihilation operators a_n^+ and a_n, we can also use the quantities $a_n^+ + a_n$ and $i(a_n^+ - a_n)$, or in other words, a set of anticommuting quantities with the commutation relations

$$[x_r,\ x_s]_+ = 2\delta_{rs},\quad r,\ s=1,\cdots 2N. \tag{7}$$

According to the theory of spinors, such quantities can be considered as making up an orthogonal basis for a $2N$-dimentional vectors space[5]. A product of two vectors $x_r x_s$ commutes with x_t if $t \neq r,\ s$. Thus we have a means for expressing commutable quantities in terms of anticommutable quantities.

As a next example we shall take the eigenvalue problem for the operator

$$P' = \sum_{n=1}^{N}(\sigma_{nx}\sigma_{n+1,x} + \sigma_{ny}\sigma_{n+1,y}) \equiv \sum(A_n + B_n), \tag{8}$$

which is a part of the operator P defined in (1). The commutation relations for A and B are

$$\begin{aligned} &[A_n,\ A_m]_- = [B_n,\ B_m]_- = 0, \\ &[A_n,\ B_m]_- = 0 \quad \text{if} \quad n=m \quad \text{or} \quad |n-m| \neq 1, \\ &[A_n,\ B_{n\pm1}]_+ = 0,\quad A_n^2 = B_n^2 = 1. \end{aligned} \tag{9}$$

That is, A_n and B_n anticommute when they are adjacent, otherwise simply commuting. Such quantities can equally be composed of totally anticommuting operators. Indeed, let us introduce $2N$ basic vectors $x_1,\ x_2, \cdots x_{2N}$ with the commutation relations (7), and define (Fig. 1), for N=even,

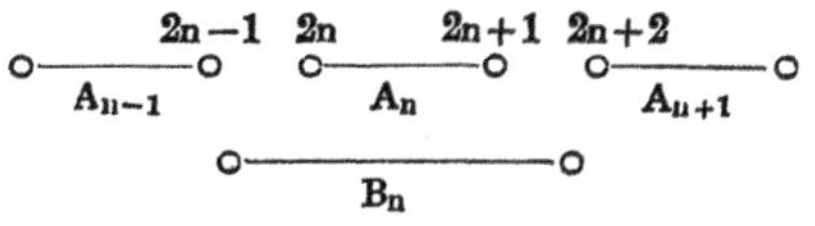

Fig. 1.

$$\begin{aligned} &A_n = i x_{2n} x_{2n+1}, \\ &\qquad n=1,\ 2,\cdots N;\ 2N+1 \equiv 1. \\ &B_n = i x_{2n-1} x_{2n+2}. \end{aligned} \tag{10}$$

Obviously these quantities satisfy all the relations. (8). The only difference is as follows. From (8) we have

$$\prod_{n=1}^{N} A_n = \prod_{n=1}^{N} B_n = 1, \tag{11}$$

while from (10), on the other hand,

$$\begin{aligned} \Pi A_n &= i^N x_2 x_3 \cdots x_{2N} x_1 = -i^N x_1 x_2 \cdots x_{2N} \equiv x = \pm 1, \\ \Pi B_n &= i^N x_1 x_4 \cdot x_3 x_6 \cdot x_5 x_8 \cdots x_{2N-1} x_2 \\ &= i^N (-1)^{N-1} x_1 x_2 \cdots x_{2N} = (-1)^N x = x, \end{aligned} \tag{11'}$$

so that the operators in (8) cover only a subspace of the operator domains for (10). We have deliberately to select those solutions which lie in the subspace $x=1$.

Now (8) can be written as

$$P'=i\sum_{n=1}^{N}(x_{2n}x_{2n+1}+x_{2n-1}x_{2n+2})=i\sum_{n=1}^{N}(x_{2n}x_{2m+1}+x_{2n-1}x_{2m+2})\delta_{nm}$$

$$=\frac{i}{N}\sum_{k=1}^{N}\sum_{n,m}(x_{2n}x_{2m+1}+x_{2n-1}x_{2m+2})e^{\frac{2\pi k}{N}i(n-m)}=2\sum_{k=1}^{N}(x_k y_{-k}+y_k x_{-k}e^{\frac{2\pi k}{N}2i})$$

$$=2i\sum_{k=1}^{N/2}\{x_k y_{-k}(1-e^{-\frac{2\pi k}{N}2i})+x_{-k}y_k(1-e^{\frac{2\pi k}{N}2i})\}$$

$$=-4\sum_{k=1}^{N/2}(x_k y_{-k}e^{-\frac{2\pi k}{N}i}-x_{-k}y_k e^{\frac{2\pi k}{N}i})\sin\frac{2\pi k}{N}\equiv-4\sum_{k=1}^{N/2}z_k\sin\frac{2\pi k}{N}, \tag{12}$$

where

$$x_k\equiv\frac{1}{\sqrt{2N}}\sum_{n=1}^{N}x_{2n}e^{\frac{2\pi k}{N}ni},\quad y_k\equiv\frac{1}{\sqrt{2N}}\sum_{n=1}^{N}x_{2n+1}e^{\frac{2\pi k}{N}ni},$$

$$[x_k,\ x_{-l}]_+=[y_k,\ y_{-l}]_+=\delta_{kl},\quad k,\ l=-N,\ -(N-1),\cdots+N, \tag{12'}$$

$$[z_k,\ z_l]_-=0,\quad [x_k,\ y_{\mp l}]_+=0.$$

Thus we see that the eigenvalue of P' is a sum of the contributions from individual terms with z_k, which have four eigenvalues 0, 0, ±1 as shown by matrix representation :

$$x_k=\begin{pmatrix} & 1\\ 0 & \end{pmatrix}\times\begin{pmatrix}1 & \\ & -1\end{pmatrix},\quad x_{-k}=\begin{pmatrix} & 0\\ 1 & \end{pmatrix}\times\begin{pmatrix}1 & \\ & -1\end{pmatrix},$$

$$y_k=\begin{pmatrix}1 & \\ & 1\end{pmatrix}\times\begin{pmatrix} & 1\\ 0 & \end{pmatrix},\quad y_{-k}=\begin{pmatrix}1 & \\ & 1\end{pmatrix}\times\begin{pmatrix} & 0\\ 1 & \end{pmatrix}, \tag{13}$$

$$\therefore\quad z_k=x_k y_{-k}e^{-\frac{2\pi k}{N}i}-x_{-k}y_k e^{\frac{2\pi k}{N}i}=-\begin{pmatrix}0 & e^{\frac{2\pi k}{N}i}\\ e^{-\frac{2\pi k}{N}i} & 0\end{pmatrix}.$$

Then

$$P'=4\sum_{k=1}^{N/2}\varepsilon_k\sin\frac{2\pi k}{N},\quad \varepsilon_k=0,\ 0,\ 1\ \text{ or }\ -1. \tag{14}$$

The restriction $x=1$ becomes as follows. We write

$$\prod_{n=1}^{N}(\cos\theta+\sin\theta x_{2n}x_{2n+1})=\Pi\exp(\theta x_{2n}x_{2n+1})=\exp[\theta\sum x_{2n}x_{2n+1}] \tag{15}$$

$$=\exp[2\theta\sum_{k=1}^{N/2}(x_k y_{-k}+x_{-k}y_k)]\equiv\prod_{k=1}^{N/2}[2\theta iR_k],$$

$$=\Pi\{1+R_k^2(\cos 2\theta-1)+i\sin 2\theta R_k\}, \tag{15'}$$

since
$$R_k^3=R_k. \tag{16}$$

Putting $\theta=\pi/2$, we have

$$(-1)^{\frac{N}{2}}x=\prod_{k=1}^{N/2}(1-2R_k^2). \tag{17}$$

R_k turns out to commute with z_k and has the eigenvalues ε_k^2. Hence we get the condition

$$(-1)^{N/2}x=\Pi(1-2\varepsilon_k^2)=(-1)^{N/2}. \tag{18}$$

which determines the number of non-zero ε_k.

Thus we see that the use of anticommuting quantities is successful for the operator (8). The problem of diagonalizing the operator has been reduced to diagonalizing a quadratic form. However, this ceases to be the case when we take instead of (8) the operator P defined in (1), which calls for a quartic form rather than a quadratic.

§ 3. Onsager's Problem.

Now let us take up the case of Onsager[2)], that is, a plane square lattice with interaction according to Ising model:

$$E=J\sum s_i s_i'+J'\sum s_i s_{i+1}, \tag{19}$$

where s_i represents the spin $(=\pm 1)$ of the ith site on a certain row and s_i' that of the neighboring one on the next row. The summation extends over all the sites in the lattice. According to Onsager, the problem of obtaining the partition function $F=Tr\exp(-E/kT)$ is reduced to asking for the largest eigenvalue of the operator

$$H=\exp\left[\beta\sum_{n=1}^{N}s_n s_{n+1}\right]\exp\left[\alpha\sum_{n=1}^{N}c_n\right],\quad N+1\equiv 1, \tag{20}$$

where
$$s_n^2=c_n^2=1,$$
$$[s_n,\ c_n]_+=0, \tag{20'}$$
$$[s_n,\ s_m]_-=[s_n,\ c_m]_-=[c_n,\ c_m]_-=0,\quad n\neq m.$$

We shall not occupy ourselves with the deduction of the formula (20), but suppose it as given, and try to solve the eigenvalue problem for H. Writing (20) simply as

$$\begin{aligned}&H=\exp[\beta\sum S_n]\exp[\alpha\sum C_n],\\&S_n\equiv s_n s_{n+1},\quad C_n\equiv c_n,\end{aligned} \tag{21}$$

we see that S_n and C_n have the properties

$$[S_n, C_{n\pm1}]_+=0, \quad \text{otherwise all commuting, and} \qquad S\equiv S_1S_2\cdots S_N=1, \quad C\equiv C_1C_2\cdots C_N=\pm1. \tag{21'}$$

In view of the relations (20) and (21′), a natural idea which suggests itself is to express H in terms of the operators of the second quantization as was illustrated in the last section. Indeed, we can put

$$s_ns_{n+1}=i(a_n^+-a_n)(a_{n+1}^++a_{n+1}), \quad c_n=2a_n^+a_n-1, \qquad a_n^++a_n=c_1c_2\cdots c_{n-1}s_n, \quad i(a_n^+-a_n)=c_1c_2\cdots c_ns_n, \tag{22}$$

with $$[a_n, a_m^+]_+=\delta_{nm}.$$

Here c_n plays just the rôle of the sign function. Such a procedure has been independently devised by Husimi. Instead of (22) we may directly start from the anticommuting orthogonal basis $\{x_n\}$. Thus we replace S_n and C_n by

$$\begin{aligned} &S_n=ix_{2n}x_{2n+1}, \quad C_n=ix_{2n-1}x_{2n},\\ &[x_r, x_s]_+=2\delta_{rs}, \quad r, s=1, 2, \cdots 2N,\\ &S=\Pi S_n=i^Nx_2x_3\cdots x_{2N}x_1=-X,\\ &C=\Pi C_n=i^nx_1x_2\cdots x_{2N}\equiv X. \end{aligned} \tag{23}$$

This substitution corresponds to the case $C=-1$ if we confine ourselves to the subspace $X=1$. The case $C=1$ obtains when one of the defining equations, say for C_1, is substituted by

$$C_1=-ix_1x_2. \tag{23'}$$

However, such minor modifications do not affect the final result for large N except some delicate problems like the boundary tension, so that we shall look apart from the eigenvalues of S and C in order to avoid unnecessary complexity. With the choice (20), the operator H can now be written as

$$H=\exp[i\beta\sum x_{2n}x_{2n+1}]\exp[i\alpha\sum x_{2n-1}x_{2n}]\equiv H_2H_1. \tag{24}$$

As mentioned before, the quantities $x_1, x_2, \cdots x_{2N}$ may be regarded as orthogonal basic vectors in a $2N$-dimensional space. In this space, a rotation of angle θ in the plane spanned by x_r and x_s is expressed by the operator

$$\begin{aligned} &x_n \to U^{-1}x_nU,\\ &U=\exp[(\theta/2)x_rx_s]=\cos(\theta/2)+\sin(\theta/2)x_rx_s. \end{aligned} \tag{25}$$

In fact it is easily verified that

$$\begin{aligned} &U^{-1}x_rU=\cos\theta x_r+\sin\theta x_s,\\ &U^{-1}x_sU=-\sin\theta x_r+\cos\theta x\ ,\\ &U^{-1}x_nU=x_n, \quad n\neq r, s. \end{aligned} \tag{26}$$

Then (24) means nothing but a product of rotations, first by angle $2\alpha i$, in the planes $x_{2n-1}x_{2n}$, and then by angle $2\beta i$, in the planes $x_{2n}x_{2n+1}$. Let us suppose that in a suitable coordinate system $\{x_n'\}$ the rotation H is brought to Jordan's standard form, so that

$$H=\exp[i\sum x_{2n}'x_{2n+1}'\gamma_n]. \tag{27}$$

Since $|ix_n'x_{n+1}'|=1$, the eigenvalue of the operator H that operates on a eigenvector Ψ is found in the expression

$$H\Psi=\exp[\sum\varepsilon_n\gamma_n]\Psi, \quad \varepsilon_n=\pm1. \tag{28}$$

We have only to solve the problem in the $2N$-dimensional subspace $\{x_n\}$ of basic vectors, instead of the whole space of Ψ with 2^N dimensions. By the first rotation H_1, a vector (x_n) is transformed into another vector (y_n), which, by the second rotation H_2, again changes into a third (z_n). Thus

$$\begin{aligned} y_{2n-1}&=\cos 2\alpha' x_{2n-1}+\sin 2\beta' x_{2n}, \\ y_{2n}&=-\sin 2\alpha' x_{2n-1}+\cos 2\beta' x_{2n}, \end{aligned} \tag{28}$$

$$\begin{aligned} z_{2n}&=\cos 2\beta' y_{2n}+\sin 2\beta' y_{2n+1}, \\ z_{2n+1}&=-\sin 2\beta' y_{2n}+\cos 2\beta' y_{2n+1}, \end{aligned} \tag{28$'$}$$

or

$$\begin{aligned} z_{2n}&=-bx_{2n-1}+dx_{2n}+ax_{2n+1}+cx_{2n+2}; \\ z&=-ax_{2n}+dx_{2n+1}+bx_{2n+2}+cx_{2n-1}. \end{aligned} \tag{28$''$}$$

Here $\alpha'=\alpha i$, $\beta'=\beta i$,

$$\begin{aligned} a&=\cos 2\alpha' \sin 2\beta', \quad b=\sin 2\alpha' \cos 2\beta', \\ c&=\sin 2\alpha' \sin 2\beta', \quad d=\cos 2\alpha' \cos 2\beta'. \end{aligned} \tag{28$'''$}$$

In matrix form,

$$H=\begin{pmatrix} \ddots & & & & & & \\ -b & d & a & c & & & \\ & & \ddots & & & & \\ c & -a & d & b & & & \\ & & \ddots & & & & \\ & & -b & d & a & c & \\ & & & & \ddots & & \\ & & c & -a & d & b & \\ & & & & & & \ddots \end{pmatrix} \begin{matrix} \\ \\ \\ \cdots 2n \\ \\ \cdots 2n+1 \\ \\ \\ \\ \end{matrix} \tag{29}$$

The eigenvalue equation for H becomes

$$\begin{aligned} -bx_{2n-1}+(d-\lambda)x_{2n}+ax_{2n+1}+cx_{2n+2}&=0, \\ cx_{2n-1}-ax_{2n}+(d-\lambda)x_{2n+1}+bx_{2n+2}&=0, \end{aligned} \tag{30}$$

or

$$\begin{aligned} ax_{2n+1}+cx_{2n+2}&=bx_{2n-1}+(\lambda-d)x_{2n}, \\ (\lambda-d)x_{2n+1}-bx_{2n+2}&=cx_{2n-1}-ax_{2n}. \end{aligned} \tag{30$'$}$$

Let us write

$$(x_{2n+1},\ x_{2n+2})=\psi_{n+1}\,,\quad (x_{2n-1},\ x_{2n})=\psi_n\,, \tag{31}$$

and

$$A=\begin{pmatrix} a, & c \\ \lambda-d, & -b \end{pmatrix},\quad B=\begin{pmatrix} b, & \lambda-d \\ c, & -a \end{pmatrix}. \tag{31'}$$

Then (30′) means

$$A\psi_{n+1}=B\psi_n\,,\quad \text{or}\quad \psi_{u+1}=A^{-1}B\psi_n\equiv C\psi_n\,, \tag{32}$$

hence $$\psi_n=C^{n-1}\psi_1\,. \tag{32'}$$

The periodicity condition requires that $\psi_{N+1}=\psi_1$, or

$$(E-C^N)\psi_1=0,\quad \det|E-C^N|=0. \tag{33}$$

If we denote the Nth roots of unity by $\xi_k=\exp(2\pi ki/N)=\exp(\varphi_k i)$, $k=1, 2,\cdots N$, then

$$\det|\xi_k E-C|=0,\quad \text{or}\quad |\xi_k A-B|=0, \tag{33'}$$

that is,

$$\begin{vmatrix} \xi_k a-b, & \xi_k c-(\lambda-d) \\ \xi_k(\lambda-d)-c, & -\xi_k b+a \end{vmatrix}=0. \tag{33''}$$

This gives a quadratic equation for λ:

$$\xi\lambda^2-\lambda[2\xi d+\xi^2 c+c]+\xi(a^2+b^2+c^2+d^2)-(1+\xi^2)(ab-cd)=0. \tag{34}$$

Substituting (28‴),

$$a^2+b^2+c^2+d^2=1,\quad ab-cd=0,$$

$$\lambda^2-\lambda[2d+(\xi+\xi^{-1})c]+1=0,$$

or $$\lambda^2-2\lambda(\cos 2\alpha'\cos 2\beta'+\sin 2\alpha'\sin 2\beta'\cos\varphi_k)+1=0. \tag{34'}$$

Putting the two roots as $\exp(\pm 2\gamma_k' i)$,

$$(1/2)(e^{2\gamma' i}+e^{-2\gamma' i})=\cos 2\gamma_k'=\cos 2\alpha'\cos 2\beta'+\sin 2\alpha'\sin 2\beta'\cos\varphi_k\,. \tag{35}$$

Returning to the original constants α and β, and writing $\gamma=\gamma' i$, we get just the relation given by Onsager:

$$\cosh 2\gamma_k=\cosh 2\alpha\cosh 2\beta-\sinh 2\alpha\sinh 2\beta\cos\varphi_k,\quad \varphi_k=2\pi k/N. \tag{35'}$$

The largest eigenvalue of H becomes then, according to (28),

$$H_{\max}=\exp[\Sigma|\gamma_k|]\,. \tag{36}$$

Such a procedure also applies, *mutatis mutandis*, to certain variants of Onsager model, e.g. the honeycomb lattice treated by Husimi and Syôzi. It fails, however, in case of those lattices with too many nearest neighbours, in particular the three-dimensional lattices. For we cannot express a quantity which anti-commutes with more than two neighbors by a product of two totally anticommuting vectors. In fact the three-dimensional model necessitates us to resort to products of four x's as illustrated below:

$$H=\exp\,[\beta\sum(s_{nm}s_{n,m+1}+s_{nm}s_{n+1,m})\,\exp\,[u\sum c_{nm}]$$
$$=\exp\,[i\beta\sum h_r]\exp\,[u\sum k_s]\,, \tag{37}$$
$$h=x_1x_3',\ x_4x_2'',\quad k=x_1x_2x_3x_4\,.\quad \text{(See Fig. 2.)}$$

Supplementary conditions are also to be considered. The occurence of quartic forms may look reasonable when we note the so-called dual transformation. Such transformations do really exist even in the three-dimensional case. In fact let us take a cubic lattice with atoms located at the middle point of each edge of the cell, and let each set of the four atoms lying on a side plane determine a four-body interaction energy $\mu s_1s_2s_3s_4$, $s_i=\pm 1$. Then it is easy to show that this lattice is dual to a simple cubic lattice made up of the body center sites of the former with ordinary two-body interactions. Perhaps the four-body force may have some bearing on the quantity k_s in (37).

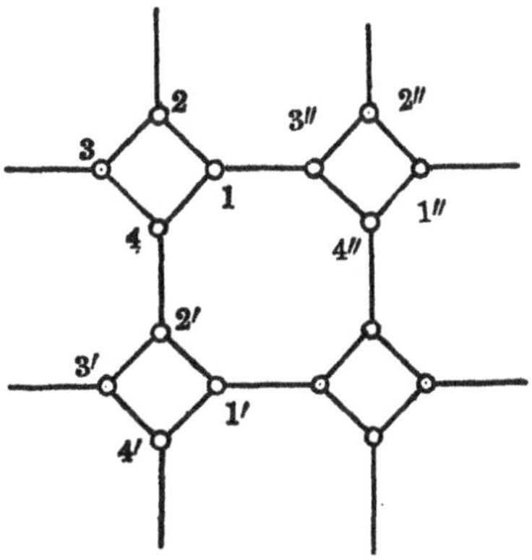

Fig. 2.

§ 4. Kramers-Wannier's method of approach.

Our next investigations concern the screw lattice model adopted by Kramers and Wannier for solving the two-dimensional case. This model seems more convenient for general purposes than that used by Onsager. Now let us try to transform the Onsager model into the K-W model as follows. Rewrite (20) (after reversing the order of H_1 and H_2) as

$$H_1H_2=\exp\,[i\alpha x_1x_2]\exp\,[i\alpha x_3x_4]\cdots\exp\,[i\beta x_2x_3]\exp\,[i\beta x_4x_5]\cdots$$
$$=\exp\,[i\alpha x_1x_2]\exp\,[i\alpha x_3x_4]\exp\,[i\beta x_2x_3]\exp\,[i\alpha x_5x_6]\exp\,[i\beta x_4x_5]\cdots$$
$$=\exp\,[i\alpha x_1x_2]\prod_{n=1}^{N-1}(\exp\,[i\alpha x_{2n+1}x_{2n+2}]\exp\,[i\beta x_{2n}x_{2n+1}]\exp\,[i\beta x_{2N}x_1]\,, \tag{38}$$

which can be modified, without grave consequences, to

$$H=\prod_{n=1}^{N}\exp\,[i\alpha x_{2n+1}x_{2n+2}]\exp\,[i\beta x_{2n}x_{2n+1}]\equiv\Pi H_n\,,$$
$$H_n\equiv\exp\,[i\alpha x_{2n+1}x_{2n+2}]\exp\,[i\beta x_{2n}x_{2n+1}]\,. \tag{38'}$$

It turns out more convenient to consider an operator which has its component factors H_n in the reverse order to (51) (which is also equivalent to a renumbering of the order). Then define a displacement operator P, which changes x_n into x_{n+1}, and put

$$\Psi_n = H_n H_{n-1} \cdots H_1 \Psi_0 \tag{39}$$

with obvious relations

$$H_n = P^n H_0 P^{-n}, \quad H_0 = \exp[i\beta x_{2N} x_1] \exp[i\alpha x_1 x_2]. \tag{39'}$$

The relation (a Schrödinger equation for discrete time variable!)

$$\Psi_{n+1} = H_n \Psi_n \tag{40}$$

is turned into

$$\Psi_{n+1}' = P^{-1} H_0 \Psi_n' \equiv A\Psi_n' \tag{40'}$$

by putting

$$P^{-\bar{n}}\Psi_n = \Psi_n' \quad \text{or} \quad \Psi_n = P^n \Psi_n'. \tag{40''}$$

But $P^{-1}H_0$ is now independent of n, hence the eigenvalue problem for Ψ_n is equivalent to

$$\lambda\Psi' = A\Psi' = P^{-1}H_0\Psi', \quad \lambda^N = E, \quad (H\Psi_0 = \Psi_N = E\Psi_0). \tag{41}$$

P can be expressed in terms of the basic vectors as

$$\begin{aligned} P = &\exp\left[\frac{\pi}{4} x_1 x_{2N-1}\right] \exp\left[-\frac{\pi}{4} x_1 x_{2N-3}\right] \cdots \exp\left[\pm\frac{\pi}{4} x_1 x_3\right] \\ &\times \exp\left[\frac{\pi}{4} x_2 x_{2N}\right] \exp\left[-\frac{\pi}{4} x_2 x_{2N-2}\right] \cdots \exp\left[\pm\frac{\pi}{4} x_2 x_4\right] \end{aligned} \tag{42}$$

because of the relations

$$\begin{aligned} &\exp\left[\frac{\pi}{4} x_1 x_2\right] x_1 \exp\left[-\frac{\pi}{4} x_1 x_2\right] \\ &= \frac{1}{2}[1 - x_1 x_2] x_1 [1 + x_1 x_3] = \frac{1}{2}(x_1 - x_1 + 2x_2) = x_2, \\ &\exp\left[\frac{\pi}{4} x_1 x_2\right] x_2 \exp\left[-\frac{\pi}{4} x_1 x^2\right] = \frac{1}{2}(x_2 - x_2 - 2x_1) = -x_1. \end{aligned} \tag{42'}$$

Thus the operator A can again be regarded as a product of rotation operators. Let us look for an eigenvector of the form

$$x = \sum_{g=1}^{N-1} \lambda^{n-1}(a x_{2n-1} + b x_{2n}) + c x_{2N} + \lambda^{N-1} a x_{2N-1}. \tag{43}$$

Substituting in the equation $AxA^{-1} = \lambda x$, we get the following result:

$$\left.\begin{array}{l}\cos 2\alpha' \cos 2\beta' x - \sin 2\alpha' \cos 2\beta' y + \sin 2\beta' z = \lambda^N x\,, \\ \sin 2\alpha' x + \cos 2\alpha' y = \lambda z\,, \\ -\sin 2\beta' \cos 2\alpha' x + \sin 2\alpha' \sin 2\beta' y + \cos 2\beta' z = \lambda^{N-1} y\,,\end{array}\right\} \quad (44)$$

or

$$\left.\begin{array}{l}\cosh 2\alpha \cosh 2\beta x + i \sinh 2\alpha \cosh 2\beta y - i \sinh 2\beta z = \lambda^N x\,, \\ -i \sinh 2\alpha x + \cosh 2\alpha y = \lambda z\,, \\ i \sinh 2\beta \cosh 2\alpha x - \sinh 2\alpha \sinh 2\beta y + \cosh 2\beta z = \lambda^{N-1} y\,.\end{array}\right\} \quad (44')$$

Eliminating z by virture of the second equation,

$$\begin{array}{l}(\cosh 2\alpha \cosh 2\beta - \sinh 2\alpha \sinh 2\beta \lambda^{-1}) x + i(\sinh 2\alpha \cosh 2\beta \\ \qquad - \sinh 2\beta \cosh 2\alpha \lambda^{-1} y = \lambda^N x\,, \\ i(\sinh 2\beta \cosh 2\alpha - \sinh 2\alpha \cos 2\beta \lambda^{-1}) x - (\sinh 2\alpha \sinh 2\beta \\ \qquad - \cosh 2\beta \cosh 2\alpha \lambda^{-1} y = \lambda^{N-1} y\,,\end{array} \quad (44'')$$

which determine λ according to

$$\begin{aligned}&\lambda^{2N-1} + \sinh 2\alpha \sinh 2\beta \lambda^N + (-\cosh 2\beta \cosh 2\alpha - \cosh 2\alpha \cosh 2\beta)\lambda^{N-1} \\ &\quad + \sinh 2\alpha \sinh 2\beta \lambda^{N-2} + (\sinh^2 2\alpha \sinh^2 2\beta + \cosh^2 2\beta \cosh^2 2\alpha \\ &\quad - \sinh^2 2\alpha \cosh^2 2\beta - \sinh^2 2\beta \cosh^2 2\alpha \lambda^{-1} \\ &\quad + (-\sinh 2\alpha \cosh 2\alpha \sinh 2\beta \cosh 2\beta + \sinh 2\alpha \cosh 2\alpha \sinh 2\beta \cosh 2\beta)\lambda^{-2} \\ &\quad - \cosh 2\alpha \sinh 2\alpha \cosh 2\beta \cosh 2\beta + \sinh 2\alpha \cosh 2\alpha \sinh 2\beta \cosh 2\beta = 0.\end{aligned}$$

This in turn reduces to

$$\lambda^{2N} + \sinh 2\alpha \sinh 2\beta (\lambda + \lambda^{-1}) - 2\cosh 2\alpha \cosh 2\beta + \lambda^{-N} = 0. \quad (45')$$

Putting $\lambda = \exp 2\gamma$,

$$\cosh 2N\gamma = \cosh 2\alpha \cosh 2\beta - \sinh 2\alpha \sinh 2\beta \cosh \gamma. \quad (46)$$

This is the equation for γ. To solve this, we assume that $2\gamma = 2\gamma_0 + \omega i$, and substitute in (46), thereby making use of the relation

$$\cosh (2\gamma_0 + \omega i) = \cosh 2\gamma_0 \cos \omega + i \sinh 2\gamma_0 \sin \omega\,. \quad (47)$$

Comparison of real and imaginary parts yields the relations

$$\begin{aligned}&\cosh 2N\gamma_0 \cos N\omega = \cosh 2\alpha \cosh 2\beta - \sinh 2\alpha \sinh 2\beta \cosh \gamma_0 \cos \omega\,, \\ &\sinh 2N\gamma_0 \sin N\omega = \sinh 2\alpha \sinh 2\beta \sinh \gamma_0 \sin \omega\,.\end{aligned} \quad (48)$$

Choose $\omega = k\pi/N$, $k = 1, 2, \cdots 2N$, so that $\sin N\omega = 0$, $\cos N\omega = (-1)^k$. In the limit $N \to \infty$ and $N\gamma_0 \to \Gamma = \text{finite}$, we have $\gamma_0 = 0$, and both equations become satisfied by this choice of ω. Thus

$$\pm \cosh 2\Gamma = \cosh 2\alpha \cosh 2\beta - \sinh 2\alpha \sinh 2\beta \cos \omega\,, \quad (49)$$

with $\omega = k\pi/N.$ (49′)

But the minus sign is impossible, hence the allowed values for ω are

$$\omega = 2k\pi/N, \quad k = 1, 2, \cdots N. \tag{49''}$$

This completely agrees with the previous results. The case of the three-dimensional lattice can be treated analogously, but we shall not give here the details.

§ 5. Additional Remarks.

The mathematical tricks employed in the preceding sections seem to allow a rather natural explanation from a somewhat more general point of view. The essential point is that we consider, instead of an ordinary eigenvalue problem, a different one of the form

$$[H, X]_{-} = \lambda X \tag{50}$$

for the " eigenoperator " X[6]. Important consequences drawn from this equation are: (a) λ is the difference of two eigenvalues of $H: \lambda = E_n - E_m$; (b) X transforms certain eigenvector Ψ_m of H with E_m into another one Ψ_n with eigenvalue $E_n = E_m + \lambda$; and (c) Product of two eigenoperators X_2X_1 is again an eigenoperator, with eigenvalue $\lambda = \lambda_1 + \lambda_2$, transforming Ψ_m into a third eigenvector Ψ_l. If we integrate the relation (50) according to

$$\frac{d}{ds} X = [H, X], \tag{51}$$

we arrive at the equation

$$e^{sH} X e^{-sH} = e^{\lambda} X. \tag{51}$$

The equation (51) just corresponds to the problem of obtaining the eigenoperator $\{x_n'\}$ for the rotation H encountered in Section 3 and 4. When the operator H has a simple structure, a general eigenoperator X will be factorized into a product of prime eigenoperators:

$$X = X_1 X_2 \cdots X_k,$$

with
$$e^{\lambda} = e^{\lambda_1 + \lambda_2 + \cdots \lambda_k}. \tag{52}$$

The largest λ is the difference of the largest and smallest eigenvalues of sH in (52). If the eigenvalues E_i are symmetrically distributed around zero, $E_{\max}$ will itself be given by $E_{\max} = \lambda_{\max}/2$, a fact which casts a sidelight on the procedure used in the previous sections.

The main part of the present work had been completed nearly two years ago. It is through the kindness of Professor Husimi a d Mr. Syôzi of Osaka University that the author enjoys the opportunity of publishing this note. He wishes to express on this occasion his cordial thanks to them and also to Professor Kubo of Tokyo University for helpful discussions.

References.

(1) H. A. Kramers and G. H. Wannier, Phys. Rev. **60** (1941), 252, 263.
E. N. Lassetre and J. P. Howe, J. Chem. Phys. **9** (1941), 747 and 801.
E. W. Montroll and J. E. Mayer, J. Chem. Phys. **9** (1941), 626.

(2) L. Onsager, Phys. Rev. **65** (1944), 117.

(3) Husimi and Syôzi, Lectures at the Annual Meeting of the Phys. Soc., 1948 and 1949; Husimi and Syozi, Prog. Theor. Phys. in press.

(4) An alternative way for solving (3) is to use commutative, rather than anticommutative, operators which corresponds to Bose statistics, together with some supplementary conditions. This leads to Bethe's rigorous solution for one-dimensional model (ZS. f. Phys. **71** (1931), 205; *Handbuch d. Phys.* XXIV/2, 604). We shall not here enter into details about it.

(5) Cf. Cartan: *Leçons sur la théorie des spineurs*, 1938.

(6) A deeper research on this subject will be found in: Y. Nambu, *On the method of the third quantization*, Prog. Theor. Phys. **4** (1949), 331 and 399.

Chapter 3: Nambu and the ω^0 Meson

On April 25, 1957, *Phys. Rev.* **106**, *1366 (1957)* received

Possible Existence of a Heavy Neutral Meson, by Yoichiro Nambu.

This is the paper, one and a half page long, where Nambu proposes the existence of the ω^0 meson to explain the Hofstadter experiment on the nucleon structure.

Background

Two theoretical discoveries in the mid-30's came to have profound importance in the development of modern particle physics. One was Hideki Yukawa's (1907–1981) 1935 proposal[29] that the strong interactions were mediated by a low-mass spinless boson, the pion. Yukawa's hypothesis of a new elementary particle was daring — a few years before Wolfgang Pauli (1900–1958) wrote his famous postcard introducing the neutrino. Most physicists were skeptical of introducing more particles than the known ones, and it was until 1947 when it finally was discovered by Cecil Powell (1903–1969) and his group. Yukawa was awarded the 1949 Nobel Prize for his prediction of the pion, and Cecil Powell a year later for its discovery.

At Dirac's suggestion, Eugene Wigner (1902–1995) published a complete study of representations of the Poincaré group. It laid the ground for how particles or fields could appear and interact in relativistic quantum field theories. Written in 1935, it was rejected by physics journals and was finally published[30] in 1939 in *Annals of Mathematics*, thanks to John von Neumann (1903–1957). In the citation to Wigner's Nobel Prize in 1963, it says "*for his contributions to the theory of the atomic nucleus and the elementary particles, particularly through the discovery and application of fundamental symmetry principles.*"

In 1932, Werner Heisenberg (1901–1976) had proposed[31] that the nuclear forces do not distinguish between the proton and the neutron. The new symmetry came to be called isospin symmetry by Wigner and protons and neutrons are the nucleons.

[29] H. Yukawa, *Proceedings of the Physico-Mathematical Society of Japan*, 3rd series, Vol. 17, 48 (1935).
[30] E. Wigner, *Annals of Mathematics* **40**(1), (1939).
[31] W. Heisenberg, *Zeit f. Phys.* **77**, 1 (1932); **78**, 156 (1932).

With Yukawa's proposal for the pion as the mediator between nucleons, a typical nuclear potential would look like

$$V = \bar{N}\Pi N,$$

where N is a nucleon field and Π the pion field. (The modern version is $V = \bar{N}\gamma^\mu\gamma^5\partial_\mu\Pi N$.) With regard to spin and isospin invariances, the nucleons are doublets under both, that is, $N \sim (\mathbf{2},\mathbf{2})$ under $SU(2)_{\text{spin}} \times SU(2)_{\text{isospin}}$. The possible representations for the Π field are to be found in the two-nucleon product,

$$(\mathbf{2},\mathbf{2}) \times (\mathbf{2},\mathbf{2}) = (\mathbf{3},\mathbf{3}) + (\mathbf{3},\mathbf{1}) + (\mathbf{1},\mathbf{3}).$$

$(\mathbf{1},\mathbf{3})$ describes three (pseudo)scalar, isovectors states, one of which is Yukawa's neutral pion. There appear in the product other possible fields — two vector fields, one isovector and one isoscalar. In a 1937 paper[32] Wigner identified those states as did Stückelberg a year later.[33] They do not propose the states as new particles, more like two-pion and three-pion states. We must remember that not even the pion had been discovered. A new particle had been discovered in 1936 but it turned out to be the muon.

It should be mentioned here that Yukawa, Soichi Sakata (1911–1970) and Mitsuo Taketani (1911–2000) in 1937 did discuss the possibility of a vector boson as the mediator.[34]

After the war, the technique with emulsions and cloud chambers had become so successful that it was possible to detect the pion as mentioned above. But also other more massive mesons were found among the tracks in the emulsions. Their mass was about half the proton mass and they were quite long-lived. They were also produced together with new baryons in associated productions. The key to understand these particles was the introduction of strangeness as a new quantum number[35] by Murray Gell-Mann and by Kazuhiko Nisjijima and Tadao Nakano, where the new particles have strangeness $+1$ or -1, while the pion and the nucleons have strangeness 0. Another strange phenomenon was that there were seemingly two different strange particles with the same mass, one decaying into the parity-even two-pion state and one into the parity-odd three-pion state, the so-called θ–τ puzzle. All this was understood when it was discovered in 1957 that the weak interactions violate parity invariance. The particle became known as the kaon. It is like the pion, a pseudoscalar particle. Much of the research in the field in the first half of the 1950s was devoted to understand these particles and there was no experimental hint of vector particles and no theoretical studies of such particles in the new field of particle physics. The history of the struggle to understand the new particles is very well-known, so we do not dwell on that any more.

[32] E. Wigner, *Phys. Rev.* **51**, 106 (1937).

[33] E. C. G. Stückelberg, *Helv. Phys. Acta* **11**, 255 (1938).

[34] H. Yukawa, S. Sakata, and M. Taketani, *Proceedings of the Physico-Mathematics Institute* **20**, 319 (1938).

[35] M. Gell-Mann, *Phys. Rev.* **92**, 833 (1953); T. Nakano and K. Nishijima, *Prog. Theor. Phys.* **10**, 581 (1953).

Also after the war, the discussion about nuclear potentials became popular again. With the success of QED, it was natural to investigate the Yukawa theory as a relativistic quantum field theory and it was a great disappointment that the field theory, even though it was renormalizable, was not a viable perturbative theory since the coupling constant was > 1. Many of the leaders of the new generation then concentrated on the new field of particle physics.

There was, however, also great progress in nuclear physics after the war. Maria Goeppert-Mayer (1906–1972) and Hans Jensen (1907–1973) and collaborators introduced the shell-model to describe the nuclei. This mimicked the atomic shells but since the force between nucleons was strong, the picture was crude. One could not, like in atomic physics, reduce the problems to two-particle problems. Aage Bohr (1922–2009) and Ben Mottelson (1926–2022) worked out the consequences of this difference and a phenomenological model of the nuclei was established. The agreement between the theory and experimental data provided support to the idea of an average potential within the nucleus.

In an impressive study[36] by M. H. Johnson and Edward Teller in 1955 of heavy nuclei with an equal number of protons and nucleons found that there would be an approximately constant meson field within nuclei and they could argue that it had to be an isoscalar. They also found that it must be a scalar and not a vector. They then discussed what it could be and said "The scalar neutral meson need not be an elementary particle in any sense of the word. It may be a virtual state composed of other mesons. It may be even a superposition of such virtual states. It may decay into π mesons so quickly that it cannot be observed." Even though the authors discussed a possible scalar neutral meson and not a vector meson, Nambu in his paper pointed to the similarity of the two proposals.

The Hofstadter experiments

In the beginning of the 1950's Robert Hofstadter and his group at Stanford started a series of experiments where electrons were scattered off nuclei. The experiments where electrons scattered off protons and deuterons turned out to be of special interest. With some ingenious techniques they managed to also extract the scattering off neutrons from the experiments on deuterons. These experiments showed clearly that the nucleons were not point-like but had a structure showing a charge distribution inside protons, and inside the neutral neutron as well.

The scatterings could be understood in terms of a theoretical scattering law developed[37] in 1950 by Marshall Rosenbluth (1927–2003). The law described the

[36]M. H. Johnson and E. Teller, *Phys. Rev.* **98**, 783 (1955).
[37]Marshall N. Rosenbluth, *Phys. Rev.* **79**, 615 (1950).

composite effect of charge and magnetic moment scattering and is given by

$$\frac{d\sigma}{d\Omega} = \sigma_{\mathrm{NS}} \left({F_1}^2 + \frac{1}{4}\frac{\hbar^2 q^2}{4M^2c^2}[2(F_1 + KF_2)^2 \tan^2 \theta/2 + K^2 {F_2}^2] \right),$$

where σ_{NS} is the scattering cross-section with a point source with no spin or magnetic moment. $F_1(q)$ is the Dirac form factor representing the nucleon charge and its associated Dirac magnetic moment, and q is the momentum transfer. $F_2(q)$ is the Pauli form factor and represents the anomalous magnetic moment of the nucleon. K is the static value (1.79) of the anomalous magnetic moment in nuclear magnetons and M the mass of the nucleon.

All the experiments at various energies could be fitted well with this formula and the form factors could be experimentally determined both for the proton as well as for the neutron. For the proton $F_1(0) = F_2(0) = 1$. For higher q's they fall off with $F_2(q)$ falling steeper than $F_1(q)$. For the neutron, the corresponding values are $F_1(0) = 0$ and $F_2(0) = 1$, consistent with the zero charge of the neutron. For higher q, again $F_2(q)$ falls off like for the proton while $F_1(q)$ fluctuates around 0.

These results have shown clearly the charge distributions within the nucleons and constitute the proof that the nucleons unlike the leptons are not point-like objects but carry an internal structure. This was of course a precursor for the quark hypothesis of Gell-Mann and George Zweig (1937–) in 1964. Hofstadter was awarded the Nobel Prize in 1961 for this discovery.

Nambu's paper

The Hofstadter experiments captured Nambu's interest. He studied it from the point of view of an effective field theory. He considered the possibility that there exists a heavy neutral meson, which is a vector field with isotopic spin zero and a mass greater than two pion masses. It should be coupled strongly to the nucleon field. He calls it ρ^0 but it came to be called ω^0 since ρ became the name for an isotopic spin-1 vector particle. An isolated ω^0 would decay through virtual nucleon pair formation according to the following schemes:

(a) $\omega^0 \to \pi^0 + \gamma, \quad 2\pi^0 + \gamma, \quad \pi^+ + \pi^- + \gamma;$
(b) $\omega^0 \to e^+ + e^-, \quad \mu^+ + \mu^-.$

He did not consider the possibility that the meson, if massive enough, could decay to three pions. Such a process would not affect the discussion of the paper. Process (b) is an electromagnetic decay while the process (a) is a strong interaction decay. Hence the lifetime for the process (a) should be some 10^{-2} to 10^{-3} s shorter.

The process (b) contributes to the electron–nucleon scattering with the exchange of ω^0 which is short range since it is a heavy meson. This is like the interchange of a "heavy photon". It will contribute to a Dirac form factor

$$F'(k^2) \sim \frac{Gg}{\mu^2 + k^2}$$

with μ the ω mass and G the nuclear coupling constant and g the effective ω-electron coupling. The electromagnetic coupling is written as e.

$$g^2 \sim G^2 e^{2^2} (\mu/M)^2$$

with M the nucleon mass. We take $\hbar = c = 1$. (Nambu is not using this convention.)

F' has the same sign for both protons and neutrons since ω is an isotopic scalar. The corresponding form factor F due to the pion cloud will change sign. Relativistic field theory shows on general grounds that $F(k^2)$ has the form

$$F(k^2) = \int_{2m_\pi}^{\infty} dm \, \frac{\rho(m)}{m^2 + k^2}.$$

It is thus possible that the two form factors cancel approximately for the neutron but is reinforced for the proton in agreement with observation.

To check that the two processes are of the same magnitude, Nambu equals tentatively the mean square radius of the proton with the one due to the ω^0. This relates the coupling constants and leads to consistent lifetimes for the two processes.

Finally he pursues (prophetically) further consequences of his assumptions

(1) ω^0 can be produced by any strong nuclear reaction.

(2) The second maximum of the pion-nucleon scattering around 1 GeV could be attributed to the process

$$\pi^- + p \to n + \omega^0,$$

if a resonance should occur for such a system.

(3) ω^0 would contribute a repulsive nuclear force of Wigner type and short range ($\lesssim 0.7 \times 10^{-13}$ cm) more or less similar to the phenomenological hard core. (In modern nuclear physics the short-range repulsive force is due to the ω meson and also to the ρ meson.)

(4) The anomalous magnetic moment of the nucleon will be affected by ω^0. The main effect seems to be that ω^0 and the usual pion make opposite contributions to the isotopic scalar part of the core momentum, thus tending to better agree between theory and experiment.

(5) If it is energetically possible the kaons and the hyperons would sometimes decay by emitting an ω^0.

Of course it turned out that kaons are less massive than the ω^0 but hyperons are not.

Further proposals of an isoscalar vector particle

In the late 1950s, the formulation[38] of the "Yang–Mills" theory by C. N. Yang (1922–) and Robert Mills (1927–1999) opened up the possibility that the strong interactions could be mediated by vector bosons. Jun John Sakurai (1933–1982) who was a junior colleague of Nambu in Chicago was a leader of this development and he argued for a strong interaction theory based on the gauge principle. He introduced both an isovector and an isoscalar vector particle. The isovector came to be called the ρ-meson. It had previously been proposed[39] as a resonance of two pions.

Also the Sakata school in Japan discussed vector particles as gauge particles in terms of the Sakata model.[40] Yasunari Fujii[41] used an analogy to electromagnetic interaction to sketch on a theory for the strong interactions. In both his and Sakurai's works, the ω-meson is called the Nambu particle.

It was Gell-Mann in his monumental work on "the eight-fold way" in 1961 who coined the name ω^0.

The discovery of the ω^0

With the new big accelerators in operation starting with the Bevatron at Berkeley in 1954 and then at Brookhaven and at CERN in the late 1950s, new mesons could be discovered in proton–proton collisions. At Berkeley, as well as at CERN and Brookhaven, one even had an antiproton beam that could look at scatterings between a proton and an antiproton which scatter into mesons only. This gave a possibility to look for resonances among the resulting pions.

The processes considered by Nambu that were important for the interpretation of the form factors measured in the Hofstadter experiments are difficult to measure experimentally since they involved photons in the final state. In 1960 Geoffrey Chew (1924–2019) argued that one could look for ω as a resonance in the three-pion spectra. A requirement for this was that the ω mass was $> 3m_\pi$.

There were a number of attempts to find the ω meson and the one that succeeded was Bogdan Maglic (1928–2017), who was a post-doc in Luis Alvarez (1911–1988) group in Berkeley. He was using the results from antiproton–proton scatterings. By scanning through several thousands of events, he could finally find a strong resonance in the three-pion spectra where the sum of charges was zero. The data set also had information about the ρ meson. The surprising fact was that the two particles were so close in mass, $m_\rho = 775$ MeV and $m_\omega = 783$ MeV. The experiments that had failed before had been looking in a lower mass interval.

[38] C. N. Yang and R. Mills, *Phys. Rev.* **96**, 191 (1954).
[39] William R. Frazer and Jose R. Fulco, *Phys. Rev. Lett.* **2**, 365 (1959).
[40] S. Sakata, *Prog. Theo. Phys.* **16**, 686 (1956).
[41] Y. Fujii, *Prog. Theo. Phys.* **21**, 232 (1958).

The discovery was noticed all the way up to the White House. President Kennedy sent a telegram to Luis Alvarez congratulating him and his three collaborators and he ended with "I am particularly pleased that one of the co-discoverers of this meson was a visiting scientist from Yugoslavia giving further evidence for scientific freedom available in this country." As far as we know, no telegram was sent to Nambu.

In his Nobel lecture in 1961 Hofstadter wrote:

"It may be noted parenthetically, that it was on the basis of the above results that Nambu postulated the existence of a new heavy neutral meson, now known as the ω-meson. Events of the past year have brilliantly confirmed the existence of this meson."

Present day nuclear potential

It is interesting to note that the present day picture of the nuclear potential consists of a short range repulsive force which is steeply falling determined by the ω- but also the ρ-meson down to a minimum, and the slow rise in the intermediate region is due to something like a scalar σ-meson, which is still a debated particle and the rising long range force is due to the pion.

Possible Existence of a Heavy Neutral Meson*

YOICHIRO NAMBU
The Enrico Fermi Institute for Nuclear Studies, The University of Chicago, Chicago, Illinois
(Received April 25, 1957)

IN an attempt to account for the charge distributions of the proton and the neutron as indicated by the electron scattering experiments,[1] we would like to consider the possibility that there may be a heavy neutral meson which can contribute to the form factor of the nucleon. We assume that this meson, ρ^0, is a vector field with isotopic spin zero and a mass two to three times that of the ordinary pion, coupled strongly to the nucleon field. An isolated ρ^0 would decay through virtual nucleon pair formation according to the following schemes:

(a) $\rho^0 \rightarrow \pi^0+\gamma, \quad 2\pi^0+\gamma, \quad \pi^+ + \pi^- + \gamma;$

(b) $\rho^0 \rightarrow e^+ + e_-, \quad \mu^+ + \mu^-;$

(c) $\rho^0 \rightarrow \pi^+ + \pi^-.$

The process (a) would have a decay probability roughly of the order of $P_a \sim (\mu c^2/\hbar)(G^2/\hbar c)(e^2/\hbar c)(\mu/M)^2$, where G is the nuclear coupling constant, μ and M the ρ^0 and the nucleon masses, respectively. For the process (b), the probability would be $P_b \sim (\mu c^2/\hbar)(G^2/\hbar c)(e^2/\hbar c)^2 \times (\mu/M)^2$. The process (c) is a forbidden transition, so that it can take place only in violation of the isotopic spin conservation, with a decay rate comparable to that for (b).

Now the process (b) gives rise to a short-range interaction between a nucleon and an electron (or a muon) by exchange of a ρ^0. This will contribute a form factor $F'(k^2) \sim Gg/(\mu^2+k^2)$ to the electron-nucleon scattering, where g is the effective ρ^0-electron coupling, $g^2/\hbar c \sim (G^2/\hbar c)(e^2/\hbar c)^2(\mu/M)^2$. Since ρ^0 is an isotopic scalar, F' has the same sign for both proton and neutron, whereas the corresponding form factor F

due to the pion cloud should change sign. Relativistic field theory shows on general grounds that $F(k^2)$ has the form

$$F(k^2)=\int_{2m_\pi}^{\infty}\frac{\rho(m)}{m^2+k^2}dm,$$

where the lower limit of integration corresponds to the threshold for pion pair creation by an external electromagnetic field. With our assumptions about ρ^0, it is thus possible that the two form factors F and F' cancel approximately for the neutron but reinforce for the proton, in agreement with observation. If we equate tentatively the mean square radius of the proton with the one due to ρ^0:

$$Gg/\mu^2 \sim e^2\langle a^2\rangle/b,$$

we get

$$(G^2/\hbar c)(g^2/\hbar c)\sim[(e^2/\hbar c)/b]^2\sim 10^{-6},$$

which checks with the previous estimate since $G^2/\hbar c$ would be of the order one. The decay lives become, very approximately,

$$\tau_a\sim 10^{-19}—10^{-20}\text{ sec},$$

$$\tau_b\sim\tau_c\sim 10^{-17}—10^{-18}\text{ sec}.$$

We can pursue further consequences of our assumption.

(1) ρ^0 could be produced by any strong nuclear reactions, but it would instantly decay mostly into a high-energy $\gamma(\gtrsim 140$ Mev) and a ρ^0. The ratio of charged to neutral components in high-energy reactions should accordingly be influenced.

(2) The second maximum of the pion-nucleon scattering around 1 Bev^2 could be attributed to the reaction

$$\pi^-+p\rightarrow n+\rho^0,$$

if a resonance should occur for such a system.

(3) ρ^0 would contribute a repulsive nuclear force of Wigner type and short range ($\lesssim 0.7\times 10^{-13}$ cm), more or less similar to the phenomenological hard core.

(4) The anomalous moment of the nucleon[3] should be affected by ρ^0. The main effect seems to be that ρ^0 and the usual pion give opposite contributions to the isotopic scalar part of the core moment, thus tending to bring better agreement between theory and experiment.

(5) If it is energetically possible, we ought to expect that K mesons and hyperons would sometimes decay by emitting a ρ^0.

It should perhaps be added that the neutral meson considered here is similar in nature to the one introduced by Teller for quite different purposes.[4]

* This work was supported by the U. S. Atomic Energy Commission.

[1] R. Hofstadter, Revs. Modern Phys. **28**, 214 (1956). Other references are given in this paper. Also E. E. Chambers and R. Hofstadter, Phys. Rev. **103**, 1456 (1956); J. A. McIntyre, Phys. Rev. **103**, 1464 (1956); R. W. McAllister, Phys. Rev. **104**, 1494 (1956); Hughes, Harvey, Goldberg, and Stafne, Phys. Rev. **90**, 497 (1953); Melkonian, Rustad, and Havens, Bull. Am. Phys. Soc. Ser. II, **1**, 62 (1956). For theoretical interpretations, see Yennie, Lévy, and Ravenhall, Revs. Modern Phys. **29**, 144 (1957).

[2] Shapiro, Leavitt, and Chen, Phys. Rev. **92**, 1073 (1953); Cool, Madansky, and Piccioni, Phys. Rev. **93**, 637 (1954); O. Piccioni and other authors, *Proceedings of the Sixth Annual Rochester Conference on High-Energy Physics 1956* (Interscience Publishers, Inc., New York, 1956). See also F. Dyson, Phys. Rev. **99**, 1037 (1955); G. Takeda, Phys. Rev. **100**, 440 (1955).

[3] H. Miyazawa, Phys. Rev. **101**, 1564 (1956); see also G. Sandri, Phys. Rev. **101**, 1616 (1956).

[4] M. H. Johnson and E. Teller, Phys. Rev. **98**, 783 (1955); H. Duerr and E. Teller, Phys. Rev. **101**, 494 (1956); E. Teller, *Proceedings of the Sixth Annual Rochester Conference on High-Energy Physics 1956* (Interscience Publishers, Inc., New York, 1956); H. Duerr, Phys. Rev. **103**, 469 (1956).

Chapter 4: Nambu and the Microscopic Theory of Superconductivity

On July 22, 1959, *Phys. Rev.* **111**, *91, (1959)* received

Quasi-Particles and Gauge Invariance in the Theory of Superconductivity, by Yoichiro Nambu.

In this paper Nambu sets up a quantum field theoretic formulation of Superconductivity to answer some open questions that arose about the BCS theory when it was published. He showed that superconductivity can be seen as a spontaneously broken phase of Quantum Electrodynamics of the electron gas with a Fermi surface, and proved that the theory is gauge invariant even though the ground-state had Cooper-pairs with charge two. Gauge invariance is non-linearly realized. This paper is also the source for his subsequent work that introduced and fully understood the relevance of spontaneous symmetry breaking in particle physics.

Background

Superconductivity is one of the most fascinating subjects in modern quantum physics. There is an enormous amount of literature on the subject and to give a full historic background would take us too far. Instead we will highlight the most important discoveries up to the Bardeen–Cooper–Schrieffer microscopic theory without going into too great details.

The history starts with the remarkable technical achievement by Heike Kamerlingh Onnes (1853–1926) in 1908 when he managed to liquify helium and to reach the temperature $4.2\,^\circ K$. On April 8, 1911, he investigated the conductivity of mercury and found[42] that it suddenly dropped to zero when this temperature was reached. This was clearly a new phenomenon, and it would take some 46 years to get a microscopic theory for it. Kamerlingh Onnes' laboratory in Leiden was the only laboratory in the world up to the 1920s with facilities to reach such low temperatures, and the rest of the world had to go there to perform extremely-low-temperature experiments. He subsequently found that also other metals could be superconducting.

Kamerlingh Onnes was awarded the Nobel Prize in physics in 1913 *"for his*

[42]Kamerlingh Onnes, *Comm. Phys. Lab.* Univ. Leiden, (1911).

investigations on the properties of matter at low temperatures which led, inter alia, to the production of liquid helium." It is noteworthy that he did not get it for the discovery of superconductivity. In a sense, it was an invention prize and as such a very well deserved one. (The Nobel Prize can be given either for a discovery or an invention.)

The next discovery[43] on superconductivity was the one by Walther Meissner (1882–1974) and Robert Ochsenfeld (1901–1993) in 1933 that a superconductor is a perfect diamagnet as well as a perfect conductor. The magnetic field vanishes in the interior of the bulk, even when it is cooled down below the transition temperature in the presence of a magnetic field. The diamagnetic currents flow in a thin penetration layer near the surface of a simply connected body to shield the interior from an externally applied field. They are stable rather than metastable.

In 1934, Cornelis Jacobus Gorter (1907–1980) and Hendrik Casimir (1909–2000) devised a two-fluid model[44] to account for the observed second order phase transition at temperature T. They proposed that the total density of electrons could be divided into two components

$$\rho = \rho_s + \rho_n.$$

The fraction ρ_s/ρ_n is then what has been condensed into the "superfluid". The fraction is really the order parameter for the second order phase transformation.

A year later, Fritz (1900–1954) and Heinz (1907–1954) London set up a phenomenological theory[45] for superconductors concentrating on the magnetic properties. They devised Maxwell-like equations for the electrical current density $\mathbf{j_s}$ carried by the superfluid which is related to the magnetic vector potential A at each point in space by

$$\mathbf{j_s} = -\frac{1}{\Lambda c}\mathbf{A},$$

where Λ is a constant depending on the material which for a free electron gas is given by $\Lambda = m/\rho_s e^2$, where m and e are the mass and the charge of an electron. From their "London equations" it follows that a magnetic field is excluded from a superconductor except within a distance

$$\lambda = \sqrt{\Lambda c^2/4\pi}$$

which is of order 10^{-6} cm in typical superconductors for T well below T_c. Even though observed values of λ are generally several times the London value, it introduced the concept of *the penetration depth.*

In the same year Fritz London suggested[46] that the diamagnetic property should follow from quantum mechanics. There should be a rigidity in the wave function

[43] W. Meissner, R. Ochsenfeld, *Naturwissenschaften* **21**, 787 (1933).
[44] C. J. Gorter, H. Casimir, *Physica* **1**, 306 (1934).
[45] F. London and H. London, *Proc. Royal Soc. A* **149**, 71 (1935).
[46] F. London, *Proc. Roy. Soc.* **152**, 24–34 (1935).

Ψ of the superconducting state such that Ψ is rigid under the influence of an externally applied magnetic field. The value $|\Psi|^2$ is really the density ratio ρ and the rigidity really means that there should be a gap in the excitation spectrum of a superconductor, which separates the energy of the superfluid electrons from the energy of the electrons in the normal electron gas. The energy gap became a very important feature of the microscopic theory of BCS.

In a book published in 1950,[47] F. London extended his theoretical conjectures by suggesting that a superconductor is a *"quantum structure on a macroscopic scale, a kind of solidification or condensation of the average momentum distribution"* of the electrons. This momentum space condensation locks the average momentum of each electron to a common value which extends over an appreciable distance in space. This became a very important ingredient of the BCS theory.

Another important theoretical discovery was the work[48] of Vitaly Ginzburg (1916–2009) and Lev Landau (1908–1968) in 1950, where they extended the wave function Ψ to be space dependent. They postulated that the free energy between the superconducting states and the normal ones at temperature T be given by

$$\Delta F = \int \left\{ \frac{\hbar^2}{2\bar{m}} | \left(\nabla + \frac{\bar{e}}{c} A(r) \right) \Psi(r)|^2 - a(T)|\Psi(r)|^2 + \frac{b(T)}{2} |\Psi(r)|^4 \right\} d^3 r,$$

where a and b are phenomenological constants, and $\bar{e}$ and $\bar{m}$ are phenomenological values for the effective charge and mass.

They applied this approach to the calculation of boundary energies between normal and superconducting phases and to many other problems. The formalism turned out to be very useful also for other types of superconductors and it is still used in modern times.

Ginzburg was awarded the Nobel Prize in 2003 *"for pioneering contributions to the theory of superconductors and superfluids."*

It was shown in experiments around 1950 that there is an isotope effect in superconductivity. The mass of the constituents of the lattice is a parameter of the theory. This made it clear that the lattice played a vital role in the interactions between the electrons. Hence the typical interactions responsible for superconductivity should be the electron–phonon interaction. Herbert Fröhlich (1905–1991) set up a theory[49] based on such an interaction that contained the isotope effect but it failed to predict the superconducting properties.

A further important concept was introduced by Brian Pippard[50] (1920–2008) in 1953; the *coherence length* ξ associated with the superconducting state such that a perturbation of the superconductor at a point necessarily influences the superfluid within a distance of that point. For pure metals $\xi \approx 10^{-4}$ cm for $T \ll T_c$.

[47]F. London, *Superfluids* Volume I, Wiley and Sons, 1950; second revised edition 1961, Dover Publications, Inc, New York.

[48]L. D. Landau, V. L. Ginzburg, *Sov. Phys. JETP* **20**, 1064 (1950).

[49]H. Frölich, *Phys. Rev.* **79**, 845 (1950).

[50]A. B. Pippard, *Proc. Roy. Soc. A* **216**, 547 (1953).

It was clear that a microscopic theory based on the energy levels of the electron was needed to get a full understanding of the problem. One major problem was that the condensation energy ΔF was typically 10^{-6} eV while the uncertainty in calculating the total energy of the electron–phonon system in even the normal state amounted to order 1 eV per electron. That meant that one had to isolate the correlations typical for the superconducting phase and treat them in detail and then assume that the remaining larger effects should be the same in the two phases and hence cancel. Comparing with Landau's theory for superfluidity, one could assume that the interactions between the electrons do not lead to discontinuous changes in the microscopic properties of the system. One can describe the two phases with the same degrees of freedom and the difference should be that in the superconducting phase some electrons form *"quasi-particles"*. The effective mass m and the Fermi velocity v_F of the quasi-particles differ from their free electron values, but aside from a weak decay rate which vanishes for states at the Fermi surface there is no essential change. It is the residual interaction between the quasi-particles which is responsible for the special correlations characterizing superconductivity. The ground state wavefunction of the superconductor Ψ is then represented by a particular superposition of these normal state configurations.

The BCS Theory

In 1955, John Bardeen (1908–1991) decided to put his efforts towards a theory for superconductivity. He invited Leon Cooper (1930–), a postdoc at the Institute for Advanced Study to join him together with his graduate student John Robert Schrieffer (1931–2019). Bardeen was one of the most respected theorists and was in the pipeline for a Nobel Prize for his share in the discovery of the transistor effect, a Prize he was awarded the year later.

Their combined effort produced the epoch-making explanation of superconductivity.[51]

They started by concentrating on the ground state Ψ_0. It should be formed as a coherent superposition of normal state configurations Φ_n

$$\Psi_0 = \sum a_n \Phi_n.$$

The normal states are the Bloch–Sommerfeld wave functions for an electron moving freely in a lattice and hence satisfying periodic boundary conditions. The only constraint is that no two states can be in the same state due to the Pauli principle. The normal ground state for a metal is where all momenta up to the Fermi surface are filled.

The energy for the ground state Ψ_0 is now

$$E_0 = (\Psi_0, H\Psi_0) = \sum_{n,n'} a^*_{n'} a_n (\Phi_{n'}, H\Phi_n).$$

[51] J. Bardeen, L. N. Cooper, J. R. Schrieffer, *Phys. Rev.* **106**, 1175 (1957).

The idea was then to check if one could restrict the coefficients so that only states with negative off-diagonal matrix elements would enter the formula. Then the energy value would be low.

When studying this problem, Cooper had found earlier that frequently a single eigenvalue was split off from the bottom of the spectrum. He studied some model cases where two electrons interacted via an attractive potential above the Fermi surface. In this way, he found that the energy eigenvalue spectrum for two electrons having zero total momentum form a bound state ("Cooper pairs") split off from the continuum of scattering states, which would lower the total energy.[52] Furthermore, he could argue that the average separation between electrons in the bound state is of order 10^{-4} cm. One problem that arose was that within the volume occupied by the bound state the centers of approximately 10^6 other pairs would be found, which showed that this is not a dilute gas of strongly bound pairs. The pairs overlap so strongly in space that the mechanism of condensation would appear to be destroyed due to the numerous pair–pair collisions interrupting the binding process of a given pair.

The problem with the argument about the matrix elements above is that summing over the enormous number of states one cannot ensure that all elements be negative. However, if one could restrict the sum to only include quasi-particles where the two electrons have the opposite momentum and spin then the matrix elements of H between such states would have a unique sign and a coherent lowering of the energy would be obtained. If these quasi-particle states are given a momentum k then there would be a net current flow. The idea here is also that all the other electrons in the electron gas behave in the same way for the normal state as for the superconducting one.

By introducing the creation and annihilation operators as $b_k^\dagger = c_{k\uparrow}^\dagger c_{-k\downarrow}^\dagger$ and $b_k = c_{k\uparrow} c_{-k\downarrow}$, where $c^\dagger$ and c are the fermionic creation and annihilation operators for electrons, one could attempt a Hamiltonian of the form

$$H_{\text{red}} = \sum_k \epsilon_k - \sum_{k\,k'} V_{k\,k'} b_{k'}^\dagger b_k.$$

The first term in this equation gives the unperturbed energy of the quasi-particles forming the pairs, while the second term is the pairing interaction. Note that the operator b is not bosonic even if it is commuting rather than anticommuting.

It is clear that for the ground state wavefunction, the average occupancy of a pair state should be one for momenta far below the Fermi surface and 0 far above. The fall off should be symmetric about the Fermi surface over a range of momenta related to the coherence length

$$\Delta k \sim \frac{1}{\xi} \sim 10^4\, cm^{-1}.$$

Hence a trial ground state wave function should look like

$$\prod_k (u_k + v_k b_k^\dagger)|0>,$$

where $u^2 + v^2 = 1$.

[52]Leon N. Cooper, *Phys. Rev.* **104**, 1189 (1956).

The leap here is bigger than meets the eye. It says that the wave function is a mixture of states with different number of electrons. What happens to gauge invariance? We will come back to this issue at length later.

By minimizing the energy for keeping the number of states fixed leads indeed to a gap, showing that with a certain electron–phonon interaction a gap can form and a superconducting current can be established.

The next step was to study the electron–phonon interactions. This was a field that had been well established during the first half of the 1950s. One could form a Hamiltonian for which one could use modern quantum field theory, use Feynman diagram techniques and renormalize. One can then compute the matrix element for phonon–electron interaction

$$M_k \sim (\hbar/2\omega_k).$$

Then the criterion for superconductivity turns out to be that the attractive phonon interaction dominates the Coulomb interaction for those matrix elements which are of importance in the superconducting wave function:

$$-V = \left\langle -\frac{|M_k|^2}{\hbar\omega_k} + \frac{4\pi e^2}{c^2}\right\rangle_{Av} < 0.$$

Then the important transition for superconductivity is such that $\Delta\epsilon < \hbar\omega_k$. The attractive Hamiltonian H from the phonon interaction now dominates the short-range screened Coulomb force between electrons so as to give a net attractive interaction. It is also consistent with the estimates above that only electrons in a narrow band around the Fermi surface contribute to the Cooper pairs.

This clearly constituted a solution to the microscopic problem of superconductivity. By further studying the theory, they could also establish the Meissner effect and make reliable computations of various parameters. Finally a microscopic theory for superconductivity was established.

Reactions

The physics community was in awe. It was clear to most people that this was a landmark result in modern quantum physics. While definitely of Nobel quality, one problem was that Bardeen had been awarded one Prize the year before. It was only in 1972 that the three discoverers were very deservedly awarded the Prize. Another problem for the Nobel committee was that Albert Einstein had "only" got one Prize!

Many physicists now joined the bandwagon and new results emerged swiftly. An important progress was the introduction of the quasi-particles of Nikolay Bogoliubov[53] (1904–1992) and J. G. Valatin[54] (1918–1978). It led to a formulation which is generally applicable to a wide range of calculations in a manner analogous

[53] N. N. Bogolyubov, *Il Nuovo Cimento* **7**, 794 (1958).

[54] J. G. Valatin, *Il Nuovo Cimento* **7**, 843 (1957).

to similar calculations in the theory of normal metals. The idea is to introduce new creation and annihilation operators with a linear transformation from fermion operators c and $c^\dagger$ as

$$\gamma^\dagger_{\mathbf{k}0} = u_{\mathbf{k}} c^\dagger_{\mathbf{K}} - v^*_{\mathbf{k}} c_{-\mathbf{K}},$$

$$\gamma^\dagger_{\mathbf{k}1} = v^*_{\mathbf{k}} c_{\mathbf{K}} + u_{\mathbf{k}} c^\dagger_{-\mathbf{K}}.$$

They satisfy canonical anticommutation relations if

$$|u_k|^2 + |v_k|^2 = 1.$$

It is now easy to see that the ground state wave function satisfies

$$\gamma_{\mathbf{k}i}\Psi_0 = 0, \quad i = 0, 1.$$

The creation operators now $\gamma^\dagger_{\mathbf{k}i}$ create excitations

$$\gamma^\dagger_{\mathbf{k}i}\gamma^\dagger_{\mathbf{l}j}\cdots\cdots\gamma^\dagger_{\mathbf{m}k}\Psi_0 = \Psi_{\mathbf{k}i,\mathbf{l}j,\ldots\ldots\mathbf{m}k},$$

and we obtain a complete set of excitations in one-to-one correspondence with the excitations of the normal metal.

However it is obvious that these relations are not gauge covariant. It was argued very strongly by Michael J. Buckingham[55] that the Meissner effect was not gauge invariant. Also M. R. Shafroth[56] was very articulate in his criticism that the BCS theory was not gauge invariant and hence its results could not be trusted. Unfortunately, he died with his wife in an airplane accident in Australia in 1959 and could not see for himself the resolution to this problem.

A very important contribution[57] came from P. W. Anderson (1923–2020) who studied the problem in great detail shortly after the BCS paper. The key was to compare superconductors with insulators and normal metals. He started by saying *"Thus, while the BCS calculation in the London gauge is probably entirely correct, and justifiable on physical grounds, it throws little light on the basic differences between the three cases — insulator, metal, and superconductor."* For example, an insulator also has a gap in the energy but shows no Meissner effect.

Bardeen had argued that the matrix elements and energy states involved in the gauge problem will bring in coherent excitations which will be strongly coupled to the plasma modes, a high-frequency phenomenon which presumably is unaltered by superconductivity. Anderson studied carefully the coherent excitations and demonstrated that for normal metals and insulators the normal excitations are the important ones which leads, for example, to diamagnetism while for superconductors there is a great difference between the longitudinal and the transverse excitations. The longitudinal ones are the ones responsible for superconductivity while the transverse ones are responsible for the Meissner effect. He then studied non-gauge invariant

[55] M. J. Buckingham, *Nuovo Cimento* **5**, 1765 (1957).
[56] M. R. Shafroth, *Phys. Rev.* **111**, 72 (1958).
[57] P. W. Anderson, *Phys. Rev.* **110**, 827 (1958).

terms and found them to be negligible in the weak coupling limit. His conclusion was that the theory is essentially gauge invariant and for practical purposes it is not a real problem. For him, this is evidence that the BSC theory is essentially correct and can be used to calculate various properties. As we shall see, from a theoretical standpoint this is quite different from the analysis of Nambu which came a year or so later.

Nambu's paper

In a short autobiographical note published in the collective volume *Broken Symmetry. Selected Papers of Y. Nambu*, edited by T. Eguchi and K. Nishijima

Nambu recalled:

"One day before publication of the BCS paper, Bob Schrieffer, still a student, came to Chicago to give a seminar on the BCS theory in progress. Gregor Wentzel, our senior professor, looked rather skeptical. I was also very much disturbed by the fact that their wave function did not conserve electron number. It did not make sense, I thought, to discuss electromagnetic properties of superconductors with such an approximation. At the same time I was impressed by their boldness, and tried to understand the problem. I ended up being captive to the BCS theory. In the meantime, Schrieffer took an assistant professor job in Chicago, so we were in close touch. But my major concern was in the purely theoretical issues like gauge invariance, and it took me two years before I was able to convince myself of its resolution and write a paper. While in the meantime, experts like N. N. Bogoliubov and P. W. Anderson were refining the BCS theory, I deliberately tried to keep my independence."

Nambu's paper contains the discovery of Spontaneous Symmetry Breaking (SSB), a name later coined by Marshall Baker (1932–) and Sheldon Glashow (1932–). This is one of the most fundamental theoretical discoveries of the last century, according to us. It allows us to eat the cake and still have it. We can allow for a variety of ground states which are not gauge invariant but the theory is still gauge invariant. As in many of Nambu's papers, he is not overstating his case. The paper is quite long with very extensive calculations. The introduction of SSB comes without hesitation and the paper is apart from calculating the usual characteristics of Superconductivity, such as the energy gap and the Meissner effect, a proof that gauge invariance is still satisfied. In more modern terms, we would say that he proves that it is nonlinearly realized. In the end, we will also discuss whether he also discovered the Brout–Englert–Higgs mechanism.

We will in the sequel give the highlights of the different sections of the paper.

Introduction

In this section, Nambu summarized the main results of the BCS paper and discussed the development after what has been highlighted above.

He started by setting up a model calculation with a Hamiltonian in the second quantization form, which is essentially a form where the phonons have been integrated out. There is still no electromagnetic interaction but the Hamiltonian is gauge invariant which here amounts to charge conservation.

$$\begin{aligned} H &= \int \sum_{i=1}^{2} \psi_i^\dagger(x) K_i \psi_i(x) d^3x + \frac{1}{2} \int\int\int \sum_{i,k} \psi_i^\dagger(x)\psi_k^\dagger(y) V(x,y) \psi_k(y)\psi_i(x) d^3x d^3y \\ &= H_0 + H_{\text{int}}. \end{aligned}$$

K is the kinetic term and we sum over the two spin states.

Nambu now considered a Hartree–Fock formulation in which one linearizes the interaction by replacing the bilinear products like $\psi_i^\dagger(x)\psi_k(y)$ with its expectation value with respect to a ground state which is determined by the interaction. However he then also considered expectation values for the bilinear $\psi_i(x)\psi_k(y)$ and $\psi_i^\dagger(x)\psi_k^\dagger(y)$. They would be zero if the ground state is an eigenstate of the number operator, but we know that we should have wave functions representing Cooper pairs. In modern language, he introduced vacuum expectation values for these operators and shifted the bilinears with these expectation values introducing new fields which are still called ψ. He was just keeping the linear terms. The linearized Hamiltonian is then

$$\begin{aligned} H_0' = &\int \sum_{I=1}^{2} \psi_i^\dagger(x) K_i \psi_i(x) + \int\int \sum_{i,k} \Big[\psi_i^\dagger(x) \chi_{ik} \psi_k(y) \\ &+ \psi_i^\dagger(x)\phi_{ik}\psi_k^\dagger(y) + \psi_i(x)\phi_{ik}^\dagger \psi_k(y) \Big] d^3x d^3y \\ = &\ H_0 + H_s, \end{aligned}$$

where

$$\begin{aligned} \chi_{ik} &= \delta_{ik}\delta^3(x-y) \int V(x,z) \sum_j \langle \psi_j^\dagger(z)\psi_j(z)\rangle d^3z - V(x,y)\langle \psi_j^\dagger(y)\psi_j(x)\rangle, \\ \phi_{ik}(x,y) &= \frac{1}{2} V(x,y) \langle \psi_k(y)\psi_i(x)\rangle, \\ \phi_{ik}^\dagger(x,y) &= \frac{1}{2} V(x,y) \langle \psi_k^\dagger(y)\psi_i^\dagger(x)\rangle. \end{aligned}$$

Nambu now interpreted this as some generalized form of Hartree–Fock equations but we can see it as a spontaneously broken form of the theory. The Hamiltonian can be written as

$$H = (H_0 + H_s) + (H_{\text{int}} - H_s) = H_0' + H_{\text{int}}'.$$

H_0' should now be diagonalized and H_{int}' shall have no matrix elements which would cause single-particle transitions.

The significance of the BCS theory is the recognition that with an essential attractive interaction V, a non-vanishing ϕ is indeed a possible solution, and the corresponding ground state has a lower energy than the normal ground state. It is also separated from the excited states by an energy gap $\sim 2\phi$.

He then discussed gauge invariance and argued that the theory should be gauge invariant. In a modern framework we could think of a path integral formulation of the problem. The starting point is certainly gauge invariant and a change of variables should not destroy it. He later checked that in great detail.

Feynman–Dyson Formulation

In this section, he started with the Lagrangian for the electron–phonon system but without the Coulomb interaction.

$$\begin{aligned}\mathcal{L} = &\sum_p \sum_i [i\psi_i^\dagger(p)\dot{\psi}_i(p) - \psi_i^\dagger(p)\epsilon_p\psi_i(p)] \\ &+\sum_k \frac{1}{2}[\dot{\phi}(k)\dot{\phi}(-k) - c^2\phi(k)\phi(-k)] \\ &-g\frac{1}{\sqrt{\mathcal{V}}}\sum_{p,k}\psi_i^\dagger(p+k)\psi_i(p)h(k)\phi(k),\end{aligned}$$

where ϕ is the phonon field with momentum k running up to a cut-off value k_m. (We are on a lattice.) ϵ_p is the electron kinetic energy relative to the Fermi surface energy (we are in a non-relativistic situation so we can fix the zero energy at will). The phonon velocity is c and $gh(k)$ is the coupling strength. $\mathcal{V}$ finally is the volume of the system.

Nambu now introduced a very useful two-component notation for the electrons.

$$\Psi(x) = \begin{pmatrix}\psi_1(x)\\ \psi_2^\dagger(x)\end{pmatrix} \quad \text{or} \quad \begin{pmatrix}\psi_1(p)\\ \psi_2^\dagger(-p)\end{pmatrix}$$

and the corresponding 2×2 Pauli matrices

$$\tau_1 = \begin{pmatrix}0&1\\1&0\end{pmatrix}, \qquad \tau_2 = \begin{pmatrix}0&-i\\i&0\end{pmatrix}, \qquad \tau_3 = \begin{pmatrix}1&0\\0&-1\end{pmatrix}.$$

The Lagrangian then becomes

$$\begin{aligned}\mathcal{L} = &\sum_p \Psi^\dagger(p)(i\frac{\partial}{\partial t} - \epsilon_p\tau_3)\Psi(p) \\ &+\sum_k \frac{1}{2}[\dot{\phi}(k)\dot{\phi}(-k) - c^2\phi(k)\phi(-k)] \\ &-g\frac{1}{\sqrt{\mathcal{V}}}\sum_{p,k}\Psi^\dagger(p+k)\Psi(p)h(k)\phi(k) + \sum_p \epsilon_p \\ = &\ \mathcal{L}_0 + \mathcal{L}_{\text{int}} + \text{constant},\end{aligned}$$

where the constant is infinite and a result of a rearrangement of the kinetic term. It will not affect the calculations except for the calculation of the total energy.

With this Lagrangian, one can set up a second quantized form and introduce canonical commutation rules for the fields and since we are interested in the energies of the system that Nambu first concentrated on, the propagators follow from the Lagrangian. With these and the three-point interaction he could now set up Feynman rules and compute the S-matrix in a perturbation series.

However, the idea was to find a different vacuum and to do so, he rewrote as in the previous section, the Lagrangian as

$$\mathcal{L} = (\mathcal{L}_0 + \mathcal{L}_s) + (\mathcal{L}_{int} - \mathcal{L}_s) = \mathcal{L}_0' + \mathcal{L}_{\text{int}}'$$

with the idea that $\mathcal{L}_0'$ is the kinetic term for new electron fields diagonal in Cooper pairs. We write

$$\mathcal{L}_0 = \sum_p \Psi_p^\dagger L_0 \Psi_p + \frac{1}{2}\sum_k \phi_k M_0 \phi_{-k}.$$

$$\mathcal{L}_s = -\sum_p \Psi_p^\dagger \Sigma \Psi_p - \frac{1}{2}\sum_k \phi_k \Pi \phi_{-k}.$$

$$L_0 - \Sigma = L \qquad M_0 - \Pi = M.$$

The idea now is to solve for Σ and Π such that L is diagonal. This can be done in a perturbative calculation. It is done by adding the second order self-energy and the first order term coming from the interaction and ensure that they cancel.

The pair creation of electrons is possible because Ψ is a two-component wave function which can have two eigenfunction with different energies for a fixed momentum. Only one of them is occupied in the ground state.

The calculation is performed by summing up an infinite series of Feynman diagrams. There will be corrections left out though in each order, which are analogous to the radiative corrections after mass renormalization in QED.

In the perturbative expansion, the Hamiltonian will still be diagonal and proportional till τ_1. There is a possibility though of a non-perturbative solution by assuming that Σ contains also a term proportional to τ_1 or τ_2. So if Σ is of the form

$$\Sigma(p) = \chi(p)\tau_3 + \phi(p)\tau_1,$$

then the form of the free Hamiltonian is

$$H_0' = (\epsilon + \chi)\tau_3 + \phi\tau_1.$$

This form bears a resemblance to the Dirac equation. Its eigenvalues are

$$E = \pm(\epsilon_p^2 + \phi_p^2)^{\frac{1}{2}}.$$

Since H_0' describes excited states by definition, we have to adopt the whole picture and conclude that the vacuum state is the state where all negative quasi-particles are occupied and no positive energy states exist. If ϕ remains finite on the Fermi surface the positive and the negative states are separated by a gap $\sim 2|\phi|$.

Nambu so set up equations for χ and ϕ and found them equivalent to the corresponding conditions of Bogoliubov and deduced a solution

$$\phi \sim \exp(-1/VN),$$

where N is the density of states on the Fermi surface.

Nambu then argued that the phonon self-energy may be studied in a similar fashion. This will determine the renormalization of the phonon field. An important correction will be given when the Coulomb interaction is also included.

It should also be noted that the choice of τ_1 instead of τ_2 which would lead to the same result as above is due to the gauge invariance where a phase change leads to the same physics.

Nonlocal (Energy-Dependent) Self-Consistency Condition

In this section, Nambu extended the self-consistency conditions to all virtual matrix elements that are not only the diagonal ones in the energy shell but also for the self-energy effects which appear in intermediate states of any process.

This means that ϕ and π will depend on energy and momentum arbitrarily. Then the self-energies can no more be incorporated in H_0' since they contain infinite orders of time derivatives. The self-consistency conditions gets a meaning in the bare particle perturbation theory as an infinite sum of a special class of diagrams. The introduction of the vacuum expectation values can be interpreted as a non-perturbative approximation to determine the dressed single particles or as the two-point function

$$\langle 0|T(\Psi(x,t)\,\Psi(x',t'))|0\rangle$$

for the true interacting system.

He then assumed that the self-consistency equations determine the approximate renormalized dispersion relations as

$$p_0^2 = E_r(p)^2, \qquad k_0^2 = \Omega_r(k)^2,$$

then he could write solutions to the full Green's function and find an expression for the energy gap equation

$$\phi(\mathbf{p},p_0) = -\frac{ig^2}{(2\pi)^4}\int \frac{\phi(\mathbf{p}',p_0')}{p_0^2 Z(\mathbf{p}'p_0')^2 - E(\mathbf{p}'p_0')^2 + i\epsilon}$$
$$\times\, h(\mathbf{p}-\mathbf{p}',p_0-p_0')^2\Delta(\mathbf{p}-\mathbf{p}',p_0-p_0')d^3p\,d^3p'.$$

We refer to the paper for further explanations for the various functions involved.

This is a fuller version of the gap equation than that in the last section, shown to give the correct volume dependence.

It is now clear that in order to really solve this equation there are a number of integrals that have to be understood. Nambu used the rest of that section to discuss these integrals.

Integral Equations for Vertex Parts

In this section, Nambu introduced the electromagnetic potential via a minimal coupling.

$$i\frac{\partial}{\partial t} \to i\frac{\partial}{\partial t} + eA_0, \qquad \mathbf{p} \to \mathbf{p} - \frac{e}{c}\,p$$

for the electron. In the two-component notation, this amounts to

$$i\frac{\partial}{\partial t} \to i\frac{\partial}{\partial t} + e\tau_3 A_0, \qquad \mathbf{p} \to \mathbf{p} - \frac{e}{c}\tau_3\,p$$

acting on Ψ. The charge and current operators now become

$$\rho(x) = \frac{e}{c}([\Psi^\dagger, \tau_3\Psi(x)] + \{\Psi^\dagger(x), \Psi(x)\}).$$

$$\begin{aligned}\mathbf{j}(x) = \frac{-ie}{4m}&([\Psi^\dagger(x), \nabla - ie\tau_3\mathbf{A})\Psi(x)] \\ &+[-(\nabla - ie\tau_3\mathbf{A})\Psi^\dagger(x), \Psi(x)] \\ &+\{\Psi^\dagger(x), (\nabla\tau_3 - ie\mathbf{A})\Psi(x)\} - \{(\tau_3\nabla - ie\mathbf{A}\Psi^\dagger(x), \Psi(x)\}).\end{aligned}$$

This expression has to be modified though when we go to the quasi-particle picture. The point is that this picture is gauge-dependent. If we want to have gauge invariance, the current must contain also terms which would cause a physically unobservable transformation, if the electromagnetic potential is replaced by the gradient of a scalar. In a perturbative scheme, one can group the diagrams into gauge invariant subsets and then let each diagram interact with a charge in all possible places along a chain of charge-carrying particle lines. The gauge invariant interaction of a quasi-particle with an electromagnetic potential should then be obtained by attaching a photon line at all possible places in the diagrams used in the previous sections.

In this way, Nambu was led to consider the modification of the vertex due to the phonon interaction in the same approximation as the self-energy is included in the quasi-particle. In fact, this corresponds to a ladder approximation for the vertex part and to an integral equation

$$\Gamma_i(p', p) = \gamma_i(p', p) - g^2 \int \tau_3 G(p'-k)\Gamma_i(p'-k, p-k)G(p-k)\tau_3 h(k)^2 \Delta(k)\, d^3k,$$

where γ_i, $I = 0, 1, 2, 3$ stand for the free particle current that follows from the expression above. This is the procedure in which a free charge-current operator γ_i of a quasi-particle is modified by a special class of radiative corrections due to H'_{int}.

He then found that there exist exact solutions to four types of vertex interactions. In this way, he showed the existence of spin and charge currents for a quasi-particle for which the continuity equations hold.

He then reflected on how surprising this is. A quasi-particle is not an eigenstate of charge. The solution must then be that an accelerated wave packet of quasi-particles, whose energy is confined to a finite region of space, continuously picks up charge from or deposits it in the surrounding medium which extends to infinity.

Gauge Invariance in the Meissner Effect

In order to understand the Meissner effect one has to calculate how the current $J(q)$ depends on the vector potential $A(q)$, where we look at the Fourier transforms.

$$J_i(q) = \sum_{j=1}^{3} K_{ij}(q)A_j(q),$$

where q is kept finite.

For free electrons, K is represented by

$$K_{ij} = K_{ij}^{(1)} + K_{ij}^{(2)}, \qquad K_{ij}^{(1)} = -\delta_{ij}ne^2/m,$$

where n is the number of electrons inside the Fermi surface. This follows from the expectation value of the current operator in the previous section. $K_{ij}^{(2)}$ is a higher order effect that can be computed from a certain loop diagram (see the paper). In the superconducting state the free electron lines will be replaced by quasi-particle lines. Because of the gap $K_{ij}^{(2)}(q)$ will tend to zero when the momentum tends to zero. The idea now is to again take care of all vertex corrections and sum up all the diagrams. Nambu was finally led to an expression

$$K_{ij}^{(2)}q_j = \frac{1}{(2\pi)^4}\frac{q_i}{m}\int Tr[\tau_3 G(p)]d^3p.$$

$K_{ij}^{(1)}$ is obtained from the expectation value of the current operator with the quasi-particles. Part of it will cancel the contribution from $K_{ij}^{(2)}$ and the remaining term can be seen to be gauge invariant. He then showed that it is really the collective excitations as described in the next section that restores the gauge invariance which P. W. Anderson had argued.

In this context, Nambu had a footnote that is not typical for him but says a lot of what he thought about some of the papers in the field. He writes "On the other hand, the way in which the collective mode accomplished this end seems to differ from one paper to another. We will attempt to analyze this situation here."

The Collective Excitation

As Nambu argued in the previous section, the collective excitations are important for the understanding of the gauge invariance of the Meissner effect. It was well known as we have said before that the essential difference between the transverse and the longitudinal vector potential in inducing a current is due to the fact that the latter can excite collective motions of quasi-particle pairs.

He started by showing that the collective excitations follow naturally from the vertex solutions. From those he could set up a homogeneous integral equation which can be interpreted as describing a pair of a particle and an antiparticle interacting with each other to form a bound state with zero momentum and energy. Since there exists a bound pair of zero momentum, there will also be pairs moving with a finite momentum giving a continuum of pair states with energies going up from zero. The rest of the section deals with the determination of their dispersion law. This is done by studying the integral equation and solving it for the relevant cases.

Calculation of the Charge-Current Vertex Functions

In this section, he continued to determine the four charge-current vertex function he had studied before where he solved for two of them. Here, he has solved the remaining two.

In a normal case one could study these vertex functions in a perturbation theory. However, there are low-lying collective excitations which do not follow from perturbation theory. Fortunately, he could solve the defining equation for momentum $q = 0$. By understanding this case fully, he could then write down the solution for the case $q \neq 0$.

He could then make the interesting observation that the solutions had some very interesting structure. The non-collective part is essentially the same as the charge current for a free quasi-particle except for the some renormalizations, whereas the collective part is spread out both in space and time.

The Plasma Oscillations

In this section, Nambu investigated the plasma oscillations. By including annihilation–creation processes in the equations for the vertex functions, one can interpret that the vertex parts get multiplied by a string of closed loops. This represents the polarization of the surrounding medium. He now solved again the integral equations. He realized that the new equations look like the ones previously solved with some substitutions.

To obtain the collective excitations, he set up another integral equation and by solving for that, he could relate it to the solution for the plasma oscillations and get a new dispersion relation law for the collective excitations.

These results are also interesting for the future development of spontaneously broken theories. Without the Coulomb interaction the dispersion is for a state with energy and momentum zero, a "massless" phonon, a *Nambu–Goldstone boson* in modern terms. When he took the Coulomb field into account, these were transformed into "massive" plasmons. In a gauge theory there are no massless Nambu–Goldstone bosons.

The section is quite brief and the calculations are quite intricate. Nambu's interest was mainly to understand the plasma oscillations. He really solved the issue and explained the difference when the Coulomb force is present or not. Like in all his papers he just described the problems to solve and how to solve them and gave the results. The reader has to interpret the result. This issue came up a few years later in the discussion whether a gauge theory could lead to short range forces and the avoidance of Nambu–Goldstone bosons in a spontaneously broken gauge theory. In an important paper[58] in 1962, P. W. Anderson studied a charged plasma by considering on the one hand, the current-potential relation from the equations of motion, and considering on the other hand the dielectric response of the media to an external field. By introducing a test particle in the plasma with its own potential, he could describe the current in two ways and compute the total field. He found that it propagates as a field for a particle with a mass related to the plasma frequency. It was indeed an example of a theory with a spontaneously broken symmetry and with no massless scalar particle or mode, although in a non-relativistic setting, this is what Nambu had already said in his paper. This was a prototype for the Brout–Englert–Higgs mechanism in spontaneously broken gauge field theories (Nobel Prize 2013). Was Nambu aware of this in his paper? We believe that he understood it in this section, but as the gentleman he was, he never advocated that he had solved the problem.

Epilog

The paper by Nambu showed that gauge invariance was indeed respected in the BCS theory of superconductivity, not approximately or in a weak limit but indeed exactly. This was, of course, a big triumph for Nambu, but also for the Bardeen, Cooper and Schrieffer theory. It was not very important for the future development in superconductivity, since the BCS theory with the input from many leading scientists was the theory to be used for the ever expanding field of superconductivity.

The biggest importance of the paper was Nambu's insight into the usefulness of spontaneously broken theories. This fact became his guiding star for the coming years. He realized that it opened up new avenues not only in condensed matter physics, but also in particle physics. As such, we think that it was one of the most important realizations in the last century.

[58] P. W. Anderson, *Phys. Rev.* **130**, 439 (1963).

PHYSICAL REVIEW VOLUME 117, NUMBER 3 FEBRUARY 1, 1960

Quasi-Particles and Gauge Invariance in the Theory of Superconductivity*

YOICHIRO NAMBU
The Enrico Fermi Institute for Nuclear Studies and the Department of Physics, The University of Chicago, Chicago, Illinois
(Received July 23, 1959)

Ideas and techniques known in quantum electrodynamics have been applied to the Bardeen-Cooper-Schrieffer theory of superconductivity. In an approximation which corresponds to a generalization of the Hartree-Fock fields, one can write down an integral equation defining the self-energy of an electron in an electron gas with phonon and Coulomb interaction. The form of the equation implies the existence of a particular solution which does not follow from perturbation theory, and which leads to the energy gap equation and the quasi-particle picture analogous to Bogoliubov's.

The gauge invariance, to the first order in the external electromagnetic field, can be maintained in the quasi-particle picture by taking into account a certain class of corrections to the charge-current operator due to the phonon and Coulomb interaction. In fact, generalized forms of the Ward identity are obtained between certain vertex parts and the self-energy. The Meissner effect calculation is thus rendered strictly gauge invariant, but essentially keeping the BCS result unaltered for transverse fields.

It is shown also that the integral equation for vertex parts allows homogeneous solutions which describe collective excitations of quasi-particle pairs, and the nature and effects of such collective states are discussed.

1. INTRODUCTION

A NUMBER of papers have appeared on various aspects of the Bardeen-Cooper-Schrieffer[1] theory of superconductivity. On the whole, the BCS theory, which leads to the existence of an energy gap, presents us with a remarkably good understanding of the general features of superconducivity. A mathematical formulation based on the BCS theory has been developed in a very elegant way by Bogoliubov,[2] who introduced coherent mixtures of particles and holes to describe a superconductor. Such "quasi-particles" are not eigenstates of charge and particle number, and reveal a very bold departure, inherent in the BCS theory, from the conventional approach to many-fermion problems. This, however, creates at the same time certain theoretical difficulties which are matters of principle. Thus the derivation of the Meissner effect in the original BCS theory is not gauge-invariant, as is obvious from the viewpoint of the quasi-particle picture, and poses a serious problem as to the correctness of the results obtained in such a theory.

This question of gauge invariance has been taken up by many people.[3] In the Meissner effect one deals with a linear relation between the Fourier components of the external vector potential A and the induced current J, which is given by the expression

$$J_i(q)=\sum_{j=1}^{3} K_{ij}(q)A_j(q),$$

with

$$K_{ij}(q)=-\frac{e^2}{m}\langle 0|\rho|0\rangle\delta_{ij}+\sum_n\left(\frac{\langle 0|j_i(q)|n\rangle\langle n|j_j(-q)|0\rangle}{E_n}+\frac{\langle 0|j_j(-q)|n\rangle\langle n|j_i(q)|0\rangle}{E_n}\right). \quad (1.1)$$

ρ and j are the charge-current density, and $|0\rangle$ refers to the superconductive ground state. In the BCS model, the second term vanishes in the limit $q\to 0$, leaving the first term alone to give a nongauge invariant result. It has been pointed out, however, that there is a significant difference between the transversal and longitudinal current operators in their matrix elements. Namely, there exist collective excited states of quasi-particle pairs, as was first derived by Bogoliubov,[2] which can be excited only by the longitudinal current.

As a result, the second term does not vanish for a longitudinal current, but cancels the first term (the longitudinal sum rule) to produce no physical effect; whereas for a transversal field, the original result will remain essentially correct.

If such collective states are essential to the gauge-invariant character of the theory, then one might argue that the former is a necessary consequence of the latter. But this point has not been clear so far.

Another way to understand the BCS theory and its problems is to recognize it as a generalized Hartree-Fock approximation.[4] We will develop this point a little further here since it is the starting point of what follows later as the main part of the paper.

* This work was supported by the U. S. Atomic Energy Commission.

[1] Bardeen, Cooper, and Schrieffer, Phys. Rev. **106**, 162 (1957); **108**, 1175 (1957).

[2] N. N. Bogoliubov, J. Exptl. Theoret. Phys. U.S.S.R. **34**, 58, 73 (1958) [translation: Soviet Phys. **34**, 41, 51 (1958)]; Bogoliubov, Tolmachev, and Shirkov, *A New Method in the Theory of Superconductivity* (Academy of Sciences of U.S.S.R., Moscow, 1958). See also J. G. Valatin, Nuovo cimento **7**, 843 (1958).

[3] M. J. Buckingam, Nuovo cimento **5**, 1763 (1957). J. Bardeen, Nuovo cimento **5**, 1765 (1957). M. R. Schafroth, Phys. Rev. **111**, 72 (1958). P. W. Anderson, Phys. Rev. **110**, 827 (1958); **112**, 1900 (1958). G. Rickayzen, Phys. Rev. **111**, 817 (1958); Phys. Rev. Letters **2**, 91 (1959). D. Pines and R. Schrieffer, Nuovo cimento **10**, 496 (1958); Phys. Rev. Letters **2**, 407 (1958). G. Wentzel, Phys. Rev. **111**, 1488 (1958); Phys. Rev. Letters **2**, 33 (1959). J. M. Blatt and T. Matsubara, Progr. Theoret. Phys. (Kyoto) **20**, 781 (1958). Blatt, Matsubara, and May, Progr. Theoret. Phys. (Kyoto) **21**, 745 (1959). K. Yosida, *ibid.* **731**.

[4] Recently N. N. Bogoliubov, Uspekhi Fiz. Nauk **67**, 549 (1959) [translation: Soviet Phys.—Uspekhi **67**, 236 (1959)], has also reformulated his theory as a Hartree-Fock approximation, and discussed the gauge invariance collective excitations from this viewpoint. The author is indebted to Prof. Bogoliubov for sending him a preprint.

Take the Hamiltonian in the second quantization form for electrons interacting through a potential V:

$$H=\int \sum_{i=1}^{2} \psi_i^+(x)K_i\psi_i(x)d^3x+\tfrac{1}{2}\int\int \sum_{i,k} \psi_i^+(x) \times\psi_k^+(y)V(x,y)\psi_k(y)\psi_i(x)d^3xd^3y \equiv H_0+H_{\text{int}}. \quad (1.2)$$

K is the kinetic energy plus any external field. $i=1, 2$ refers to the two spin states (e.g., spin up and down along the z axis).

The Hartree-Fock method is equivalent to linearizing the interaction H_{int} by replacing bilinear products like $\psi_i^+(x)\psi_k(y)$ with their expectation values with respect to an approximate wave function which, in turn, is determined by the linearized Hamiltonian. We may consider also expectation values $\langle\psi_i(x)\psi_k(y)\rangle$ and $\langle\psi_i^+(x)\psi_k^+(y)\rangle$ although they would certainly be zero if the trial wave function were to represent an eigenstate of the number of particles, as is the case for the true wave function.

We write thus a linearized Hamiltonian

$$H_0'=\int \sum_i \psi_i^+K_i\psi_i d^3x+\int\int \sum_{i,k} [\psi_i^+(x)\chi_{ik}(xy)\psi_k(y) +\psi_i^+(x)\phi_{ik}(xy)\psi_k^+(y) +\psi_k(x)\phi_{ki}^+(xy)\psi_i(y)]d^3xd^3y \equiv H_0+H_s, \quad (1.3)$$

where

$$\chi_{ik}(xy)=\delta_{ik}\delta^3(x-y)\int V(xz)\sum_j\langle\psi_j^+(z)\psi_j(z)\rangle d^3z -V(xy)\langle\psi_k^+(y)\psi_i(x)\rangle, \quad (1.4)$$

$$\phi_{ik}(xy)=\tfrac{1}{2}V(xy)\langle\psi_k(y)\psi_i(x)\rangle,$$

$$\phi_{ik}^+(xy)=\tfrac{1}{2}V(xy)\langle\psi_k^+(y)\psi_i^+(x)\rangle.$$

We diagonalize H_0' and take, for example, the ground-state eigenfunction which will be a Slater-Fock product of individual particle eigenfunctions. The defining equations (1.4) then represent just generalized forms of Hartree-Fock equations to be solved for the self-consistent fields χ and ϕ.

The justification of such a procedure may be given by writing the original Hamiltonian as

$$H=(H_0+H_s)+(H_{\text{int}}-H_s)\equiv H_0'+H_{\text{int}}',$$

and demanding that H_{int}' shall have no matrix elements which would cause single-particle transitions; i.e., no matrix elements which would effectively modify the starting H_0': to put it more precisely, we demand our approximate eigenstates to be such that

$$\langle n|H_{\text{int}}'|0\rangle=\langle n|H|0\rangle=0, \quad (1.5)$$

if in $|n\rangle$ more than one particle change their states from those in $|0\rangle$. This condition is contained in Eq. (1.4).[5] Since in many-body problems, as in relativistic field theory, we often take a picture in which particles and holes can be created and annihilated, the condition (1.5) should also be interpreted to include the case where $|n\rangle$ and $|0\rangle$ differ only by such pairs. The significance of the BCS theory lies in the recognition that with an essentially attractive interaction V, a non-vanishing ϕ is indeed a possible solution, and the corresponding ground state has a lower energy than the normal state. It is also separated from the excited states by an energy gap $\sim 2\phi$.

The condition (1.5) was first invoked by Bogoliubov[2] in order to determine the transformation from the ordinary electron to the quasi-particle representation. He derived this requirement from the observation that H_{int}' contains matrix elements which spontaneously create virtual pairs of particles with opposite momenta, and cause the breakdown of the perturbation theory as the energy denominators can become arbitrarily small. Equation (1.5), as applied to such pair creation processes, determines only the nondiagonal part (in quasi-particle energy) of H_s in the representation in which H_0+H_s is diagonal. The diagonal part of H_s is still arbitrary. We can fix it by requiring that

$$\langle 1'|H_{\text{int}}'|1\rangle=0, \quad (1.6)$$

namely, the vanishing of the diagonal part of H_{int} for the states where one more particle (or hole) having a Hamiltonian H_0' is added to the ground state. In this way we can interpret H_0' as describing single particles (or excitations) moving in the "vacuum," and the diagonal part of H_s represents the self-energy (or the Hartree potential) for such particles arising from its interaction with the vacuum.

The distinction between Eqs. (1.5) and (1.6) is not so clear when applied to normal states. On the one hand, particles and holes (negative energy particles) are not separated by an energy gap; on the other hand, there is little difference when one particle is added just above the ground state.

In the above formulation of the generalized Hartree fields, χ and ϕ will in general depend on the external field as well as the interaction between particles. There is a complication due to the fact that they are gauge dependent. This is because a phase transformation $\psi_i(x)\rightarrow e^{i\lambda(x)}\psi_i(x)$ applied on Eq. (1.3) will change χ and ϕ according to

$$\chi(xy)\rightarrow e^{-i\lambda(x)+i\lambda(y)}\chi(xy),$$
$$\phi(xy)\rightarrow e^{i\lambda(x)+i\lambda(y)}\phi(xy), \quad (1.6)$$
$$\phi^+(xy)\rightarrow e^{-i\lambda(x)-i\lambda(y)}\phi^+(xy).$$

[5] Equation (1.5) refers only to the transitions from occupied states to unoccupied states. Transitions between occupied states or unoccupied states are given by Eq. (1.6). These two together then are equivalent to Eq. (1.4). For the analysis of the Hartree approximation in terms of diagrams, see J. Goldstone, Proc.

It is especially serious for ϕ (and ϕ^+) since, even if $\phi(xy)=\delta^3(x-y)$ times a constant in some gauge, it is not so in other gauges. Therefore, unless we can show explicitly that physical quantities do not depend on the gauge, any calculation based on a particular ϕ is open to question. It would not be enough to say that a longitudinal electromagnetic potential produces no effect because it can be transformed away before making the Hartree approximation. A natural way to reconcile the existence of ϕ, which we want to keep, with gauge invariance would be to find the dependence of ϕ on the external field explicitly. If the gauge invariance can be maintained, the dependence must be such that for a longitudinal potential $A=-\text{grad}\lambda$, it reduces to Eq. (1.6). This should not be done in an arbitrary manner, but by studying the actual influence of H_{int} on the primary electromagnetic interaction when ϕ is first determined without the external field.

After these preliminaries, we are going to study the points raised here by means of the techniques developed in quantum electrodynamics. We will first develop the Feynman-Dyson formulation adapted to our problem, and write down an integral eauation for the self-energy part which corresponds to the Hartree approximation. It is observed that it can possess a nonperturbational solution, and the existence of an energy gap is immediately recognized.

Next we will introduce external fields. Guided by the well-known theorems about gauge invariance, we are led to consider the so-called vertex parts, which include the "radiative corrections" to the primary charge-current operator. When an integral equation for the general vertex part is written down, certain exact solutions are obtained in terms of the assumed self-energy part, leading to analogs of the Ward identity.[6] They are intimately related to inherent invariance properties of the theory. Among other things, the gauge invariance is thus strictly established insofar as effects linear in the external field are concerned, including the Meissner effect.

Later we look into the collective excitations. A very interesting result emerges when we observe that one of the exact solutions to the vertex part equations becomes a homogeneous solution if the external energy-momentum is zero, and expresses a bound state of a pair with zero energy-momentum. Then by perturbation, other bound states with nonzero energy-momentum are obtained, and their dispersion law determined. Thus the existence of the bound state is a logical consequence of the existence of the special self-energy ϕ and the gauge invariance, which are seemingly contradictory to each other.

When the Coulomb interaction is taken into account, the bound pair states are drastically modified, turning into the plasma modes due to the same mechanism as in the normal case. This situation will also be studied.

Roy. Soc. (London) **A239**, 267 (1957). Compare also T. Kinoshita and Y. Nambu, Phys. Rev. **94**, 598 (1953).

[6] J. C. Ward, Phys. Rev. **78**, 182 (1950).

2. FEYNMAN-DYSON FORMULATION

We start from the Lagrangian for the electron-phonon system, which is supposed to be uniform and isotropic.[7]

$$\mathcal{L}=\sum_p\sum_i[i\dot{\psi}_i^+(p)\psi_i(p)-\psi_i^+(p)\epsilon_p\psi_i(p)]+\sum_k\tfrac{1}{2}[\dot{\varphi}(k)\dot{\varphi}(-k)-c^2\varphi(k)\varphi(-k)]-g\frac{1}{\sqrt{\mathcal{V}}}\sum_{p,k}\psi_i^+(p+k)\psi_i(p)h(k)\varphi(k). \quad (2.1)$$

φ is the phonon field, with the momentum k (energy $\omega_k=ck$) running up to a cutoff value $k_m(\omega_m)$; c is the phonon velocity. ϵ_p is the electron kinetic energy relative to the Fermi energy; $gh(k)$ represents the strength of coupling.[8] ($\mathcal{V}$ is the volume of the system.)

The Coulomb interaction between the electrons is not included for the moment in order to avoid complication. Later we will make remarks whenever necessary about the modifications when the Coulomb interaction is taken into account.

It will turn out to be convenient to introduce a two-component notation[9] for the electrons

$$\Psi(x)=\begin{pmatrix}\psi_1(x)\\ \psi_2^+(x)\end{pmatrix} \quad\text{or}\quad \Psi(p)=\begin{pmatrix}\psi_1(p)\\ \psi_2^+(-p)\end{pmatrix}, \quad (2.2)$$

and the corresponding 2×2 Pauli matrices

$$\tau_1=\begin{pmatrix}0&1\\1&0\end{pmatrix},\quad \tau_2=\begin{pmatrix}0&-i\\i&0\end{pmatrix},\quad \tau_3=\begin{pmatrix}1&0\\0&-1\end{pmatrix}. \quad (2.3)$$

The Lagrangian then becomes:

$$\mathcal{L}=\sum_p\Psi^+(p)\left(i\frac{\partial}{\partial t}-\epsilon_p\tau_3\right)\Psi(p)+\sum_k\tfrac{1}{2}[\dot{\varphi}(k)\dot{\varphi}(-k)-c^2\varphi(k)\varphi(-k)]-g\frac{1}{\sqrt{V}}\sum_{p,k}\Psi^+(p+k)\tau_3\Psi(p)h(k)\varphi(k)+\sum_p\epsilon_p$$
$$=\mathcal{L}_0+\mathcal{L}_{\text{int}}+\text{const}.$$

The last infinite c-number term comes from the rearrangement of the kinetic energy term. This is certainly uncomfortable, but will not be important except for the calculation of the total energy.

The fields obey the standard commutation relations.

[7] We use the units $\hbar=1$.

[8] For convenience, we have included in $h(k)$ the frequency factor: $h(k)=h_1(k)k_0$.

[9] P. W. Anderson [Phys. Rev. **112**, 1900 (1958)], has also introduced this two-component wave function.

Especially for Ψ, we have

$$\begin{aligned}\{\Psi_i(x),\Psi_j^+(y)\} &\equiv \Psi_i(x)\Psi_j^+(y)+\Psi_j^+(y)\Psi_i(x)\\ &=\delta_{ij}\delta^3(x-y), \qquad (2.5)\\ \{\Psi_i(p),\Psi_j^+(p')\} &=\delta_{ij}\delta_{pp'}.\end{aligned}$$

We may now formally treat H_{int} as perturbation, using the formulation of Feynman and Dyson.[10] The unperturbed ground state (vacuum) is then the state where all individual electron states $\epsilon_p<0(>0)$ are occupied (unoccupied) in the representation where $\psi_i^+(p)\psi_i(p)$ is the occupation number.

Having defined the vacuum, the time-ordered Green's functions for free electrons and phonons

$$\begin{aligned}\langle T(\Psi_i(xt),\Psi_j^+(x't'))\rangle &=[G_0(x-x',t-t')]_{ij},\\ \langle T(\varphi(xt),\varphi(x't'))\rangle &=\Delta_0(x-x',t-t')\end{aligned} \qquad (2.6)$$

are easily determined. We get for their Fourier representation (in the limit $\mathcal{V}\to\infty$)[10a]

$$G_0(xt)=(1/(2\pi)^4)\int G_0(pp_0)e^{ip\cdot x-ip_0t}d^3p\,dp_0,$$

$$\Delta_0(xt)=\frac{1}{(2\pi)^4}\int_{|k|<k_m}\Delta_0(kk_0)e^{ik\cdot x-ik_0t}d^3k\,dk_0,$$

$$\begin{aligned}G_0(pp_0)&=i\left[P\frac{1}{p_0-\epsilon_p\tau_3}-i\pi\,\text{sgn}(\tau_3\epsilon_p)\delta(p_0-\tau_3\epsilon_p)\right] \qquad (2.7)\\ &=i(p_0+\epsilon_p\tau_3)/(p_0^2-\epsilon_p^2+i\epsilon),\\ \Delta_0(kk_0)&=i\left[P\frac{1}{k_0^2-c^2k^2}-i\pi\delta(k_0^2-c^2k^2)\right]\\ &=i/(k_0^2-c^2k^2+i\epsilon).\end{aligned}$$

With the aid of these Green's functions, we are able to calculate the S matrix and other quantities according to a well-defined set of rules in perturbation theory.

We will analyze in particular the self-energies of the electron and the phonon. In the many-particle system, these energies express (apart from the self-interaction of the electron) the average interaction of a single particle or phonon placed in the medium. Because the phonon spectrum is limited, there will be no ultraviolet divergences, unlike the case of quantum electrodynamics.

These self-energies may be obtained in a perturbation expansion with respect to H_{int}. We are, however, interested in the Hartree method which proposes to take account of them in an approximate but nonperturbational way. It is true that the self-energies are in general complex due to the instability of single particles. But to the extent that the single-particle picture makes physical sense, we will ignore the small imaginary part of the self-energies in the following considerations.

FIG. 1. Second order self-energy diagrams. Solid and curly lines represent electron and phonons, respectively, themselves being under the influence of the self-energies Σ and Π. All diagrams are to be interpreted in the sense of Feynman, lumping together all topologically equivalent processes.

Let us thus introduce the approximate self-energy Lagrangian $\mathcal{L}_s$, and write

$$\begin{aligned}\mathcal{L}&=(\mathcal{L}_0+\mathcal{L}_s)+(\mathcal{L}_{\text{int}}-\mathcal{L}_s)\\ &\equiv\mathcal{L}_0'+\mathcal{L}_{\text{int}}',\\ \mathcal{L}_0&=\textstyle\sum_p\Psi_p^+L_0\Psi_p+\sum_k\frac{1}{2}\varphi_kM_0\varphi_{-k}, \qquad (2.8)\\ \mathcal{L}_s&=-\textstyle\sum_p\Psi_p^+\Sigma\Psi_p-\sum_k\frac{1}{2}\varphi_k\Pi\varphi_{-k},\\ L_0-\Sigma&\equiv L,\quad M_0-\Pi\equiv M.\end{aligned}$$

The free electrons with "spin" functions u and phonons obey the dispersion law

$$L_0(\mathbf{p},p_0=\epsilon_p)u_p=0,\quad M_0(\mathbf{k},k_0=\omega_k)=0, \qquad (2.9)$$

whereas they obey in the medium

$$L(\mathbf{p},p_0=E_p)u_p=0,\quad M(\mathbf{k},k_0=\Omega_k)=0. \qquad (2.9')$$

Σ will be a function of momentum p and "spin." Π will consist of two parts: $\Pi(k_0k)=\Pi_1(k)k_0^2+\Pi_2(k)$ in conformity with the second order character (in time) of the phonon wave equation.[11]

The propagators corresponding to these modified electrons and phonons are

$$\begin{aligned}G(pp_0)&=i/(L(pp_0)+i\,\text{sgn}(p_0)\epsilon),\\ \Delta(kk_0)&=i/(M(kk_0)+i\epsilon).\end{aligned} \qquad (2.10)$$

We now determine Σ and Π self-consistently to the second order in the coupling g. Namely the second order self-energies coming from the phonon-electron interaction have to be cancelled by the first order effect of $\mathcal{L}_{\text{int}}$.

These second order self-energies are represented by the nonlocal operators[12] (Fig. 1)

$$\begin{aligned}\mathcal{S}((t+t')/2)&=\int\int\int\Psi^+(xt)S(x-x',t-t')\\ &\qquad\times\Psi(x't')d^3x\,d^3x'\,d(t-t'),\\ \mathcal{P}((t+t')/2)&=\tfrac{1}{2}\int\int\int\varphi(xt)P(x-x',t-t')\\ &\qquad\times\varphi(x't')d^3x\,d^3x'\,d(t-t'),\end{aligned} \qquad (2.11)$$

[10] F. J. Dyson, Phys. Rev. 75, 486, 1736 (1949); R. P. Feynman, Phys. Rev. 76, 769 (1949); J. Schwinger, Phys. Rev. 74, 1439 (1948). Although we followed here the perturbation theory of Dyson, there is no doubt that the relations obtained in this paper can be derived by a nonperturbational formulation such as J. Schwinger's: Proc. Natl. Acad. Sci. U. S. 37, 452, 455 (1951).

[10a] P stands for the principal value; $i\epsilon$ in the denominator is a small positive imaginary quantity.

[11] In the same spirit Σ should actually be in the form $\Sigma_1(p)p_0+\Sigma_2(p)$. Here we neglect the renormalization term Σ_1 since the two conditions (2.13) can be met without it.

[12] We use the word nonlocal here for nonlocality in time.

where S and P have the Fourier representation

$$S(pp_0)=-ig^2\tau_3\delta^3(p)\delta(p_0)h^2(0)\Delta(0)$$
$$\times\int \mathrm{Tr}[\tau_3 G(p'p_0)]d^3p dp_0$$
$$-ig^2\int \tau_3 G(p-k,\, p_0-k_0)\tau_3 h^2(kk_0)$$
$$\times\Delta(kk_0)d^3k dk_0, \quad (2.12)$$
$$P(kk_0)=ig^2h(kk_0)^2\int \mathrm{Tr}[\tau_3 G(pp_0)$$
$$\times G(p+k,\, p_0+k_0)]d^3p dp_0.$$

In Eq. (2.11) we have chosen more or less arbitrarily $(t+t')/2$ as the fixed time to which we refer the nonlocal operators $\mathcal{S}$ and $\mathcal{P}$. The self-consistency requirements (1.5) and (1.6) mean in the present case that Σ, Π must be identical with S, P (*a*): for the diagonal elements [on the energy shell, Eq. (2.9)], and (*b*): for the nondiagonal matrix elements for creating a pair out of the vacuum.

The pair creation of electrons is possible because Ψ, being a two-component wave function, can have in general two eigenfunctions u_{ps} $(s=1, 2)$ with different energies E_{ps} for a fixed momentum, p, only one of which is occupied in the ground state.

Thus taking particular plane waves $u_{ps}{}^*e^{-i\mathbf{p}\cdot\mathbf{x}+ip_0t}$, $u_{p's'}e^{i\mathbf{p}'\cdot\mathbf{x}'-ip_0't'}$ for Ψ^+ and Ψ in (2.11), we easily find that the diagonal matrix element of Σ corresponds to $u_{ps}{}^*S(p,E_{ps})u_{ps}$, while the nondiagonal part corresponds to $u_{ps}{}^*S(p,0)u_{ps'}$, $s\neq s'$ $(p_0'=-p_0)$.

A similar situation holds also for the photon self-energy Π. Since Π consists of two parts, the diagonal and off-diagonal conditions will fix these.

With this understanding, the self-consistency relations may be written

$$\begin{aligned}\Sigma(\mathbf{p}E_p)_{\mathrm{D}}&=S(\mathbf{p},E_p)_{\mathrm{D}}, \quad \Sigma(\mathbf{p}0)_{\mathrm{ND}}=S(\mathbf{p}0)_{\mathrm{ND}},\\ \Pi(\mathbf{k}\Omega_k)&=P(\mathbf{k}\Omega_k), \quad \Pi(\mathbf{k}0)=P(\mathbf{k}0),\end{aligned} \quad (2.13)$$

where D, ND signify the diagonal and nondiagonal parts in the "spin" space. As stated before, we have agreed to omit possible imaginary parts in S and P. (The nondiagonal components, however, will turn out to be real.)

Before discussing the general solutions, let us consider the meaning of Eq. (2.13) in terms of perturbation theory. Suppose we expand G occurring in Eq. (2.12), with respect to Σ:

$$G=G_0-iG_0\Sigma G_0-G_0\Sigma G_0\Sigma G_0+\cdots,$$

and expand Σ itself with respect to g^2, then we easily realize that Eq. (2.13) defines an infinite sum of a particular class of diagrams, which are illustrated in Fig. 2. The first term in S of Eq. (2.12) corresponds to the ordinary Hartree potential which is just a constant,

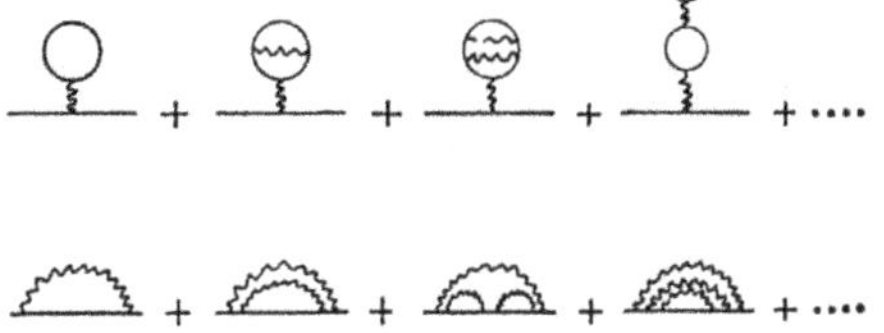

FIG. 2. Expansion of the self-consistent self-energy $\Sigma\sim S$ in terms of bare electron diagrams.

whereas the second term gives an exchange effect. In the latter, the approximation is characterized by the fact that no phonon lines cross each other.

It must be said that the Hartree approximation does not really sum the series of Fig. 2 completely since we equate in Eq. (2.13) only special matrix elements of both sides. For in the perturbation series the Σ obtained to any order is a function of p_0, whereas in Eq. (2.13) it is replaced by a p_0-independent quantity. Hence there will be a correction left out in each order (analogous to the radiative correction after mass renormalization in quantum electrodynamics).

In this perturbation expansion, S in Eq. (2.13) is always proportional to τ_3 on the energy shell since $H_0\propto\tau_3$. Accordingly Σ will be $\propto\tau_3$ and commute with H_0, so that no off-diagonal part exists.[11]

It is important, however, to note the possibility of a nonperturbational solution by assuming that Σ contains also a term proportional to τ_1 or τ_2. Thus, take

$$\begin{aligned}\Sigma(p)&=\chi(p)\tau_3+\phi(p)\tau_1,\\ H_0'&=(\epsilon+\chi)\tau_3+\phi\tau_1\\ &\equiv\bar{\epsilon}\tau_3+\phi\tau_1.\end{aligned} \quad (2.14)$$

This form bears a resemblance to the Dirac equation. Its eigenvalues are

$$E=\pm E_p\equiv\pm(\bar{\epsilon}_p{}^2+\phi_p{}^2)^{\frac{1}{2}}. \quad (2.15)$$

Since H_0' describes by definition excited states, we have to adopt the hole picture and conclude that the ground state (vacuum) is the state where all negative energy "quasi-particles" $(E<0)$ are occupied and no positive energy particles exist. If ϕ remains finite on the Fermi surface, the positive and negative states are separated by a gap $\sim 2|\phi|$. The corresponding Green's function G now has the representation

$$G(pp_0)=i\frac{p_0+\bar{\epsilon}_p\tau_3+\phi_p\tau_1}{p_0{}^2-E_p{}^2+i\epsilon}. \quad (2.16)$$

In order to extract the diagonal and nondiagonal parts in spin space, we will use the trick

$$\begin{aligned}O_{\mathrm{D}}&=\tfrac{1}{2}\,\mathrm{Tr}\,(\Lambda O),\\ O_{\mathrm{ND}}&=-(i/2)\,\mathrm{Tr}\,(\Lambda O\tau_2),\\ \Lambda&=[E_p+H_0'(p)]/2E_p.\end{aligned} \quad (2.17)$$

Applying this to Eq. (2.13a) with Eqs. (2.12), (2.14), and (2.15), we finally obtain the following equations for χ and ϕ

$$\frac{\epsilon_p\chi_p+\phi_p{}^2}{E_p}=\frac{g^2\pi}{(2\pi)^4}P\int\left[\frac{E_p}{\Omega_k}+\frac{E_{p-k}+\Omega_k}{E_pE_{p-k}\Omega_k}\right.$$
$$\left.\times(\bar{\epsilon}_p\bar{\epsilon}_{p-k}-\phi_p\phi_{p-k})\right]\frac{h(k)^2}{E_p{}^2-(E_{p-k}-\Omega_k)^2}d^3k,$$

$$\epsilon_p\phi_p=\frac{g^2\pi}{(2\pi)^4}\int(\bar{\epsilon}_{p-k}\phi_p+\bar{\epsilon}_p\phi_{p-k})$$
$$\times\frac{h(k)^2d^3k}{E_{p-k}\Omega_k(E_{p-k}+\Omega_k)}. \quad (2.18)$$

The second equation, coming from the nondiagonal condition, has a trivial solution $\phi=0$. If a finite solution ϕ exists, it cannot follow from perturbation treatment since there is no inhomogeneous term to start with.

Equation (2.18) is equivalent to, but slightly different from, the corresponding conditions of Bogoliubov because of a slightly different definition of the nondiagonal part of the self-energy operator, which is actually due to an inherent ambiguity in approximating nonlocal operators by local ones. (This is the same kind of ambiguity as one encounters in the derivation of a potential from field theory. The difference between the local operator Σ and the nonlocal one S shows up in a situation like that in Fig. 3, and the compensation between Σ and S is not complete.) We may avoid this unpleasant situation, by extending the Hartree self-consistency conditions to all virtual matrix elements, but this would mean that ϕ (and χ) must be treated as nonlocal. We will discuss this situation in a separate section since such a generalization brings simplification in dealing with the problem of gauge invariance and collective excitations.

For the moment we consider the second equation of (2.18) and rewrite it

$$\phi_p=A_p\frac{g^2\pi}{(2\pi)^4}\int\frac{\phi_{p-k}}{E_{p-k}}\frac{h(k)d^3k}{\Omega_k(E_{p-k}+\Omega_k)},$$
$$A_p=\bar{\epsilon}_p\Big/\left(\epsilon_p-\frac{g^2\pi}{(2\pi)^4}\int\frac{\bar{\epsilon}_ph(k)^2d^3k}{E_{p-k}\Omega_k(E_{p-k}+\Omega_k)}\right). \quad (2.19)$$

This is essentially the energy gap equation of BCS if $g^2A_ph(k)^2/\Omega_k(E_{p-k}+\Omega_k)$ is identified with the effective interaction potential V, and if $\bar{\epsilon}_p\sim\epsilon_p(\chi_p\sim 0)$. It has a solution

$$\phi\sim\Omega_m\exp(-1/VN),$$

if $VN\ll 1$, N being the density of states: $N=dn/d\epsilon_p$ on the Fermi surface.

The phonon self-energy Π may be studied similarly from Eq. (2.13), which should determine the renormalization of the phonon field. It does not play an essential role in superconductivity, though it gives rise to an important correction when the Coulomb effect is taken into account. (See the following section.)

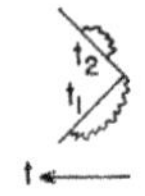

FIG. 3. An example of the situation where the cancellation of Σ_{ND} *versus* S_{ND} is not complete. The two self-energy parts overlap in time, and their centers of time t_1 and t_2 are such that $t_1>t_2$. If calculated according to the usual perturbation theory, this process will not be eliminated by the condition $\Sigma_{ND}=(S_1)_{ND}$.

From the nature of Eq. (2.12), it is clear that $\tau_1\phi$ can actually be pointed in any direction in the 1–2 plane of the τ space: $\tau_1\phi_1+\tau_2\phi_2$. It was thus sufficient to take $\phi_1\neq 0$, $\phi_2=0$. Any other solution is obtained by a transformation

$$\Psi\rightarrow\exp(i\alpha\tau_3/2)\Psi,$$
$$(\phi,0)\rightarrow(\phi\cos\alpha,\phi\sin\alpha). \quad (2.20)$$

In view of the definition of Ψ, Eq. (2.20) is a gauge transformation with a constant phase. Thus the arbitrariness in the direction of ϕ is the 1–2 plane is a reflection of the gauge invariance.

For later use, we also mention here the particle-antiparticle conjugation C of the quasi-particle field Ψ. This is defined by

$$C:\quad \Psi\rightarrow\Psi^C=C\Psi^+=\tau_2\Psi^+,$$

or

$$\begin{pmatrix}\psi_1{}^C\\ \psi_2{}^{+C}\end{pmatrix}=\begin{pmatrix}-i\psi_2\\ i\psi_1{}^+\end{pmatrix}, \quad (2.21)$$

and changes quasi-particles of energy-momentum (p_0,p) into holes of energy-momentum $(-p_0,-p)$, or interchanges up-spin and down-spin electrons. Under C, the τ operators transform as

$$C:\quad \tau_i\rightarrow C^{-1}\tau_iC=-\tau_i{}^T,\quad i=1,2,3 \quad (2.22)$$

where T means transposition.

As a consequence, we have also

$$C:\quad L(p)\rightarrow L^C(-p)=-L(-p)^T. \quad (2.23)$$

Finally we make a remark about the Coulomb interaction. When this is taken into account, the phonon interaction factor $g^2h(k)^2\Delta(k,k_0)$ in Eq. (2.12a) has to be replaced by

$$[g^2h(k)^2\Delta(kk_0)+ie^2/k^2]/$$
$$\{1-i\Pi(kk_0)[\Delta(kk_0)+ie^2/g^2h(k)^2k^2]\}.$$

As is well known, the denominator represents the screening of the Coulomb interaction. Discussion about this point will be made later in connection with the plasma oscillations.

3. NONLOCAL (ENERGY-DEPENDENT) SELF-CONSISTENCY CONDITIONS

In the last section we remarked that the self-consistency conditions Eq. (2.13) may be extended to all virtual matrix elements, namely, not only on the energy shell (diagonal) and for the virtual pair creation out of the vacuum, but also for the self-energy effects which appear in intermediate states of any process.

This simply means that ϕ and π are now nonlocal; i.e., depend both on energy and momentum arbitrarily, and are to be completely equated with S and P, respectively,

$$\Sigma(pp_0)=S(pp_0),\quad \Pi(kk_0)=P(kk_0). \tag{3.1}$$

Actually, these self-energies can no more be incorporated in H_0' as the zeroth order Lagrangian since they contain infinite orders of time derivatives.[13] Nevertheless, Eq. (3.1) has a precise meaning in the bare particle perturbation theory. It defines the (proper) self-energy parts (in the sense of Dyson) as an infinite sum of the special class of diagrams illustrated in Fig. 2.

The earlier condition of Eq. (2.13) represented, as was noted there, only an approximation to this sum. In other words, Eqs. (2.13) and (3.1) are not exactly identical even on the energy shell.

The Hartree-Fock approximation based on Eq. (3.1) could be interpreted as a nonperturbation approximation to determine the "dressed" single particles (together with the "dressed vacuum") or the Green's function $\langle 0|T(\Psi(xt),\Psi^+(x't'))|0\rangle$ for the true interacting system. Such single particles will satisfy

$$L(p,p_0)u\cong 0,\quad M(k,k_0)\cong 0. \tag{3.2}$$

We use the approximate equality since a really stable single particle may not exist.

Let us assume that these determine the approximate renormalized dispersion law

$$p_0^2=E_r(p)^2,\quad k_0^2=\Omega_r(k)^2. \tag{3.3}$$

If we write for Σ

$$\Sigma(pp_0)=p_0\zeta(pp_0)+\chi(pp_0)\tau_3+\phi(pp_0)\tau_1, \tag{3.4}$$

where ζ, χ, ϕ are even functions of p_0, then

$$\begin{aligned}E_r^2(p)&=[\bar{\epsilon}(pp_0)^2+\phi(pp_0)^2]/[1-\zeta(pp_0)]^2|_{p_0^2=E_r(p)^2}\\&\equiv E(pp_0)^2/Z(pp_0)^2|_{p_0^2=E_r(p)^2}.\end{aligned} \tag{3.5}$$

The Green's functions G and Δ will be given by

$$\begin{aligned}G(pp_0)&=i/L(pp_0)\\&=i\int_0^\infty \frac{dx}{p_0^2-x+i\epsilon}\\&\qquad\times\mathrm{Im}\frac{p_0Z(px)+\bar{\epsilon}(px)\tau_3+\phi(px)\tau_1}{x^2Z(px)^2-E(px)^2},\end{aligned} \tag{3.6}$$

$$\begin{aligned}\Delta(kk_0)&=i/M(kk_0)\\&=i\int_0^\infty \frac{dx}{k_0^2-x+i\epsilon}\,\mathrm{Im}\frac{1}{M(kx)}.\end{aligned}$$

This representation assumes that $G(p_0)[\Delta(k_0)]$ is analytic except for a branch cut on the real axis. The imaginary part in the integrand is expected to have a delta function or a sharp peak at $x=E_r^2(p)$ $[\Omega_r(k^2)]$. These properties are necessary in order that the vacuum is stable and the quasi-particles and phonons have a valid physical meaning as excitations.[14] In the following, we will generally consider this quasi-particle peak only, and write

$$G(pp_0)=i\frac{p_0Z(pp_0)+\bar{\epsilon}(pp_0)\tau_3+\phi(pp_0)\tau_1}{p_0Z(pp_0)^2-E(pp_0)^2+i\epsilon},\quad \text{etc.}$$

The Hartree equations now take the form

$$\Sigma(pp_0)=-i\frac{g^2}{(2\pi)^4}\int \tau_3G(p-k,\,p_0-k_0)\tau_3h(kk_0)^2d^3kdk_0,$$

$$\begin{aligned}\Pi(kk_0)=i\frac{g^2}{(2\pi)^4}\int \mathrm{Tr}\,[&\tau_3G(k-p,\,k_0-p_0)\\&\times\tau_3G(pp_0)]d^3pdp_0.\end{aligned} \tag{3.7}$$

This equation for Σ is much simpler than the previous one (2.18) since we may just equate the coefficients of 1, τ_3, τ_1 on both sides. In particular, we get the energy gap equation

$$\begin{aligned}\phi(pp_0)=-\frac{ig^2}{(2\pi)^4}\int &\frac{\phi(p'p_0')}{p_0^2Z(p'p_0')^2-E(p'p_0')^2+i\epsilon}\\&\times h(\mathbf{p}-\mathbf{p}',\,p_0-p_0')^2\Delta(\mathbf{p}-\mathbf{p}',\,p_0-p_0')d^3p'dp_0',\end{aligned} \tag{3.8}$$

which is to be compared with Eq. (2.19).

Although the existence of a solution to Eq. (3.6) may be difficult to establish, the solution, if it exists, should not be much different from the older solution to Eq. (2.19). At any rate, our assumption about the analyiticity of G and Π is consistent with Eq. (3.6) or (3.7) which implies that Σ and Π are also analytic except for a cut on the real axis.

In later calculations we shall encounter various integrals which we may classify into three types regarding their sensitivity to the energy gap. First, a normal self-energy part, for example, represents the effect of the bulk of the surrounding electrons on a particular electron, and is insensitive to the change of the small fraction $\sim\phi/E_F$ of the electrons near the Fermi surface in a superconductor. Such a quantity is

[13] It would seem then that we lose the advantage of the generalization since we cannot find the Bogoliubov transformation. However, we could still start from the older solution (2.13) as the zeroth approximation to Eq. (3.1), and then calculate the correction; namely, the "radiative" correction to the Bogoliubov vacuum and the Bogoliubov quasi-particle. These corrections would take account of the single-particle transitions which remain after the Bogoliubov condition (2.13) is imposed.

[14] This is a representation of the Lehmann type [H. Lehmann, Nuovo cimento **11**, 342 (1954)] which can be derived by defining the Green's functions in terms of Heisenberg operators. See also V. M. Galizkii and A. B. Migdal, J. Exptl. Theoret. Phys. U.S.S.R. **34**, 139 (1958) [translation: Soviet Phys. JETP **7**, 96 (1958)].

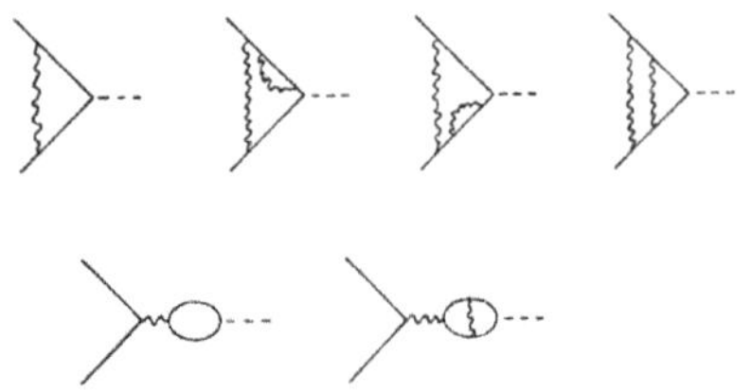

FIG. 4. Construction of the vertex part Γ in bare particle picture. The second line represents the polarization diagrams.

given by an integral like

$$g^2\int \frac{\epsilon_k}{E_k}f(\mathbf{p}-\mathbf{k})d^3k, \tag{3.9}$$

where the region $\epsilon_k \lesssim E_k = (\epsilon_k{}^2+\phi^2)^{\frac{1}{2}}$ makes little contribution if $f(p-k)$ is a smooth function.

Second, the energy gap itself is determined from an equation of the form

$$g^2\int \frac{d^3k}{E_k}f(\mathbf{p}-\mathbf{k}) \sim g^2\int_{E_k\lesssim\omega_m} \frac{d^3k}{E_k}f(\mathbf{p}-\mathbf{k}) \sim 1, \tag{3.10}$$

which means that even if g^2 is small, such an expression is always of the order 1.

Finally we meet with integrals like

$$g^2\int \frac{\epsilon_k\phi}{E_k{}^3}f(\mathbf{p}-\mathbf{k})d^3k,\quad g^2\int \frac{\phi^2}{E_k{}^3}f(\mathbf{p}-\mathbf{k})d^3k,\quad \text{etc.} \tag{3.11}$$

They have an extra cutoff factor $\sim 1/E$, $1/E^2$, etc., in the integrand which restricts the contribution to an energy interval $\sim 2\phi$ near the Fermi surface. The integrals are thus of the order

$$g^2N\phi/\omega_m,\quad g^2N,\quad \text{etc.}$$

In the following, we will not be primarily concerned with the ordinary self-energy effects. We will assume that proper renormalization has been carried out, or else simply disregard it unless essential. When we carry out perturbation type calculations, we will arrange things so that quantities of the second type are taken into account rigorously, and treat quantities of the third type as small, and hence negligible ($g^2N \ll 1$).

4. INTEGRAL EQUATIONS FOR VERTEX PARTS[15]

In the presence of an electromagnetic potential, the original Lagrangian $\mathfrak{L}$ has to be modified according to the rule

$$i\frac{\partial}{\partial t} \rightarrow i\frac{\partial}{\partial t}+eA_0,\quad \mathbf{p}\rightarrow \mathbf{p}-\frac{e}{c}\mathbf{A}$$

for the electron. Going to the two-component representation, this corresponds to the prescription

$$i\frac{\partial}{\partial t} \rightarrow i\frac{\partial}{\partial t}+e\tau_3A_0,\quad \mathbf{p}\rightarrow \mathbf{p}-\frac{e}{c}\tau_3\mathbf{A} \tag{4.1}$$

acting on Ψ. It can also be inferred from the gauge transformation $\Psi \rightarrow \exp(i\alpha\tau_3)\Psi$ as was observed previously. So the ordinary charge-current operator turns out to be in our form given by

$$\rho(x) = \frac{e}{2}([\Psi^+(x),\tau_3\Psi(x)]+\{\Psi^+(x),\Psi(x)\}),$$

$$\begin{aligned}\mathbf{j}(x) = \frac{-ie}{4m}([\Psi^+(x),\ (\nabla-ie\tau_3\mathbf{A})\Psi(x)]\\ +[(-\nabla-ie\tau_3A)\Psi^+(x),\ \Psi(x)]\\ +\{\Psi^+(x),\ (\nabla\tau_3-ieA)\Psi(x)\}\\ -\{(\tau_3\nabla-ie\mathbf{A})\Psi^+(x),\ \Psi(x)\}).\end{aligned} \tag{4.2}$$

The second terms on the right-hand side, being infinite C numbers, arise from the rearrangement of ψ and ψ^+, and will actually be compensated for by the first terms.

This expression, however, has to be modified when we go to the quasi-particle picture.

For we have seen that the self-energy ϕ of a quasi-particle is a gauge-dependent quantity. If we want to have the quasi-particle picture and gauge invariance at the same time, then it is clear that the electromagnetic current of a quasi-particle must contain, in addition to the normal terms given by Eq. (4.2), terms which would cause a physically unobservable transformation of ϕ if the electromagnetic potential is replaced by the gradient of a scalar. In other words, the complete charge current of a quasi-particle has to satisfy the continuity equation, which Eq. (4.2) does not, since

$$\partial\rho/\partial t+\nabla\cdot\mathbf{j} = 2\Psi^+\phi\tau_2\Psi.$$

In order to find such a conserving expression for charge current, it is instructive to go back to the bare electron picture, in which the self-energy is represented by a particular class of diagrams discussed in the previous sections.

It is well known[16] in quantum electrodynamics that, in any process involving electromagnetic interaction, perturbation diagrams can be grouped into gauge-invariant subsets, such that the invariance is maintained by each subset taken as a whole. Such a subset can be constructed by letting each photon line in a diagram interact with a charge of all possible places along a chain of charge-carrying particle lines. The gauge-invariant interaction of a quasi-particle with an electromagnetic potential should then be obtained by attaching a photon line at all possible places in the diagrams of Fig. 2. The result is illustrated in Fig. 4,

[15] Hereafter we will often use the four-dimensional notation $x=(\mathbf{x},t)$, $p=(\mathbf{p},p_0)$, $d^4p=d^3pdp_0$.

[16] Z. Koba, Progr. Theoret. Phys. (Kyoto) **6**, 322 (1951).

which consists of the "vertex" part Γ and the self-energy part Σ.

In this way we are led to consider the modification of the vertex due to the phonon interaction in the same approximation as the self-energy effect is included in the quasi-particle. It is not difficult to see that it corresponds to a "ladder approximation" for the vertex part, and we get an integral equation[17]

$$\Gamma_i(p',p)=\gamma_i(p',p)-g^2\int\tau_3 G(p'-k)\Gamma_i(p'-k,\,p-k)\times G(p-k)\tau_3 h(k)^2\Delta(k)d^4k, \quad (4.3)$$

where γ_i, $i=0,1,2,3$ stand for the free particle charge current $[\tau_3, (1/2m)(\mathbf{p}+\mathbf{p}')]$ which follows from Eq. (4.2). Similar equations may be set up for any type of vertex interactions.

Equation (4.3) is the basis of the rest of this paper. It expresses a clear-cut approximation procedure in which the "free" charge-current operator γ_i of a quasi-particle is modified by a special class of "radiative corrections" due to H_{int}'.

As the next important step, we observe that there exist exact solutions to Eq. (4.3) for the following four types of vertex interactions

$$\begin{aligned}
(a)\quad &\gamma^{(a)}(p',p)=L_0(p')-L_0(p)\\
&\qquad=(p_0'-p_0)-\tau_3(\epsilon_{p'}-\epsilon_p),\\
&\Gamma^{(a)}(p',p)=L(p')-L(p)\\
&\qquad=\gamma_a(p')-[\Sigma(p')-\Sigma(p)],\\
(b)\quad &\gamma^{(b)}(p',p)=L_0(p')\tau_3-\tau_3L_0(p)\\
&\qquad=(p_0'-p_0)\tau_3-(\epsilon_{p'}-\epsilon_p),\\
&\Gamma^{(b)}(p',p)=L(p')\tau_3-\tau_3L(p),\\
(c)\quad &\gamma^{(c)}(p',p)=L_0(p')\tau_1+\tau_1L_0(p'),\\
&\Gamma^{(c)}(p',p)=L(p')\tau_1+\tau_1L(p),\\
(d)\quad &\gamma^{(d)}(p',p)=L_0(p')\tau_2+\tau_2L(p),\\
&\Gamma^{(d)}(p',p)=L(p')\tau_2+\tau_2L(p).
\end{aligned}\quad (4.4)$$

The verification is straightforward by noting that $G(p)=i/L(p)$, and making use of Eq. (3.7).

The fact that there are simple solutions is not accidental. These solutions express continuity equations and other relations following from the four types of operations, which do not depend on the presence or absence of the interaction:

$$\begin{aligned}
(a)\quad &\Psi(x)\to e^{i\alpha(x)}\Psi(x), &&\Psi^+(x)\to\Psi^+(x)e^{-i\alpha(x)},\\
(b)\quad &\Psi\to e^{i\tau_3\alpha}\Psi, &&\Psi^+\to\Psi^+e^{-i\tau_3\alpha},\\
(c)\quad &\Psi\to e^{\tau_1\alpha}\Psi, &&\Psi^+\to\Psi^+e^{\tau_1\alpha},\\
(d)\quad &\Psi\to e^{\tau_2\alpha}\Psi, &&\Psi^+\to\Psi^+e^{\tau_2\alpha},
\end{aligned}\quad (4.5)$$

where $\alpha(x)$ is an arbitrary real function.

(*a*) and (*b*) correspond, respectively, to the spin rotation around the *z* axis, and the gauge transformation. The entire Lagrangian is invariant under them, so that we obtain continuity equations for the *z* component of spin and charge, respectively:

$$\begin{aligned}
&\frac{\partial}{\partial t}\boldsymbol{\Psi}^+\boldsymbol{\Psi}+\nabla\cdot\boldsymbol{\Psi}^+\frac{\mathbf{p}}{m}\tau_3\boldsymbol{\Psi}=0,\\
&\frac{\partial}{\partial t}\boldsymbol{\Psi}^+\tau_3\boldsymbol{\Psi}+\nabla\cdot\boldsymbol{\Psi}^+\frac{\mathbf{p}}{m}\boldsymbol{\Psi}=0,
\end{aligned}\quad (4.6)$$

where $\boldsymbol{\Psi}$ is the true Heisenberg operator.

These equations are identical with

$$\boldsymbol{\Psi}^+\gamma^{(a)}\boldsymbol{\Psi}=0,\quad \boldsymbol{\Psi}^+\gamma^{(b)}\boldsymbol{\Psi}=0. \quad (4.7)$$

Taken between two "dressed" quasi-particle states, the left-hand side of Eq. (4.7) will become

$$\begin{aligned}
e^{-i(p'-p)\cdot x}\langle p'|\boldsymbol{\Psi}^+(x)\gamma^{(n)}\boldsymbol{\Psi}(x)|p\rangle
&=u_{p'}^*\Gamma^{(n)}(p',p)u_p\\
&=0,\quad (n=a,b)
\end{aligned}\quad (4.7')$$

where u_p, $u_{p'}$ are single-particle wave functions satisfying $L(p)u_p=u_{p'}^*L(p')=0$.

In this way we have shown the existence of spin and charge currents $\Gamma_i^{(a)}(p',p)$ and $\Gamma_i^{(b)}(p',p)$ for a quasi-particle, for which the continuity equations

$$(p_0'-p_0)\cdot\Gamma_0^{(n)}-\sum_{i=1}^{3}(p'-p)_i\cdot\Gamma_i^{(n)}=\Gamma^{(n)}(p',p)=0$$

will hold.

The last two transformations of Eq. (4.4) are not unitary, but mix ψ_1 and ψ_2^+ in such a way as to keep $\Psi^+\tau_3\Psi$ invariant. From infinitesimal transformations of these kinds we get

$$\begin{aligned}
&i\boldsymbol{\Psi}^+\tau_1\left(\frac{\overrightarrow{\partial}}{\partial t}-\frac{\overleftarrow{\partial}}{\partial t}\right)\boldsymbol{\Psi}+\nabla\cdot\boldsymbol{\Psi}^+\tau_2\left(\frac{\overrightarrow{\mathbf{p}}}{m}+\frac{\overleftarrow{\mathbf{p}}}{m}\right)\boldsymbol{\Psi}=0,\\
&-i\boldsymbol{\Psi}^+\tau_2\left(\frac{\overrightarrow{\partial}}{\partial t}-\frac{\overleftarrow{\partial}}{\partial t}\right)\boldsymbol{\Psi}+\nabla\cdot\Psi^+\tau_1\left(\frac{\overrightarrow{\mathbf{p}}}{m}+\frac{\overleftarrow{\mathbf{p}}}{m}\right)\boldsymbol{\Psi}=0,
\end{aligned}\quad (4.8)$$

which bear the same relations to $\gamma^{(c),(d)}$ and $\Gamma^{(c),(d)}$ as Eq. (4.6) did to $\gamma^{(a),(b)}$ and $\Gamma^{(a),(b)}$. Note that the above equations are unaffected by the presence of the phonon interaction.

The fact that we can find a conserved charge-current

[17] This equation may also be derived simply by considering the self-energy equation (3.7) in the presence of an external field, and expanding Σ in *A*. Σ should be now a function of initial and final momenta, and we define

$$\Sigma^{(A)}(p',p)\equiv\Sigma(p)\delta^4(p'-p)+\sum_{i=0}^{3}(\Gamma_i(p',p)-\gamma_i(p',p))\times A^i(p'-p)+O(A^2).$$

In the limit $p'-p=0$, $\Gamma_i-\gamma_i=\partial\Sigma/\partial A^i$, which is the content of the Ward identity.[8] Investigation of the higher order terms in *A* is beyond the scope of this paper.

FIG. 5. The diagram for the kernel $K^{(2)}$.

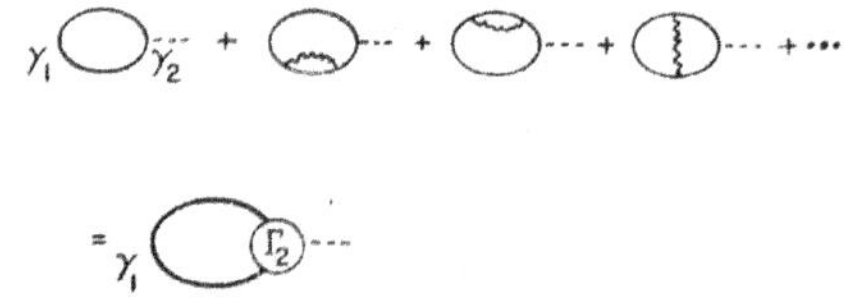

FIG. 6. Graphical derivation of Eq. (5.5). The thick lines represent quasi-particles.

for a quasi-particle is rather surprising. A quasi-particle cannot be an eigenstate of charge since it is a linear combination of an electron and a hole, tending to an electron well above the Fermi surface, and to a hole well below. We must conclude than that an accelerated wave packet of quasi-particles, whose energy is confined to a finite region of space, continuously picks up charge from, or deposits it with, the surrounding medium which extends to infinity. This situation will be studied in Sec. 7, where we will derive the charge current operators Γ_i explicitly.

5. GAUGE INVARIANCE IN THE MEISSNER EFFECT

We will next discuss how the gauge invariance is maintained in the problem of the Meissner effect when the external magnetic field is static. We calculate the Fourier component of the current $J(q)$ induced in the superconducting ground state by an external vector potential $A(q)$:

$$J_i(q)=\sum_{j=1}^{3} K_{ij}(q)A_j(q), \tag{5.1}$$

where q is kept finite.

For free electrons, K is represented by

$$K_{ij}=K_{ij}^{(1)}+K_{ij}^{(2)}, \quad K_{ij}^{(1)}=-\delta_{ij}ne^2/m, \tag{5.2}$$

where n is the number of electrons inside the Fermi sphere. $K^{(1)}$ comes from the expectation value of the current operator Eq. (4.2), whereas $K^{(2)}$ corresponds to the diagram in Fig. 5. [Compare also Eq. (1.1).] It is well known that in this case K_{ij} is of the form

$$K_{ij}(q)=(\delta_{ij}q^2-q_iq_j)K(q^2), \tag{5.3}$$

so that for a longitudinal vector potential $A_i(q)\sim q_i\lambda(q)$, we have

$$J_i(q)=K_{ij}q_j\lambda(q)=0, \tag{5.4}$$

establishing the unphysical nature of such a potential.

In the case of a superconducting state, the free electron lines in Fig. 5 will be replaced by quasi-particle lines. But then we have $K^{(2)}(q)\rightarrow 0$ as $q\rightarrow 0$ since the intermediate pair formation is suppressed due to the finite energy gap, whereas $K^{(1)}$ is essentially unaltered. Thus Eq. (5.2) takes the form of the London equation, except that even a longitudinal field creates a current.

According to our previous argument, this lack of gauge invariance should be remedied by taking account of the vertex corrections. Starting again from the free electron picture, and inserting the phonon interaction effects, as indicated in Fig. 6, we arrive at the conclusion that either one of the vertices γ in Fig. 5 has to be replaced by the full Γ^{10}. In addition, there is the polarization correction represented by a string of bubbles. Let us, however, first neglect this correction. $K_{ij}^{(2)}$ is then

$$K_{ij}^{(2)}(q)=\frac{-ie^2}{(2\pi)^4}\int \mathrm{Tr}[\gamma_i(p-q/2,\,p+q/2)G(p+q/2)$$
$$\times\Gamma_j(p+q/2,\,p-q/2)G(p-q/2)]d^4p. \tag{5.5}$$

Although we do not know $\Gamma_j(p,p)$ explicitly, we can establish Eq. (5.4) easily. For

$$-\sum_{j=1}^{3}\Gamma_j(p+q,\,q)q_j$$

is exactly the solution $\Gamma^{(b)}(p+q,\,p)$ of Eq. (4.4) where q_0 is equal to zero. Substituting this solution in Eq. (4.5) we find

$$K_{ij}^{(2)}(q)q_j$$
$$=\frac{-1}{(2\pi)^4}\int \mathrm{Tr}\{\gamma_i(p-q/2,\,p+q/2)$$
$$\times[\tau_3G(p+q/2)-G(p-q/2)\tau_3]\}d^4p$$
$$=\frac{-1}{(2\pi)^4}\int \mathrm{Tr}\{[\gamma_i(p-q,\,p)-\gamma_i(p,\,p+q)]$$
$$\times\tau_3G(p)\}d^4p$$
$$=\frac{1}{(2\pi)^4}\frac{q_i}{m}\int \mathrm{Tr}[\tau_3G(p)]d^4p, \tag{5.6}$$

where the properties of γ_i and G under particle conjugation and a translation in p space were utilized in going from the first to the second line.

On the other hand, the part $K^{(1)}$ is, according to Eq. (4.2) given by

$$K_{ij}^{(1)}=-\delta_{ij}\frac{e^2}{2m}(\langle 0|[\Psi^+(x),\tau_3\Psi(x)]|0\rangle+\{\Psi^+(x),\Psi(x)\})$$
$$=K_{ij}^{(1a)}+K_{ij}^{(1b)}. \tag{5.7}$$

The first term becomes further

$$-\delta_{ij}\frac{e^2}{2m}\langle 0|[\Psi^+(x),\tau_3\Psi(x)]|0\rangle$$

$$=-\delta_{ij}\frac{e^2}{m}\tfrac{1}{2}\lim_{\epsilon\to 0}\sum_{\pm}\langle 0|T(\Psi^+(x,t\pm\epsilon)\tau_3\Psi(x,t))|0\rangle$$

$$=-\delta_{ij}\frac{e}{m}\mathrm{Tr}[\tau_3 G(xt=0)]$$

$$=-\delta_{ij}\frac{e}{m}\frac{1}{(2\pi)^4}\int \mathrm{Tr}[\tau_3 G(p)]d^4p. \quad (5.8)$$

Thus

$$[K_{ij}^{(1a)}(q)+K_{ij}^{(2)}(q)]q_j=0. \quad (5.9)$$

The second term $K^{(1b)}$ comes from the c-number term of the current operator (4.2), and is just the anticommutator of the electron field, which does not depend on the quasi-particle picture, nor on the presence of interaction. Therefore we may write for this contribution

$$K_{ij}^{(1b)}(q)A_j(q)=\frac{-ie}{2m}\frac{1}{(2\pi)^3}\int e^{-iqx}d^3x$$
$$\times\{\Psi^+(x),(\tau_3\nabla-ieA(x))_i\Psi(x)\} \quad (5.10)$$

to show its formal gauge invariance since $\tau_3\nabla-ieA(x)$ is certainly a gauge-invariant combination for free electron field.

As for the polarization correction, we can easily show in a similar way that it vanishes for the static case ($q_0=0$) because

$$\int \mathrm{Tr}\,\Gamma_i(p-q/2,p+q/2)G(p+q/2)$$
$$\times\gamma_0(p+q/2,p-q/2)G(p-q/2)d^4p=0.$$

Thus the above proof is complete and independent of the Coulomb interaction which profoundly influences the polarization effect. Although the proof is thus rigorous, it is still somewhat disturbing since $K^{(1a)}$, $K^{(1b)}$ and $K^{(2)}$ are all infinite. Actually there is a certain ambiguity in the evaluation of $K^{(2)}$, Eq. (5.6), which is again similar to the one encountered in quantum electrodynamics.[18] An alternative way would be to expand quantities in q without making translations in p space. In this case we may write

$$-\Gamma^{(b)}(p+q/2,p-q/2)=\bar{\epsilon}(p+q/2)-\bar{\epsilon}(p-q/2)$$
$$-i\tau_2[\phi(p+q/2)+\phi(p-q/2)]$$
$$\approx \mathbf{p}\cdot\mathbf{q}/m-2i\tau_2\phi. \quad (5.11)$$

The first term then gives

$$\frac{e^2}{(2\pi)^3}\int\frac{\phi^2}{4E_p{}^5}\left(\frac{\mathbf{p}\cdot\mathbf{q}}{m}\right)^2 p_i\mathbf{p}\cdot\mathbf{q}d^3p\propto q^2q_i, \quad (5.12)$$

which is convergent and the same as the one obtained from Eq. (1.1) using the bare quasi-particle states. The second term also is finite and equal to

$$\frac{e^2}{(2\pi)^3}\int\frac{\phi^2}{E_p{}^3}\frac{p^2}{3m^2}q_id^3p+O(q^2)q_i\approx N\alpha^2q_i=n(e^2/m)q_i. \quad (5.13)$$

The last line follows from Eqs. (6.11) and (6.11′) below.

The calculation of $K^{(1)}$ from Eqs. (5.7) and (5.10), gives, on the other hand, the same value as Eq. (5.2), so that we get $(K_{ij}^{(1)}+K_{ij}^{(2)})q_j=0$ in the limit of small q. (The polarization correction is again zero.)

Since Eq. (5.13) is a contribution from the collective intermediate state (see Secs. 6 and 7), we may say that the collective state saves gauge invariance, as has been claimed by several people.[3,19]

It goes without saying that the effect of the vertex correction on K_{ij} will be felt also for real magnetic field. But as we shall see later, it is a small correction of order g^2N (except for the renormalization effects), and not as drastic as for the longitudinal case.

6. THE COLLECTIVE EXCITATIONS

In order to understand the mechanism by which gauge invariance was restored in the calculation of the Meissner effect, and also to solve the integral equations for general vertex interactions, it is necessary to examine the collective excitations of the quasi-particles. In fact, people[3] have shown already that the essential difference between the transversal and longitudinal vector potentials in inducing a current is due to the fact that the latter can excite collective motions of quasi-particle pairs.

We see that the existence of such collective excitations follows naturally from our vertex solutions Eq. (4.4). For taking $p=p'$, the second solution $\Gamma^{(b)}(p',p)$ becomes

$$\Gamma^{(b)}(p,p)=L(p)\tau_3-\tau_3L(p)$$
$$=2i\tau_2\phi, \quad (6.1)$$
$$\gamma^{(b)}=0.$$

In other words $\tau_2\phi(p)\equiv\Phi_0(p)$ satisfies a homogeneous integral equation:

$$\Phi_0(p)=-\frac{g^2}{(2\pi)^4}\int\tau_3G(p')\Phi(p')G(p')$$
$$\times\tau_3h(p-p')^2\Delta(p-p')d^4p. \quad (6.2)$$

We interpret this as describing a pair of a particle and an antiparticle interacting with each other to form a bound state with zero energy and momentum $q=p'-p=0$.

[18] H. Fukuda and T. Kinoshita, Progr. Theoret. Phys. (Kyoto) **5**, 1024 (1950).

[19] On the other hand, the way in which the collective mode accomplishes this end seems to differ from one paper to another. We will not attempt to analyze this situation here.

In fact, by defining

$$F(p,-p)\equiv -G(p)\Phi_0(p)G(p), \qquad (6.3)$$

Eq. (6.2) becomes

$$L(p)F(p,-p)L(p)=-\frac{g^2}{(2\pi)^4}\int \tau_3F(p',-p')\times\tau_3 h(p-p')^2\Delta(p-p')d^4p',$$

or

$$\sum_{j,l=1}^{2} L(p)_{ij}L^C(-p)_{kl}F(p,-p)_{jl}=\frac{-g^2}{(2\pi)^4}\int\sum_{j,l}(\tau_3)_{ij}(\tau_3)_{kl}F(p',-p')_{jl}\times h(p-p')^2\Delta(p-p')d^4p. \qquad (6.4)$$

The particle-conjugate quantity L^C was defined in Eq. (2.23).

Equations (6.2) and (6.4) are the analog of the so-called Bethe-Salpeter equation[20] for the bound pair of quasi-particles with zero total energy-momentum. $F_{ij}(p,-p)$ is the four-dimensional wave function with the spin variables i, j and the relative energy-momentum $(p_0,\mathbf{p})$.

Since there, thus, exists a bound pair of zero momentum, there will also be pairs moving with finite momentum and kinetic energy. In other words, there will be a continuum of pair states with energies going up from zero. We have to determine their dispersion law.

For a finite total energy-momentum q, the homogeneous integral equation takes the form

$$\Phi_q(p)\equiv L(\tfrac{1}{2}q+p)F(\tfrac{1}{2}q+p,\tfrac{1}{2}q-p)L(p-\tfrac{1}{2}q)=-g^2\frac{1}{(2\pi)^4}\int\tau_3F(\tfrac{1}{2}q+p',\tfrac{1}{2}q-p')\times\tau_3h(p-p')^2\Delta(p-p')d^4p. \qquad (6.5)$$

From here on we carry out perturbation calculation. Let us expand F and L in terms of the small change $L(p\pm q/2)-L(p)$, thus

$$\begin{aligned}F(\tfrac{1}{2}q+p,\tfrac{1}{2}q-p)&=F^{(0)}(p)+F^{(1)}(p,q/2)+\cdots,\\ L(p\pm q/2)&=L(p)+\Delta L(p,\pm q/2).\end{aligned} \qquad (6.6)$$

Collecting terms of the first order, we get

$$\begin{aligned}L(p)F^{(1)}(p,q/2)L(p)+U^{(1)}(p,q/2)&=-g^2\frac{1}{(2\pi)^4}\int\tau_3F^{(1)}(p',q/2)\times\tau_3h(p-p')^2\Delta(p-p')d^4p',\\ U^{(1)}(p,q/2)&=\Delta L(p,q/2)F^{(0)}(p)L(p)+L(p)F^{(0)}(p)\Delta L(p,-q/2).\end{aligned} \qquad (6.7)$$

[20] E. E. Salpeter and H. A. Bethe, Phys. Rev. **84**, 1232 (1951).

This is an inhomogeneous integral equation for $F^{(1)}$. In order that it has a solution, the inhomogeneous term $U(p)$ must be orthogonal to the solution $\Phi_0(p)$ of the homogeneous equation. This condition can be derived as follows:

We multiply Eq. (6.7) by $F^{(0)}(p)=-G(p)\Phi_0(p)G(p)$, and integrate thus:

$$\int \mathrm{Tr}\,F^{(0)}(p)L(p)F^{(1)}(p,q/2)L(p)d^4p+\int\mathrm{Tr}\,F^{(0)}(p)U^{(1)}(p,q/2)d^4p=-g^2\frac{1}{(2\pi)^4}\int\int\mathrm{Tr}\,F^{(0)}(p)\tau_3F^{(1)}(p',q/2)\times\tau_3h(p-p')^2\Delta(p-p')d^4pd^4p'.$$

In view of Eq. (6.5) the last line is

$$=\int\mathrm{Tr}\,L(p')F^{(0)}(p')L(p')F^{(1)}(p',q/2)d^4p',$$

so that

$$(F^{(0)},U^{(1)})\equiv\int\mathrm{Tr}\,F^{(0)}(p)U^{(1)}(p,q/2)d^4p=0. \qquad (6.8)$$

This is the desired condition.

For the evaluation of Eq. (6.8), we will neglect the p dependence of the self-energy terms. Thus

$$\begin{aligned}F^{(0)}(p)&=\tau_2\phi/(p_0^2-E_p^2+i\epsilon),\quad E_p^2=\epsilon_p^2+\phi^2,\\ \Delta L(p,q/2)&=q_0/2-\tau_3(\mathbf{p}\cdot\mathbf{q}/2m+(q/2)^2/2m).\end{aligned} \qquad (6.9)$$

We then obtain

$$(F^{(0)},U^{(1)})=2\pi i\int\frac{\phi^2}{E_p^3}\left[\left(\frac{q_0}{2}\right)^2-\left(\frac{\mathbf{p}\cdot\mathbf{q}}{2m}\right)^2-\frac{\epsilon_p}{m^2}\left(\frac{q}{2}\right)^2\right]d^3p=0,$$

or

$$\left(\frac{q_0}{2}\right)^2-\left(\frac{q}{2}\right)^2\left[\frac{1}{3}\frac{\bar{p}^2}{m^2}-\frac{\bar{\epsilon}_p}{m}\right]=0, \qquad (6.10)$$

where the average $\bar{f}$ is defined

$$\bar{f}=\int f(p)\frac{\phi^2}{E^3}d^3p\Big/\int\frac{\phi^2}{E^3}d^3p. \qquad (6.10')$$

The weight function $\phi^2/E_p^3=\phi^2/(\epsilon_p^2+\phi^2)^{\frac{3}{2}}$ peaks around the Fermi momentum, so that $p^2\sim p_F^2$, $\epsilon_p\sim 0$. Thus

$$q_0^2\approx q^2\frac{1}{3}\frac{\bar{p}^2}{m^2}\equiv\alpha^2q^2,\quad \alpha^2\approx p_F^2/3m^2, \qquad (6.11)$$

which is the dispersion law for the collective excitations.[2,3] We also note, incidentally, that

$$\frac{1}{(2\pi)^3}\int\frac{\phi^2}{E^3}d^3p\approx N=mp_F/\pi^2, \tag{6.11'}$$

$$\alpha^2N\approx p_F{}^3/3\pi^2m=n.$$

We would like to emphasize here that these collective excitations are based on Eq. (6.2), which takes account of the phonon-Coulomb scattering of the quasi-particle pairs, but does not take into account the annihilation-creation process of the pair due to the same interaction.

It is well known that this annihilation-creation process is very important in the case of the Coulomb interaction, and plays the role of creating the plasma mode of collective oscillations. We will consider it in a later section.

As for the wave function $F^{(1)}$ itself, we have still to solve the integral equation (6.7). But this can be done by perturbation because on substituting $U^{(1)}$ in the integrand, we find that all the terms are of the type (3.11). In other words, to the zeroth order we may neglect the integral entirely and so

$$F^{(1)}(p,q/2)=-G(p)U^{(1)}(p,q/2)G(p). \tag{6.12}$$

The original function

$$\Phi_q(p)=-L(p+q/2)F(p,q/2)L(p-q/2)$$

is even simpler. We get

$$\Phi_q(p)\approx\Phi_0(p) \tag{6.13}$$

to this order.

7. CALCULATION OF THE CHARGE-CURRENT VERTEX FUNCTIONS

In this section we determine explicitly the charge-current vertex functions Γ_i, ($i=0$, 1, 2, 3) from their integral equations. Only the particular combination $\Gamma^{(b)}$ of these was given before.

Let us first go back to the integral equation for Γ_0 generated by τ_3:

$$\begin{aligned}\Gamma_0(p+q/2,\ p-q/2)\\ =\tau_3-g^2\int\tau_3G(p'+q/2)\Gamma_0(p'+q/2,\ p'-q/2)\\ \times G(p'-q/2)\tau_3h(p-p')^2\Delta(p-p')d^4p',\end{aligned}$$

or

$$\begin{aligned}L(p+q/2)F_0(p+q/2,\ p-q/2)L(p-q/2)\\ =\tau_3+g^2\int\tau_3F_0(p'+q/2,\ p'-q/2)\\ \times\tau_3h(p-p')^2\Delta(p-p')d^4p'.\end{aligned} \tag{7.1}$$

For small g^2, the standard approach to solve the equation would be the perturbation expansion in powers of g^2. We know, however, that there are low-lying collective excitations, discussed before, to which τ_3 can be coupled, and these excitations do not follow from perturbation.[21]

Fortunately, if we assume $q=0$, $q_0\neq0$, then we have an exact solution to Eq. (7.1) in terms of $\Gamma^{(b)}$ of Eq. (4.4). Namely,

$$\begin{aligned}\Gamma_0(p+q/2,\ p-q/2)=\Gamma^{(b)}(p+q/2,\ p-q/2)/q_0\\ =\tau_3\{[Z(p+q/2)+Z(p-q/2)\}/2\\ +(p_0/q_0)[Z(p+q/2)-Z(p-q/2)]\}\\ -[\chi(p+q/2)-\chi(p-q/2)]/q_0\\ +i\tau_2[\phi(p+q/2)+\phi(p-q/2)]/q_0,\end{aligned} \tag{7.2}$$

which can readily be verified.

The second term is the result of the coupling of τ_3 to the collective mode. This can be understood in the following way. Γ_0 contains matrix elements for creation or annihilation of a pair out of the vacuum. These processes can go through the collective intermediate state with the dispersion law (6.11), so that Γ will contain terms of the form

$$R_\pm/(q_0\pm\alpha q).$$

The residues $R_\pm$ can be obtained by taking the limit

$$R_\pm=\lim_{q_0\pm\alpha q\to0}\Gamma_0(p+q/2,\ p-q/2)(q_0\pm\alpha q). \tag{7.3}$$

Applying this procedure to the integral equation (7.1) for Γ_0, we find that $R_\pm$ must be a solution of the homogeneous equation; namely,

$$R_\pm=C_\pm\Phi_q(p), \tag{7.4}$$

under the condition $q_0\pm\alpha q=0$.

For the particular case $q=0$, $\Phi_q(p)$ reduces to $\tau_2\phi(p)$, which in fact agrees with Eq. (7.2) if

$$C_\pm=-2i. \tag{7.5}$$

This observation enables us to write down Γ_0 for $q\neq0$. According to the results of Sec. 6, $\Phi_q(p)=\Phi_0(p)$ in the zeroth order in g^2N. Since corrections to the non-collective part of Γ_0 also turn out to be calculable by perturbation, we may now put

$$\begin{aligned}\Gamma_0(p+q/2,\ p-q/2)\approx\tau_3\bar{Z}+2i\tau_2\bar{\phi}q_0/(q_0{}^2-\alpha^2q^2),\\ \bar{\phi}\equiv[\phi(p+q/2)+\phi(p-q/2)]/2,\\ \bar{Z}\equiv[Z(p+q/2)+Z(p-q/2)\}/2\end{aligned}$$

to the extent that terms of order g^2N and/or the p-dependence of the renormalization constants are neglected.

In quite a similar way the current vertex $\mathbf{\Gamma}$ may be constructed. This time we start from the longitudinal

---

[21] If we proceeded by perturbation theory, we would find in each order terms of order 1.

component for $q_0=0$, $q\neq 0$, which has the exact solution

$$\Gamma(p+q/2,\,p-q/2)\cdot\mathbf{q}/q=-\Gamma^{(b)}(p+q/2,\,p+q/2)/q$$
$$=\frac{\mathbf{p}\cdot\mathbf{q}}{mq}\left\{1+\frac{\chi(p+q/2)-\chi(p-q/2)}{\mathbf{p}\cdot\mathbf{q}/m}\right\}$$
$$-\tau_3 p_0[\zeta(p+q/2)-\zeta(p-q/2)]/q$$
$$-2i\tau_2\frac{\phi(p+q/2)+\phi(p-q/2)}{2q}. \quad (7.7)$$

For $q_0\neq 0$, then, we get

$$\Gamma(p+q/2,\,p-q/2)\cdot\mathbf{q}/q$$
$$\approx(\mathbf{p}\cdot\mathbf{q}/q)\bar{Y}+2i\tau_2\phi\alpha^2 q/(q_0^2-\alpha^2q^2), \quad (7.8)$$
$$\bar{Y}\equiv 1+[\chi(p+q/2)-\chi(p-q/2)]/(\mathbf{p}\cdot\mathbf{q}/m).$$

Combining (7.6) and (7.8), the continuity equation takes the form

$$q_0\Gamma_0-\mathbf{q}\cdot\Gamma=q_0\tau_3\bar{Z}+(\mathbf{p}\cdot\mathbf{q}/m)\bar{Y}+2i\tau_2\bar{\phi}$$
$$\approx\Gamma^{(b)},$$

which is indeed zero on the energy shell.

The transversal part of Γ, on the other hand, is not coupled with the collective mode because the latter is a scalar wave.[22] We may, therefore, write instead of Eq. (7.8)

$$\Gamma(p+q/2,\,p-q/2)\approx(\mathbf{p}/m)\bar{Y}$$
$$+2i\tau_2\bar{\phi}\alpha^2 q/(q_0^2-\alpha^2q^2). \quad (7.10)$$

Equations (7.6) and (7.10) for Γ_i have a very interesting structure. The noncollective part is essentially the same as the charge current for a free quasi-particle except for the renormalization $\bar{Z}$ and $\bar{Y}$, whereas the collective part is spread out both in space and time. Neglecting the momentum dependence of $\bar{Z}$, $\bar{Y}$, and ϕ, we may thus write the charge-current density (ρ,j) as

$$\rho(x,t)\cong e\Psi^+\tau_3 Z\Psi(x,t)+\frac{1}{\alpha^2}\frac{\partial f(x,t)}{\partial t}\equiv\rho_0+\frac{1}{\alpha^2}\frac{\partial f}{\partial t}, \quad (7.11)$$
$$\mathbf{j}(x,t)\cong e\Psi^+(\mathbf{p}/m)Y\Psi(x,t)-\nabla f(x,t)\equiv\mathbf{j}_0-\nabla f,$$

where f satisfies the wave equation

$$\left(\Delta-\frac{1}{\alpha^2}\frac{\partial^2}{\partial t^2}\right)f\approx-2e\Psi^+\tau_2\phi\Psi. \quad (7.12)$$

$(\rho_0,\mathbf{j}_0)$ is the charge-current residing in the "core" of a quasi-particle. The latter is surrounded by a cloud of the excitation field f. In a static situation, for example, f will fall off like $1/r$ from the core. When the particle is accelerated, a fraction of the charge is exchanged between the core and the cloud.

The total charge residing in a finite volume around a core is not constant because the current $-\nabla f$ reaches out to infinity.

[22] There may be transverse collective excitations (Bogoliubov, reference 2), but they do not automatically follow from the self-energy equation nor affect the energy gap structure.

8. THE PLASMA OSCILLATIONS

The inclusion of the annihilation-creation processes in the equations of the previous sections means that the vertex parts get multiplied by a string of closed loops, which represent the polarization (or shielding effect) of the surrounding medium. We will call the new quantities Λ, which now satisfy the following type of integral equations

$$\Lambda(p',p)=\gamma-i\int\tau_3G(p'-k)\Lambda(p'-k,\,p-k)$$
$$\times G(p-k)\tau_3D(k)d^4k$$
$$+iD(p'-p)\tau_3\int\mathrm{Tr}[\tau_3G(p'-k) \quad (8.1)$$
$$\times\Lambda(p'-k,\,p-k)G(p-k)]d^4k,$$
$$D(q)\equiv-ig^2h(q)^2\Delta(q)+e^2/q^2.$$

$D(q)$ includes the effect of the Coulomb interaction [see Eq. (2.24)]. Putting

$$\bar{X}(p'-p)\equiv i\int\mathrm{Tr}[\tau_3G(p'-k)\Lambda(p'-k,\,p'-k)$$
$$\times G(p-k)]d^4k, \quad (8.2)$$

Eq. (8.1) takes the same form as Eq. (4.3) for Γ with the inhomogeneous term replaced by $\gamma+\tau_3D\bar{X}$, so that Λ is a linear combination of the Γ corresponding to γ and Γ_0:

$$\Lambda=\Gamma+\Gamma_0D\bar{X}. \quad (8.3)$$

Substitution in Eq. (8.2) then yields

$$\bar{X}(p'-p)=i\int\mathrm{Tr}[\tau_3G(p'-k)$$
$$\times\Gamma(p'-k,\,p-k)G(p-k)]d^4k$$
$$+iD(p'-p)\bar{X}(p'-p)\int\mathrm{Tr}[\tau_3G(p'-k)$$
$$\times\Gamma_0(p'-k,\,p-k)G(p-k)]d^4k,$$

or

$$\bar{X}(p'-p)=i\int\mathrm{Tr}[\tau_3G(p'-k)$$
$$\times\Gamma(p'-k,\,p-k)G(p-k)]d^4k$$
$$\times\left\{1-iD(p'-p)\int\mathrm{Tr}[\tau_3G(p'-k)\right.$$
$$\left.\times\Gamma_0(p'-k,\,p-k)G(p-k)]d^4k\right\}^{-1}$$
$$\equiv X(p'-p)/[1-D(p'-p)X_0(p'-p)]. \quad (8.4)$$

Especially for $\gamma=\tau_3$, we get

$$\bar{X}_0(p'-p)=X_0(p'-p)/[1-D(p'-p)X_0(p'-p)],\quad \Lambda_0(p',p)=\Gamma_0(p',p)/[1-D(p'-p)X_0(p',p)]. \tag{8.5}$$

To obtain the collective excitations, let us next write down the homogeneous integral equation:

$$\Theta_q(p)=-i\int \tau_3 G(p'+q/2)\Theta_q(p') \times G(p'-q/2)\tau_3 D(p-p')d^4p' +i\tau_3 D(q)\int \mathrm{Tr}\,[\tau_3 G(p'+q/2)\Theta_q(p') \times G(p'-q/2)]d^4p', \tag{8.6}$$

which means

$$\Theta_q(p)=\Gamma_0(p+q/2,\,p-q/2)D(q)\chi(q),$$

$$\chi(q)\equiv i\int \mathrm{Tr}[\tau_3 G(p'+q/2) \times \Theta_q(p')G(p'-q/2)\tau_3]d^4p'. \tag{8.7}$$

Substituting Θ_q in the second equation from the first, we get

$$1=D(q)X_0(q), \tag{8.8}$$

where $X_0(q)$ is defined in Eq. (8.4).

The solutions to Eq. (8.8) determine the new dispersion law $q_0=f(q)$ for the collective excitations.

With the solution (7.6), the quantity X_0 in Eq. (8.8) can be calculated. After some simplifications using Eq. (6.11), we obtain

$$X_0=\frac{1}{(2\pi)^3}\left[\frac{\alpha^2q^2}{q_0^2-\alpha^2q^2}\int\frac{\phi^2d^3p}{E_p(E_p^2-\alpha^2q^2/4)} +\frac{q^2}{4}\int\frac{\phi^2d^3p}{E_p(q_0^2/4-E_p^2)}\left(\frac{p^2}{3m^2E_p^2} -\frac{\alpha^2}{E_p^2-\alpha^2q^2/4}\right)\right]+O(q^4). \tag{8.9}$$

For $\alpha q\ll\phi$, and $q_0\gg\phi$ or $\ll\phi$, the second integral may be dropped and

$$X_0\cong\alpha^2q^2N/(q_0^2-\alpha^2q^2). \tag{8.10}$$

For small q^2, the dominant part of $D(q)$ in Eq. (8.8) is the Coulomb interaction e^2/q^2. Equation (8.8) then becomes

$$q_0^2=e^2\alpha^2N=e^2n\quad(q^2\to 0), \tag{8.11}$$

where n is the number of electrons per unit volume. This agrees with the ordinary plasma frequency for free electron gas.

We see thus that the previous collective state with $q_0^2=\alpha^2q^2$ has shifted its energy to the plasma energy as a result of the Coulomb interaction.

On the other hand, if Coulomb interaction is neglected, Eq. (8.8) leads to[23]

$$q_0^2=\alpha^2q^2[1-ig^2\Delta(q,q_0)h(q,q_0)^2N]. \tag{8.12}$$

The correction term, however, is of the order g^2N, hence should be neglected to be consistent with our approximation.

We can also study the behavior of X_0 in the limit $q_0\to 0$ for small but finite q^2:

$$X_0\approx\frac{1}{(2\pi)^3}\int\frac{\phi^2}{E^3}d^3p\approx N, \tag{8.13}$$

which comes entirely from the noncollective part of Γ_0, but again agrees with the free electron value.

Another observation we can make regarding $\bar{X}_0(q,q_0)$ is the following. $\bar{X}_0$ represents the charge density correlation in the ground state:

$$\bar{X}_0(q,q_0)=\int\langle 0|T(\rho(xt),\rho(0))|0\rangle e^{-i\mathbf{q}\cdot\mathbf{x}+iq_0t}d^3xdt.$$

If $|0\rangle$ is an eigenstate of charge, $\bar{X}_0$ should vanish for $q\to 0$, $q_0\neq 0$ since the right-hand side then consists of the nondiagonal matrix elements of the total charge operator Q:

$$\bar{X}_0(0,q_0)\propto\sum_n\left(\frac{1}{q_0-E_n}-\frac{1}{q_0+E_n}\right)|\langle n|Q|0\rangle|^2.$$

The converse is also true if $E_n>|q_0|$, $n\neq 0$ for some $q_0\neq 0$. Our result for $\bar{X}_0$, as is clear from Eqs. (8.5) and (8.9), has indeed the correct property in spite of the fact that the "bare" vacuum, from which we started, is not an eigenstate of charge.

9. CONCLUDING REMARKS

We have discussed here formal mathematical structure of the BCS-Bogoliubov theory. The nature of the approximation is characterized essentially as the Hartree-Fock method, and can be given a simple interpretation in terms of perturbation expansion. In the presence of external fields, the corresponding approximation insures, if treated properly, that the gauge invariance is maintained. It is interesting that the quasi-particle picture and charge conservation (or gauge invariance) can be reconciled at all. This is possible because we are taking account of the "radiative corrections" to the bare quasi-particles which are not eigenstates of charge. These corrections manifest themselves primarily through the existence of collective excitations.

There are some questions which have been left out. We would like to know, for one thing, what will happen if we seek corrections to our Hartree-Fock approximation by including processes (or diagrams) which have not been considered here. Even within our ap-

[23] Compare Anderson, reference 7.

proximation, there is an additional assumption of the weak coupling ($g^2N \ll 1$), and the importance of the neglected terms (of order g^2N and higher) is not known.

Experimentally, there has been some evidence[24] regarding the presence of spin paramagnetism in superconductors. This effect has to do with the spin density induced by a magnetic field and can be derived by means of an appropriate vertex solution. However, this does not seem to give a finite spin paramagnetism at 0°K.[25] The collective excitations do not play an important role here as they are not excited by spin density. [$\Gamma^{(a)}$, Eq. (4.4), does not have the characteristic pole.]

It is desirable that both experiment and theory about spin paramagnetism be developed further since this may be a crucial test of the fundamental ideas underlying the BCS theory.

ACKNOWLEDGMENT

We wish to thank Dr. R. Schrieffer for extremely helpful discussions throughout the entire course of the work.

[24] Knight, Androes, and Hammond, Phys. Rev. **104**, 852 (1956); F. Reif, Phys. Rev. **106**, 208 (1957); G. M. Androes and W. D. Knight, Phys. Rev. Letters **2**, 386 (1959).

[25] K. Yosida, Phys. Rev. **110**, 769 (1958).

Chapter 5: Nambu and the Pion and Spontaneously Broken Chiral Symmetry

On February 23, 1960, *Phys. Rev. Letters* **4**, *No. 7, 380 (1960)* received

Axial Vector Current Conservation in Weak Interactions by Yoichiro Nambu.

In 1960 at a Purdue University conference:

A 'Superconductor' Model of Elementary Particles and its Consequences by Yoichiro Nambu. (No proceedings exist, but Nambu's talk has been retrieved.)

On October 27, 1960, *Phys. Rev.* **122**, *No. 1, 345 (1960)* received

Dynamical Model of Elementary Particles Based on an Analogy with Superconductivity I by Yoichiro Nambu and Giovanni Jona-Lasinio.

On May 10, 1961, *Phys. Rev.* **124**, *No. 1, 246 (1961)* received

Dynamical Model of Elementary Particles Based on an Analogy with Superconductivity II by Yoichiro Nambu and Giovanni Jona-Lasinio.

This is the series of papers around the year 1960 in which Nambu made an analogy from his understanding of superconductivity to introduce spontaneous symmetry breaking in particle physics, laying the ground for pion physics, modern nuclear physics and in the end, Quantum Chromodynamics.

Background

In 1956 Tsung-Dao Lee (1926–) and Chen-Ning Yang (1922–) proposed[59] that parity is not conserved in the weak interactions which was subsequently experimentally verified by Chien-Shiung Wu (1912–1997) and collaborators with a paper[60] submitted on January 15, 1957. Several other experiments also showed the same result and another revolution in fundamental physics took place. By then, Lee and Yang had shared the physics Nobel Prize in 1957. Mme Wu was not included. All this is well known history and we will only mention a few important papers that came to

[59]T. D. Lee, C. N. Yang, *Phys. Rev.* **104**, 254 (1956).
[60]C. S. Wu *et al.*, *Phys. Rev.* **105**, 141 (1957).

be very important for Nambu when he turned his interest fully to particle physics after his momentous work on superconductivity.

The $(V-A)$ theory of the weak interactions

Once the parity violation of weak interactions was established, the race was on for a theory of the weak interactions. Since the most typical process, the β-decay of the nucleons and the decay of the muon only involve fermions, it was natural to look for a Hamiltonian with only fermions. The natural attempt would be the four-point couplings since the typical decays involve four particles. Already in 1934 Enrico Fermi (1901–1954) used the "criterion of simplicity" to propose a Hamiltonian as a product of two vector currents $V \cdot V$,

$$H = -G\bar{n}(x)\gamma^{\mu}p(x)\bar{e}(x)\gamma_{\mu}\nu(x) + h.c. \,.$$

In September 1957 the $(V-A)$ theory of weak interactions was proposed by Robert Marshak (1916–1992) and his student George Sudarshan (1931–2018). Journal publication was delayed due to a conference proceeding[61] because it contradicted a number of experiments by renowned experimentalists.[62]

The same month, R. P. Feynman and M. Gell-Mann submitted[63] the same $(V-A)$ arrangement in the Fermi interaction.

In both papers, the neutrino only exists with left-handed chirality.

The Feynman–Gell-Mann (FG) paper is very comprehensive and full of carefully conceived new ideas, and we discuss its highlights to set the stage for Nambu's papers.

There are five Lorentz-invariant types of Fermi's four-fermion interactions,

$$\bar{n}(x)O_i p(x)\bar{e}(x)O_i\nu(x)$$

where O_i represent the Lorentz covariant couplings, scalar, pseudoscalar, vector, axial vector and tensor $i = S, P, V, A, T$.

Left-handed neutrinos are represented through chiral projection $(1+\gamma_5)/2$, suggesting two possible combinations $S-P$ and $V-A$. Earlier that year, in a review[64] with Arthur Rosenfeld (1926–2017), Gell-Mann had proposed that the interaction should involve only V and A.

The combination $(V-A)$ was the right one. It was obviously a bold proposal in the light of the various experiments that contradicted it, but Feynman and Gell-Mann insisted that those experiments should be redone which they were, and eventually all agreed with their theory.

---

[61] E. C. G. Sudarshan, R. E. Marshak, *Proc. of the Padua-Venice Conf. on Mesons and Newly Discovered Particles*, Sept. 1957.

[62] Years later at Sudarshan's Festschrift, Robert Marshak apologized for the delay.

[63] R. P. Feynman, M. Gell-Mann, *Phys. Rev.* **109**, 193 (1958).

[64] M. Gell-Mann, A. H. Rosenfeld, *Ann. Rev. Nuc. Sc.* **25**, 555 (1957).

The weak leptonic current[65] J_l^μ as $\bar{e}(x)(\gamma_\mu - i\gamma^\mu\gamma^5)\nu(x)$ determined its hadronic counterpart J_h^μ as $\bar{n}(x)(\gamma^\mu - i\gamma^\mu\gamma^5)p(x)$, augmented with other pairs of hadrons and leptons lead to more interactions. An immediate success was a Hamiltonian that gave the lifetime of the muon within 2% within the experimental limit.

The theory was non-renormalizable, but they could *"imagine that the interaction is due to some intermediate (electrically charged) vector meson of very high mass M_0"*. Gell-Mann later called it X, Schwinger who had similar ideas called it W, a name that stuck as the charged intermediate vector boson of the Standard Model. Its coupling to the weak currents could be small since its large propagator mass would dominate small momentum transfers.

Considering the hadronic vector current, it is obvious that it is conserved in the classical Hamiltonian if we disregard the mass differences among the hadronic pairs. It is one of the generators, T^+ of the isotopic symmetry $SU(2)$ of the strong interactions, and hence it would be natural that it be conserved, the *Conserved Vector Hypothesis, CVC.*

On the other hand, the axial vector current is seemingly not conserved: in the Dirac Lagrangian, its divergence, proportional to the sum of its fermion masses, does not vanish. Another opposing reason is that, while the vector current generates the isotopic symmetry via Amalie Emmy Noether's (1882–1935) theorem, with the hadrons lying in its representations, there were no similar sets of "pseudohadrons".

Feynman and Gell-Mann pictured charged pion decays as a neutron-proton pair[66] that coupled weakly either to a muon and an antineutrino or to an electron and an antineutrino. Without recourse of the detailed strong interaction, they could calculate the branching ratio of the two decay modes $\pi^- \to e^- + \bar{\nu}, \pi^- \to \mu^- + \bar{\nu}$, to be 10^{-4}; the electron decay mode had yet to be observed, which they saw as a *"very serious discrepancy"* to their proposal. It was discovered a year later at CERN in accord with their prediction.

This paper set the stage for a lot of developments in the coming years when Nambu was so heavily involved in his search for a theoretical understanding of the BCS theory.

The Goldberger–Treiman Relation and PCAC

In a parallel work to the $V - A$ theory, Marvin Goldberger (1922–2014) and Sam Treiman (1925–1999) were studying[67] pion decay from the point of view of the strong interactions. As in the Fermi–Yang picture, they thought of pion decay as a nucleon–nucleon loop with the neutron and the proton in a four-point interaction

[65]The matrix γ^5 is defined slightly differently from one paper to another. Here we use the FG convention, but there can be differences from one subsection to another.
[66]E. Fermi, C. N. Yang, *Phys. Rev.* **76**, 1739 (1949).
[67]M. Goldberger, S. Treiman, *Phys. Rev.* **110**, 1178 (1958).

with a muon and antineutrino with a 99% branching ratio. They worked with an underlying quantum field theory in the back of their minds and a Fermi coupling for the weak part. They set up a general amplitude in terms of the ongoing fields,

$$\langle \mu\nu|\pi\rangle = i(2\pi)^4\delta(p_\mu + p_\nu - p_\pi)\langle\mu|\tilde{f}_\nu(0)|\pi\rangle(1+\gamma_5)u(p_\nu),$$

where $\tilde{f}_\nu(0)$ is the source of the neutrino field. Goldberger and Treiman then concentrated on the spinor coefficient of $(1+\gamma_5)u(p_\nu)$ with appropriate normalization,

$$\mathcal{M} \equiv \left(\frac{p_{\mu 0}}{m_\mu}\right)^{\frac{1}{2}} (2p_{\pi 0})^{\frac{1}{2}}\langle\mu|\tilde{f}_\nu|\pi\rangle.$$

By using the sources for the fields and making some appropriate translations and dropping terms that would not contribute to pion decay they could write it as,

$$\mathcal{M} = i\left(\frac{p_{\mu 0}}{m_\mu}\right)^{\frac{1}{2}} \int dx e^{-ip_\nu\cdot x}\theta(x)\langle\mu|[\tilde{f}_\nu, J(0)_+]|0\rangle,$$

where J is the source of the charged meson field.

It follows from invariance principles that $\mathcal{M}$ must be of the form

$$\mathcal{M} = F(p_\pi^2)\bar{u}(p_\mu)\gamma_5\gamma_\lambda {p_\pi}^\lambda.$$

All strong interaction parts here are in the unknown function F, like $J(0)$ in the above formula.

Much of the work lies now in setting up a dispersion relation for F. The imaginary part can be obtained by returning to the formula before and the whole exercise is now to find an expression for $F(0)$ in terms of some unknown function which can be regarded as a phase. They make strong assumptions about the unknown functions which cannot really be justified but end up with the following expressions

$$F(0) = -\frac{m}{2\pi^2}\sqrt{2}Gg_A\frac{J}{1+(G^2/4\pi)(2J/\pi)}$$

with

$$J = \int_0^\infty dk\frac{k^2}{(k^2+m^2)}\cos\varphi(k)\exp\left(\frac{2}{\pi}\int_0^\infty dk'\varphi(k')\left(\frac{1}{k'^2-k^2}-\frac{1}{k'^2+m^2}\right)\right),$$

where $\varphi(k)$ is the unknown phase. G is the pion–nucleon coupling constant ($G^2/4\pi \approx 15$) and $g_A \approx 1.26$ is the beta decay Gamow–Teller coupling constant, the strength of the axial vector coupling.

This is a remarkably complex expression and it seems impossible to make further progress. Yet, Goldberger and Treiman did when they applied these formulae to the pion decay rate, and rewrote it in terms of one unknown factor,

$$\left(\frac{J}{1+G^2J/2\pi^2}\right),$$

and then extracted its value from the data,

$$\frac{J}{1+G^2J/2\pi^2} = 0.13.$$

They observed that for a large range of values, $J \gg 0.1$, the one in the denominator can be neglected and J drops out, yielding

$$\left(\frac{J}{1+G^2J/2\pi^2}\right) \approx \frac{2\pi^2}{G^2} \approx 0.10,$$

a value surprisingly in good agreement with the experiments.

This is a very impressive result in the light of the long calculations and all the approximations and assumptions made. In a sense, this is a bit of great luck since the final expression is rather insensitive to the unknown functions, but a very important result emerged.

A fact that was not emphasized in their paper is that if we go back to the expression for $F(0)$, it will also be simplified. If we again neglect the unity in the denominator, the expression becomes again independent of the unknown function J and we find the relation,

$$F(0)G = -\sqrt{2}mg_A.$$

A series of steps, translating $F(0)$ into the pion scattering length f_π, taking into account the different normalizations of $\pi^+(x) = \frac{1}{\sqrt{2}}(\pi^1 + i\pi^2)$ and the axial vector $A^+_\mu = A^1_\mu + iA^2_\mu$, and that G is defined up to a sign, yielded

$$f_\pi G = m_n g_A,$$

known today as the *Goldberger–Treiman* relation. It was a most surprising result that involved only the pion decay constant, the mass of the nucleons and the axial vector weak coupling. It was satisfied to within some 10%.

In another important development, John C. Taylor (1930–) had found[68] that if the axial vector is strictly conserved, the pion cannot decay to muons and neutrinos, so now the general conclusion is that the axial vector could only be partially conserved. Even so the missing sector of "pseudohadrons" made the non-conservation natural. So what was the divergence of the axial vector?

Actually Goldberger and Treiman gave a clue. Going back to the expression for $\mathcal{M}$, we find that it is proportional to the pion momentum. It is then natural to state in the form,

$$\langle 0|A^\mu_a(0)|\pi_b(p)\rangle = if_{ab}\,p^\mu$$

with constant f_{ab}. Assuming isospin symmetry, $f_{ab} = \delta_{ab}f_\pi$

$$\langle 0|A^\mu_a(0)|\pi_b(p)\rangle = i\delta_{ab}\,f_\pi\,p^\mu$$

where f_π is the pion decay constant. The divergence follows,

$$\langle 0|\partial_\mu A^\mu_a|\pi_b(p)\rangle = \delta_{ab}f_\pi {m_\pi}^2.$$

Normalizing the pion field ϕ to $\langle 0|\phi_a(0)|\pi_b(p)\rangle = \delta_{ab}$, yields the operator relation,

$$\partial^\mu A^a_\mu = f_\pi m^2_\pi \phi^a.$$

This came to be known as PCAC, the Partial Conservation of the Axial Current. Note that it is proportional to the pion mass squared. If it is zero, the axial current is conserved, and since the mass is much smaller than the scale of strong interactions, the axial current is indeed "partially" conserved.

[68] J. C. Taylor, *Phys. Rev.* **110**, 1216 (1958).

Gell-Mann and Lévy

In 1960, four days before Nambu's first paper on the subject, Gell-Mann and Maurice Lévy (1922–2022) published a very influential paper on the subject. It is clear that they were bothered by the success of the Goldberger–Treiman relation and wanted to find some stronger theoretical ground to link it to PCAC in the operator form,

$$\partial^\mu A_\mu = \frac{ia}{\sqrt{2}}\pi^- .$$

Before that, they discussed the Fermi and the Gamow–Teller coupling constants G_V and G_A. In Gell-Mannesque fashion, they offered different hypotheses whether $G = G_\mu$, the coupling constant in muon decay, is a universal phenomena because of CVC. In a modern sense they are both connected. They discussed the decay of ^{14}O, where experiments had found that G_V/G_μ is less than one (.97). Where did this discrepancy come from? In a strike of genius they added a note in proof where they suggested that the vector current also contains a $\Delta S = 1$ strangeness-changing piece in addition to the expected $\Delta S = 0$. They computed the ratio of the two contributions to get a universal value and discovered in this way a small mixing angle that came to be named after Nicola Cabibbo (1935–2010).

They next discussed three models that lead to PCAC, none of them very realistic. This is a bit of a search in the dark. They started by showing that PCAC in the form above leads to the Goldberger–Treiman result, as we sketched above. It is done by also considering renormalization constants and unknown form factors to have their backs free. They pinpoint the PCAC result as the key to understand the Goldberger–Treiman result. It is interesting to see that since the Goldberger–Treiman paper was so complicated and full of assumptions PCAC was in fact hidden in it. This did not seem to be known at the time.

Another interesting aspect of the paper is when they discussed the symmetries from the vector and axial vector currents — they treated them like gauge symmetries, not particularly distinguishing between local and global symmetries. We will see that Nambu also did the same.

They proceeded to investigate the standard Yukawa pseudoscalar type Lagrangian but got an axial current whose divergence is not of the type above. They then introduced three models for pion–nucleon interaction.

(i) They start with the gradient coupling theory,

$$\mathcal{L}_1 = -\bar{N}(\gamma \cdot \partial + m_0 + if_0 \boldsymbol{\tau} \cdot \partial^\mu \boldsymbol{\pi} \gamma_\mu \gamma_5)N - \frac{(\partial \boldsymbol{\pi})^2}{2} - \frac{\mu_0^2 \boldsymbol{\pi}^2}{2}.$$

It has a chiral symmetry of the form

$$\delta N = 0, \qquad \delta \boldsymbol{\pi} = \frac{\mathbf{v}}{g_0},$$

and axial current,

$$\mathbf{P}_\mu = \bar{N}\boldsymbol{\tau}\gamma_\mu\gamma_5 N - \frac{i}{f_0}\partial_\mu\boldsymbol{\pi}.$$

Its divergence is not of the expected form, but they could show that they can derive the Goldberger–Treiman relation.

The model although much used in nuclear physics with a gradient coupling, leads to a nuclear potential of the expected form that is not renormalizable with no simple way to make it.

(ii) In the second example, they took up an idea of Schwinger and John Polkinghorne (1930–2021) and introduced a new scalar field σ with isotopic spin zero. Such a particle had not been found but being much heavier than the pion, it could decay strongly to two pions. They considered the following Lagrangian:

$$\begin{aligned}\mathcal{L}_2 = &- \bar{N}[\gamma\cdot\partial + m_0 - g_0(\sigma + i\boldsymbol{\tau}\cdot\boldsymbol{\pi}\gamma_5)]N - \left(\frac{\partial\boldsymbol{\pi}}{2}\right)^2 - \left(\frac{\partial\sigma}{2}\right)^2 - \frac{\mu_0^2\boldsymbol{\pi}^2}{2}\\ &- \left(\mu_0^2 + \frac{2\lambda_0}{f_0^2}\right)\frac{\sigma^2}{2} - \lambda_0[\boldsymbol{\pi}^2 + \sigma^2) - \frac{2}{f_0}\sigma(\sigma^2 + \boldsymbol{\pi}^2)],\end{aligned}$$

where $f_0 = g_0/2m_0$. The nucleon seems to have a mass term but by shifting the σ-field

$$\sigma \to \sigma - \frac{1}{2f_0}$$

they arrived at the Lagrangian (up to an additive constant)

$$\begin{aligned}\mathcal{L}_2 = &- \bar{N}[\gamma\cdot\partial + -g_0(\sigma + i\boldsymbol{\tau}\cdot\boldsymbol{\pi}\gamma_5)]N - \left(\frac{\partial\boldsymbol{\pi}}{2}\right)^2 - \left(\frac{\partial\sigma}{2}\right)^2 - \frac{\mu_0^2}{2}(\boldsymbol{\pi}^2 + \sigma^2)\\ &-\lambda_0\left[\boldsymbol{\pi}^2 + \sigma^2 - \frac{1}{4f_0}\right]^2 - \frac{\mu_0^2}{2f_0}\sigma.\end{aligned}$$

It is easy to see that apart from the last term, this Lagrangian is invariant under the chiral symmetry

$$\delta N = i\boldsymbol{\tau}\cdot\boldsymbol{v}\gamma_5 N, \quad \delta\boldsymbol{\pi} = -2\boldsymbol{v}\sigma, \quad \delta\sigma = 2\boldsymbol{v}\cdot\boldsymbol{\pi}.$$

This theory has a partially conserved axial symmetry, with the axial current

$$\mathbf{P}_\mu = \bar{N}\boldsymbol{\tau}\gamma_\mu\gamma_5 N + 2i(\sigma\partial_\mu\boldsymbol{\pi} - \boldsymbol{\pi}\partial_\mu\sigma),$$

and divergence,

$$\partial^\mu\mathbf{P} = -i\frac{\mu_0^2}{f_0}\boldsymbol{\pi},$$

which is the desired form for PCAC. We will compare this with Nambu's theory for the pion and PCAC later in the chapter. The field theory is renormalizable but introduces a scalar isoscalar field. Note that if the pion mass is zero, the symmetry is exact. This is also true if the nucleon mass is zero. The fact that the σ-coupling is responsible for the nucleon mass is a curious property of the model and not very satisfactory. There are strong renormalization effects on the theory.

The model is semi-realistic but predicts a scalar isoscalar meson. Such a particle has been searched for since then. It has been appearing and disappearing. The present situation is that there is some agreement that such a particle exists called $f_0(500)$ with some curious properties. It gives a very broad resonance in the pion–pion spectrum and in other channels and does not seem to have the properties predicted by the model. In modern terms, we do not see a quark–antiquark bound state.

(iii) In the third model, they introduce "the nonlinear σ-model". Since they are uncomfortable with the introduction of a new particle, they get rid of it by making it a function of the pion field. In order to save the symmetry they introduce the constraint

$$\pi^2 + \sigma^2 = \frac{1}{4f_0^2}.$$

This leads to a positive mass term for the nucleon but the theory is not renormalizable. Yet it has the same symmetries as the linear σ-model and is therefore an example of PCAC. It was used very much later in more general contexts.

Nambu's first paper

Nambu's attempt to understand the Goldberger–Treiman relation grew out of his ideas to use his knowledge of the BCS theory, spontaneous symmetry breaking and the Nambu–Goldstone boson applied to particle physics. In this short paper he was not referring to that knowledge, but it is certainly behind his thinking. He started with an expression for a conserved axial current, which is certainly different from the results of Gell-Mann and Lévy. He called it Γ_μ^A. We change notations a bit to conform with the previous discussions,

$$\Gamma_\mu^A(p', p) = i\bar{N}(p')\gamma_5\gamma_\mu N(p) - 2M\bar{N}(p')\gamma_5 q_\mu/q^2 N(p).$$

M is the nucleon mass and q is the momentum difference. This expression has two appealing points to Nambu. One is the modest renormalization effect on the axial vector beta decay constant ($g_A/g_V \approx 1.25$). He acknowledged that it is wrong from different points of view: it forbids the decay $\pi \to e + \nu$, since the divergence is zero, even though it had been observed, and it is not compatible with Goldberger–Treiman, so it had to be discarded. Note that this signals a massless pole, which must be the pion.

Now he makes the ingenious conclusion that if one introduces a small pion mass and allow for new form factors to make it more general, one can attempt an expression like,

$$g_A\Gamma_\mu^A(p', p) = g_V\left[iF_1(q^2)\,\bar{N}(p')\gamma_5\gamma_\mu N(p) - 2mF_2(q^2)\,\bar{N}(p')\gamma_5\,\frac{q_\mu}{q^2 + m_\pi^2}\,N(p)\right],$$

$$F_1(0) \approx g_A/g_V \approx F_2(0), \qquad F_1(q^2) \sim F_2(q^2) \quad q^2 \gg m_\pi^2.$$

This leads to PCAC and the pion is still the Nambu–Goldstone boson although with a small mass different from the strong interaction scale. He is now setting up dispersion relations for the form factors around $q^2 = -m_\pi^2$,

$$F_i(q^2) = F_i(m_\pi^2) - (q^2 + m_\pi^2) \int_{(3m_\pi)^2}^{\infty} \frac{\rho_i(m^2)}{(q^2 - m^2)(m^2 - m_\pi^2)} dm^2,$$

with the assumption that the first open channel is 3π.

Since we are mostly interested in very low momentum transfers, we can expect $F_1(0) = g_A/g_V$.

From here Nambu argued that the second term in the axial current is directly related to the pion decay via the pion propagator, and it is only required to be considered when discussing pion decay, even though it is small compared to the first one. He found,

$$\sqrt{2}G_\pi g_\pi = 2mg_V F_2(-m_\pi^2) \approx 2Mg_A,$$

where G_π is the pion–nucleon coupling constant and g_π is the pion decay constant previously and usually denoted f_π. Hence Nambu could directly read off the Goldberger–Treiman relation. This is certainly a much more direct derivation than Goldberger and Treiman's. It is also fairly model-independent, in contrast to the Gell-Mann and Lévy models.

Taken as such, the result was like pulling it out of a hat. There is no flaw in the argument, except perhaps some assumptions about weak dependence of q^2 of the form factors, and this can look a bit farfetched. As we will see in his next paper, Nambu had a lot more up his sleeve.

In the rest of this short paper he extended the discussion to the strangeness-changing beta decays, which was natural at the time but not so important for further development.

The Purdue talk

There was a conference at Purdue University in the late spring/early summer of 1960. It did not seem to have proceedings, and the written version of the talk by Nambu was missing for many years but was discovered in the 1990s when the first volume of Nambu's selected works was to be published. A similar talk at the Rochester conference later that year can be found today on the web. Giovani Jona-Lasinio (1922–) delivered the talk, and his two later papers with Nambu show that he was fully involved in Nambu's ideas and work. It was a very useful talk for understanding his ideas which he spelled them out very clearly leaving stricter proofs and applications in their two papers that we will discuss in the next section.

Nambu had said that he felt much freer to discuss his ideas when he gave a talk and then wrote it up rather than publishing it in a journal. That is why the notes from his talks are gold mines in order to dig deeper into his ideas.

(i) This talk described his ideas for applying the structure of his BCS work to the strong interactions. His first observation was that the Bogolyubov formulation of BCS was a good starting point for a discussion of the fermions in the Dirac equation. As we saw before, the electrons near the Fermi surface in the BCS theory satisfy the following equations,

$$\begin{aligned} E\psi_{p+} &= \epsilon_p \psi_{p+} + \phi \psi^{\dagger}_{-p-}, \\ E\psi^{\dagger}_{-p-} &= -\epsilon_p \psi^{\dagger}_{-p-} + \phi \psi_{p+}, \end{aligned}$$

where ψ_{p+} is the wave function of an electron of momentum p and spin up and $\psi^{\dagger}_{-p-}$ is that for a hole of momentum p and spin up, i.e. the absence of an electron of momentum $-p$ and spin down. Solving for the energy, one gets

$$E_p = \pm\sqrt{\epsilon_p^2 + \phi^2}.$$

Hence ϕ is the gap, the energy needed to lift a quasi-electron from the lower to the upper state. As Nambu showed in the superconductivity paper, this is a non-perturbative solution that cannot be reached by perturbation calculations. The gap ϕ can be calculated from the theory, and is an inherent consequence of the theory.

Nambu now compared this to the Dirac equation written in the Weyl representation.

$$\begin{aligned} E\psi_1 &= \boldsymbol{\sigma} \cdot \boldsymbol{p}\, \psi_1 + m\psi_2, \\ E\psi_2 &= -\boldsymbol{\sigma} \cdot \boldsymbol{p}\, \psi_2 + m\psi_1, \\ E_p &= \pm\sqrt{p^2 + m^2}, \end{aligned}$$

where ψ_1 and ψ_2 are the two chirality states.

The idea here is that the mass of the fermion can also be a feature of the interacting theory and not a free parameter. This is certainly a novel idea!

He now developed his program and made the following comparisons,

Superconductivity	Elementary Particle
Free electrons	Bare fermions (zero or small mass)
QED	Some unknown strong interaction
Energy gap	Observed fermion mass
Collective excitations	Meson, bound fermion pair
Charge	Chirality
Gauge Invariance	Chiral invariance

The real ingenuity is to proceed without knowing the interaction. In the end, the idea is to extract information in a fairly model independent way. Nambu was discussing a pure fermion theory similar to Heisenberg's ideas but also a non-abelian gauge theory. He found the last alternative interesting but chose not to use it, since *"we do not know whether a finite observed mass (of the vector particle) can be compatible with the invariance assumption."*

(ii) The second attempt was to work with a pure fermion field theory. The aim was to show that a bound state meson can be created in the theory. He discussed a field theory of the type Heisenberg advocated in the late 1950s. In modern terms we can think of a consistent, renormalizable theory (say QCD, even though he did not think of quarks but nucleons as the basic fermion constituents), where we integrate out all fields except the fermion fields, and then throw away all terms but the lowest four-point coupling. Such an effective field theory is, of course, non-renormalizable but one can use a cut-off to discuss higher-order terms. We can assume that such a procedure can be a good approximation at low momenta. The parameters of the theory is then the coupling constant g and the cut-off K. The starting point is the Lagrangian

$$\mathcal{L} = -\bar{\psi}\gamma_\mu \partial^\mu \psi - g[\bar{\psi}\psi\,\bar{\psi}\psi - \bar{\psi}\gamma_5\psi\,\bar{\psi}\gamma_5\psi].$$

Note that the field is massless. Then the theory is invariant under two $U(1)$ symmetries. One is the charge invariance and the other is the chiral invariance

$$\psi \to e^{i\alpha}\psi \qquad \bar{\psi} \to \bar{\psi}e^{-i\alpha},$$
$$\psi \to e^{i\alpha'\gamma_5}\psi \quad \bar{\psi} \to \bar{\psi}e^{+i\alpha'\gamma_5}.$$

There are two conserved currents

$$V^\mu = \bar{\psi}\gamma^\mu\psi, \qquad A^\mu = \bar{\psi}\gamma_5\gamma^\mu\psi.$$

Nambu now treated this in the Hartree–Fock approximation. In modern language, he assumes that there is a spontaneous symmetry breaking with nonzero vacuum expectation values of $\bar{\psi}\psi$ and $\bar{\psi}\gamma_5\psi$ and changes of variables to

$$\bar{\psi}\psi \to \bar{\psi}\psi + \langle\bar{\psi}\psi\rangle,$$

$$\bar{\psi}\gamma_5\psi \to \bar{\psi}\gamma_5\psi + \langle\bar{\psi}\gamma_5\psi\rangle.$$

When we introduce this back into the Lagrangian we find that we have generated a mass term for the fermion field.

$$\begin{aligned} m &= 2g[\langle\bar{\psi}\psi\rangle - \gamma_5\langle\bar{\psi}\gamma_5\psi\rangle] \\ &= [TrS^{(m)}(0) - \gamma_5 Tr(\gamma_5 S^{(m)}(0))], \end{aligned}$$

where $S^{(m)}(x)$ is the fermion Green's function. In momentum space we arrive at

$$m = \frac{g}{(2\pi)^3}\int \frac{m\,d^3p}{\sqrt{p^2+m^2}}.$$

The perturbative solution $m = 0$ is obvious, but with a cut-off K, one finds another solution

$$\frac{\pi^2}{|g|\,K^2} = \sqrt{1+m^2/K^2} - m^2/K^2 sinh^{-1}\,|K/m|,$$

provided that $g < 0$ and $\pi^2 < |g|K^2$.

The important thing here is not the exact value of m but the fact that there is a solution with a positive mass in a theory that starts out classically to be massless. This is the same phenomena as is the BCS theory where the gap is a truly quantum effect.

Like in the superconductivity paper, Nambu showed an infinite series of Feynman diagrams with this solution that is hence a truly non-perturbative effect.

Nambu then discussed how the axial vector current must be corrected to still be conserved. Note that we have not tampered with the symmetries, so the axial current must still be conserved even though it cannot be seen in the leading order. A generalized form for the current is now suggested to be

$$A^{\mu} = \bar{\psi}\left(\gamma_5\gamma^{\mu} + \frac{2im\gamma_5 q^{\mu}}{q^2}\right)\psi F(q^2).$$

The key fact here is that there is a pole signaling a massless particle, the meson, the collective excitation. The axial symmetry is spontaneously broken and the net result is a massive fermion and a massless meson, the Nambu–Goldstone boson. It comes out as a bound state of the fermions and must be there.

However, we know that the current cannot be conserved (PCAC), so Nambu broke chiral invariance by adding a small fermion mass, much smaller than the strong interaction scale, just enough to have a partially conserved axial current. He suggested the form,

$$A^{\mu} = \bar{\psi}\left(\gamma_5\gamma^{\mu} + \frac{2im\gamma_5 q^{\mu}}{q^2 + \mu^2}\right)\psi F(q^2),$$

with μ the tiny meson (pion) mass. This formula is the one he had used in the *Phys. Rev. Lett.* papers above to derive the Goldberger–Treiman relation, and the background for that paper is the basis of this talk. Note the difference with the BCS theory where the gauge invariance is not broken even though the plasma mode is massive. Here, we see another side of Nambu's genius in that he is not following the BCS slavishly. In the Standard Model, PCAC comes from the small masses of the u- and d-quark.

Nambu's theory for the emergence of the pion as a bound state and as a quasi-Nambu–Goldstone boson with a tiny mass was a completely new way to understand the pion. As we saw earlier, the pion as a nucleon–antinucleon bound state had been suggested in 1949 by Fermi and Yang, but it was not really the mainstream idea, especially after the multitude of new particles were discovered when the big accelerators came into operation. The remarkably small mass of the pion compared to the strong interaction scale is one of the key elements of nuclear physics and the whole formalism of spontaneous symmetry breaking came to be one of the key features of the Standard Model when it was developed.

Nambu–Jona–Lasinio I and II

(I) In this paper[69] they treated the ideas and calculations of the Purdue talk in more detail and with more discussions. Here they only discussed a theory with symmetry under the axial current.

Once they concluded that there are both a massless and massive fermion solution to the self-energy solution they repeated a calculation that Nambu did in the superconductivity paper, namely they expanded the two solutions in creation and annihilation operators and defined a vacuum for each, $\Omega^{(0)}$ and $\Omega^{(m)}$. They then calculated the overlap between them,

$$(\Omega^{(0)}, \Omega^{(m)})$$

and found that it diverges as $e^{\int d^3p}$. From this, they concluded that the overlap must be zero.

When they looked deeper into the γ_5 invariance, they found an infinity of choices signaling a $U(1)$ invariance. This will be more important in the future development of spontaneous symmetry breaking.

They also discussed further the collective excitations. The pion is only the lowest lying excitation. They also found a scalar state with mass $2m$ as well as a pseudovector state and a state with nucleon number 2. It is clear that here they are out on thin ice. At the time of the work, the large accelerator had only started to collide protons on protons at high energies and the multitude of resonances decaying under the strong interactions had not yet been discovered. That happened during the 1960s. This paper was published slightly before the paper[70] by Geoffrey Chew (1924–2019) and Steven Frautschi (1933–), where they conjectured that strongly interacting particles lie on straight Regge trajectories when the spin is plotted against the mass squared. This introduced a new parameter of the strong interaction, the "Regge slope". This is clearly a consequence of QCD which could not be discovered by the analysis of Nambu and Jona-Lasinio.

(II) In the accompanying paper[71] which was submitted half a year later they extended the analysis to involve both isospin and a small bare fermion mass. They repeated the calculation to compute the emerging fermion mass and showed that it is essentially the same as in the previous calculation.

They again went through the emerging mesons that come out as bound states of a fermion–antifermion pair. The pion is an isovector and is the Nambu–Goldstone boson although with a small mass because of the small bare fermion mass.

They further studied a scalar isoscalar meson and found that an unbound pseudo scalar meson. They also discussed the case of one scalar, like the σ-particle of Schwinger and Gell-Mann and Lévy, that should be checked experimentally.

[69] Y. Nambu and G. Jona-Lasinio, *Phys. Rev.* **122**, 345 (1961).
[70] Geoffrey F. Chew, S. C. Frautschi, *Phys. Rev. Lett.* **7**, 394 (1961).
[71] Y. Nambu, G. Jona-Lasinio, *Phys. Rev.* **124**, 246 (1961).

As we already saw in the section about the ω-particle, we should expect an isovector vector particle as well as an isoscalar vector particle, and both are discussed. As said before, these discussions are premature in the light of what we know now from QCD in terms of three families of quarks where still the meson masses is a deep problem. They discussed strange particles and attempted to derive results for the kaon and other particles, but again that discussion is premature but extremely interesting. This shows the deep knowledge they have of the situation of that day and how much of that was possible due to the knowledge.

One fascinating calculation which still holds is the estimate of how big the bare fermion mass can be to account for the small pion mass. They found,

$$m_0 \lesssim 5 \text{ MeV}.$$

In retrospect, we know that the masses of the u- and d-quarks are in this range.

The conclusion we can draw is that calculations of the fermion and pion parameters at low momentum are quite well described in their analysis while the absence of a theory for the strong interactions (QCD) makes calculations at higher momenta a bit dubious. The authors were well aware of this.

Epilog

A similar model was investigated by Valentin Vaks (1932–) and Anatoly Larkin (1932–2005) but published in a Soviet journal,[72] it was not much noticed in the West.

Nambu's work was regarded as quite complicated when it was published, but directly when the first preprint came out Jeffrey Goldstone (1933–) got interested and wrote a very influential paper.[73] He started by first checking a renormalizable quantum field theory with a fermion interacting with a pseudoscalar boson with negative mass squared. After lengthy calculations he found a non-perturbative solution with a positive boson mass squared. He then went on to study a pure bosonic theory, first for a real field and then for a complex field with a $U(1)$ symmetry, again with a negative mass square. In the latter case he found a solution where one component of the field is massless and the other massive.

This was a result that caught people's eyes. Soon after, together with Abdus Salam and Steven Weinberg, he formalized[74] it to a proof that in a field theory with spontaneously broken global symmetry, there is always a massless boson. Now the problem is rather that there is no massless scalar field in Nature except the nearly massless pion. When non-abelian gauge fields were tried for the weak interactions and the symmetry was broken spontaneously there would be a remnant of a massless scalar. All this was solved in 1964 in the works of Robert Brout (1928–2011) and

[72] V. G. Vaks, A. I. Larkin, *Sov. Phys. JETP* **13**(1), 192.

[73] J. Goldstone, *Nuovo Cimento* **19**, 154 (1961).

[74] J. Goldstone, A. Salam and S. Weinberg, *Phys. Rev.* **127**, 965 (1962).

François Englert (1932–), Peter Higgs (1929–), and also Gerald Guralnik (1936–2014), Carl Hagen (1937–) and Tom Kibble (1932–2016) used a complex scalar field to break the symmetry. Alexander Migdal (1945–) and Alexander Polyakov (1945–) used dynamical breaking[75] like Nambu's. The history is so well-known, culminating in the 2013 Nobel to Englert and Higgs, as Brout had died two year earlier.

The massless scalar was named the Goldstone boson from the beginning and only later it was called Nambu–Goldstone boson, even though Goldstone clearly remarked that he got the idea from Nambu's work.

It took quite some time before the Nambu and Jona-Lasinio model became popular. In the mid-1960's Weinberg and others introduced "chiral perturbation theory" in which he instead considered an effective theory with only pions. Thus he could look at possible terms in the interaction, consistent with the chiral symmetry, and compute a manifold of results. He proposed and used it as an alternative to the current algebra that was very popular at that time.

Instead came the development of the Standard Model with its electroweak sector and the strong one in terms of QCD in the 1970s. With the discovery of asymptotic freedom by David Politzer, David Gross and Frank Wilczek (Nobel Prize in 2004), it became clear that perturbation expansions could only be used to study processes with large momenta. It then became popular to use the Nambu–Jona-Lasinio framework as an effective field theory to study quark dynamics at small momenta where the coupling constant is large and not amenable to perturbation expansions. This has turned out to be very useful in order to study such questions as bound states and the meson spectrum. It is still a very popular method and a wealth of results from QCD has been achieved as a result.

In 2008, Nambu was awarded the Nobel Prize in physics "*for the discovery of the mechanism of spontaneous broken symmetry in subatomic physics.*"

[75]A. A. Migdal, A. M. Polyakov, *Soviet Physics JETP* **24**, 91 (1967).

VOLUME 4, NUMBER 7 PHYSICAL REVIEW LETTERS APRIL 1, 1960

AXIAL VECTOR CURRENT CONSERVATION IN WEAK INTERACTIONS*

Yoichiro Nambu
Enrico Fermi Institute for Nuclear Studies and Department of Physics
University of Chicago, Chicago, Illinois
(Received February 23, 1960)

In analogy to the conserved vector current interaction in the beta decay suggested by Feynman and Gell-Mann, some speculations have been made about a possible conserved axial vector current.[1-3] One can formally construct an axial vector nucleon current, which satisfies a continuity equation,

$$\Gamma_\mu^A(p',p) = i\gamma_5\gamma_\mu - 2M\gamma_5 q_\mu/q^2, \quad q = p' - p, \tag{1}$$

where p and p' are the initial and final nucleon momenta. Such an attempt has some appeal in view of the apparently modest renormalization effect on the axial vector beta decay constant ($g_A/g_V \approx 1.25$), although the second appealing point,[1] namely, the possible forbidding of $\pi \to e + \nu$, has now lost its relevance.

The expression (1), unfortunately, can be easily ruled out experimentally, as was pointed out by Goldberger and Treiman,[3] since it introduces a large admixture of pseudoscalar interaction.

On the other hand, Eq. (1) arouses theoretical curiosity as to the origin of the second term if it really exists; according to our conventional field theory, we would have to interpret the denominator q^2 as implying a massless, pseudoscalar, and charged quantum bridging the nucleon and lepton currents.

We would like to suggest that there may not be a strict pseudovector current conservation, but that we may have an approximate conservation which becomes rigorous in the limit $q^2 >> m_\pi^2$, m_π being the pion mass. Specifically, we propose that the axial vector part of the nucleon beta decay vertex has the following form and properties:

$$g_A \Gamma_\mu^A(p',p)$$

$$= g_V\left[i\gamma_5\gamma_\mu F_1(q^2) - \frac{2M\gamma_5 q_\mu}{q^2+m_\pi^2} F_2(q^2)\right],$$

$$F_1(0) = g_A/g_V \approx F_2(0);$$

$$F_1(q^2) \sim F_2(q^2) \quad \text{for} \quad q^2 >> m_\pi^2. \qquad (2)$$

The pion is then the analog of the massless quantum mentioned above. This is consistent with the dispersion relations expected for Γ_μ^A. Namely, F_1 and F_2 should have in general the form

$$F_i(q^2) = F_i(-m_\pi^2)$$

$$-(q^2+m_\pi^2)\int_{m_0^2}^{\infty} \frac{\rho_i(m^2)dm^2}{(q^2+m^2)(m^2-m_\pi^2)}$$

$$(i = 1, 2), \qquad (3)$$

where $m_0 = 3m_\pi$ unless there are new particles of low mass. Thus the F's will be slowly varying for $|q^2| << m_0^2$. The conditions in Eq. (2) imply that $F_1/F_2 \approx 1$ for all q^2. If $m_\pi = 0$ and $F_1/F_2 \equiv 1$, then we restore exact current conservation,[2] and we also expect $F_1(0) = g_A/g_V = 1$.

If we adopt Eq. (2), the second term of Γ_μ^A immediately gives a relation between g_A, the pion decay (pseudovector) constant g_π, and the pion-nucleon (pseudoscalar) coupling G_π:

$$2Mg_A \approx 2Mg_V F_2(-m_\pi^2) = \sqrt{2}\, G_\pi g_\pi. \qquad (4)$$

With $g_A = 1.25\, g_V = 1.75\times10^{-49}$ erg cm^3,[4] $G_\pi^2/4\pi = 13.5$, this gives a π-μ decay life of 2.7×10^{-8} sec as compared with the observed value 2.56 $\times10^{-8}$ sec.

Goldberger and Treiman[5] have arrived at the same relation Eq. (4) (in the limit of their self-energy integral $J\to\infty$) from an entirely different approach. In our opinion, this is not a coincidence, as will be explained elsewhere.

We are tempted to extend this approximate conservation of the axial vector (and naturally also the vector current) to the strangeness-nonconserving beta decays. We take, for example, the ΛN axial vector in the form

$$\Gamma_\mu^A(p_N', P_\Lambda) \approx i\gamma_5\gamma_\mu - \frac{(M_\Lambda + M_N)\gamma_5 q_\mu}{q^2+m_K^2}, \qquad (5)$$

and attribute the second term to the pseudoscalar K meson.[6] The degree of accuracy of the relation (5) will be poorer than in the previous case in view of the Λ-N mass difference (which destroys vector conservation) and the large K-meson mass. At any rate, we obtain an analog of Eq. (4):

$$(M_\Lambda + M_N)\, g_A' \approx G_K g_K, \qquad (6)$$

which relates the Λ beta decay axial vector coupling g_A', the ΛNK coupling G_K, and the K_μ decay coupling g_K.

With the observed $K_{\mu 2}$ lifetime 2.1×10^{-8} sec and a tentative value $G_K^2/4\pi = \frac{1}{4}G_\pi^2/4\pi$, we get

$$g_A'/g_A \approx 1/10. \qquad (7)$$

This is not inconsistent with the observed beta decay of Λ which seems an order of magnitude less than predicted from a universal coupling scheme $g_V' = g_A' = g_V$.[7]

We can still go further, though the argument becomes more arbitrary. Let us assume that a fundamental weak coupling $(\bar{N}N\bar{N}\Lambda)$ gives rise to an effective V-A interaction (or at least part of it) of the form

$$g''(\Gamma_\mu^V - \Gamma_\mu^A)_{\bar{N}N}(\Gamma_\mu^V - \Gamma_\mu^A)_{\bar{N}\Lambda}. \qquad (8)$$

Here $\Gamma_\mu^V = i\gamma_\mu$ which is approximately conserved by itself, and Γ_μ^A stands for Eq. (2) or (5). We see easily that Eq. (8) contains information about the $\Lambda\to N+\pi$ decay matrix element:

$$(2M_N g''/\sqrt{2}\,G_\pi) q_\mu (\Gamma_\mu^V - \Gamma_\mu^A)_{\bar{N}\Lambda}. \qquad (9)$$

Combined with the assumption of $\Delta T = \frac{1}{2}$ selection rule, this gives a lifetime of 2.5×10^{-10} sec

VOLUME 4, NUMBER 7 PHYSICAL REVIEW LETTERS APRIL 1, 1960

for $g'' = g_V$ as compared with the observed value 2.8×10^{-10} sec.

It is possible to apply this kind of consideration to other hyperons. Moreover, if the Feynman–Gell-Mann coupling scheme such as $(\pi\pi e\nu)$ is formally extended to $(K\pi e\nu)$, etc. as has been tried by some people, all the observed decay processes may be covered. Here we would like to point out that if all baryons should satisfy Eqs. (4) and (6), the ratios g_A/G_π and g_A'/G_K must be approximately common constants.

Our final remark concerns the theoretical basis for the assumptions made here. If the baryons are derived from some fundamental field ψ which possesses an invariance under a transformation of the type $\psi\to\exp(i\vec{\alpha}\cdot\vec{\tau}\gamma_5)\psi$,[8] then there will be a conservation of the pseudovector charge-current. A finite observed mass can be compatible with the conservation if the particle is coupled with a boson as was noted in Eq. (1).

This situation may be understood by making an analogy to the theory of superconductivity originated by Bardeen, Cooper, and Schrieffer,[9] and refined by Bogoliubov.[10] There gauge invariance, the energy gap, and the collective excitations are logically related to each other as was shown by the author.[11] In the present case we have only to replace them by γ_5 invariance, baryon mass, and the mesons. In fact, the mathematical method used in superconductivity may be taken over to study the self-energy problem of elementary particles. It is interesting that pseudoscalar mesons automatically emerge in this theory as bound states of baryon pairs. The nonzero meson masses and baryon mass splitting would indicate that the γ_5 invariance of the bare baryon field is not rigorous, possibly because of a small bare mass of the order of the pion mass.

The above-mentioned model of elementary particles will be studied in a separate paper.

*This work was supported by the U. S. Atomic Energy Commission.

[1]J. C. Taylor, Phys. Rev. 110, 1216 (1958).

[2]J. C. Polkinghorne, Nuovo cimento 8, 179 and 781 (1958).

[3]M. L. Goldberger and S. B. Treiman, Phys. Rev. 110, 1478 (1958).

[4]A. I. Alikhanov, Ninth Annual International Conference on High-Energy Physics, Kiev, 1959 (unpublished).

[5]M. L. Goldberger and S. B. Treiman, Phys. Rev. 110, 1178 (1958); M. L. Goldberger, Revs. Modern Phys. 31, 797 (1959).

[6]It is also possible to associate a scalar K meson with the ΛN vector current conservation, while leaving the axial vector unaccounted for.

[7]Again Eq. (5) and the subsequent conclusions are essentially the same as those of C. H. Albright, Phys. Rev. 114, 1648 (1959) and B. Sakita, Phys. Rev. 114, 1650 (1959), which are based on the Goldberger-Treiman method. For the Λ-decay case below, see L. Tenaglia, Nuovo cimento 14, 499 (1959).

[8]F. Gürsey (private communication) has recently obtained similar results on the π decay based on this γ_5 invariance. We do not here specify the interaction of the ψ field, which may be of the nonlinear Heisenberg type, or due to an intermediate boson (different from π or K).

[9]J. Bardeen, L. N. Cooper, and J. R. Schrieffer, Phys. Rev. 106, 162 (1957).

[10]N. N. Bogoliubov, V. V. Tolmachev, and D. V. Shirkov, A New Method in the Theory of Superconductivity (Academy of Sciences of USSR, Moscow, 1958).

[11]Y. Nambu, Phys. Rev. 117, 648 (1960).

A 'SUPERCONDUCTOR' MODEL OF ELEMENTARY PARTICLES AND ITS CONSEQUENCES by Y. Nambu (University of Chicago)†

(In absence of the author the paper was presented by G. Jona-Lasinio.)

1

In recent years it has become fashionable to apply field-theoretical techniques to the many-body problems one encounters in solid state physics and nuclear physics. This is not surprising because in a quantized field theory there is always the possibility of pair creation (real or virtual), which is essentially a many-body problem. We are familiar with a number of close analogies between ideas and problems in elementary particle theory and the corresponding ones in solid state physics. For example, the Fermi sea of electrons in a metal is analogous to the Dirac sea of electrons in the vacuum, and we speak about electrons and holes in both cases. Some people must have thought of the meson field as something like the shielded Coulomb field. Of course, in elementary particles we have more symmetries and invariance properties than in the other, and blind analogies are often dangerous.

At any rate, we should expect a close interaction of the two branches of physics in terms of concepts and mathematical techniques, which make up the content of quantum field theory. In this talk we are going to show another possibility of such an interaction, but this time in the opposite direction to what has been the general trend. Namely, the model of elementary particles we are going to talk about is motivated by the mathematical theory of superconductivity which was first worked out with great success by Bardeen, Cooper and Schrieffer[1]. The characteristic feature of the theory is that the ground state of a superconductor is found to be separated by a gap from the excited states, which, of course, has been confirmed experimentally. The gap is caused by the fact that the attractive phonon interaction between electrons produce correlated pairs of electrons with opposite momenta near the Fermi surface, and it takes a finite amount of energy to break the correlation.

The BCS theory was given an elegant mathematical basis by Bogoliubov[2], who introduced a coherent mixture of electrons and holes to discuss the el-

† **Retypeset by M. Okai, Nov. 1993**
Courtesy of : Purdue University and the Purdue Research Foundation

ementary excitations (quasi-particles) in a superconductor. It is easy to see that such a particle has a finite "rest energy," which corresponds to the finite energy gap. Let us assume the following equations for electrons near the top of the Fermi surface:

$$
\begin{aligned}
E\psi_{p+} &= \epsilon_p \psi_{p+} + \phi \psi^{\dagger}_{-p-} \ , \\
E\psi^{\dagger}_{-p-} &= -\epsilon_p \psi^{\dagger}_{-p-} + \phi \psi_{p+} \ .
\end{aligned} \tag{1}
$$

ψ_{p+} is the wave function for an electron of momentum p and spin $+$ (up), and $\psi^{\dagger}_{-p-}$ is one for a hole of momentum p and spin $+$, which means the absence of an electron of momentum $-p$ and spin $-$ (down). ϵ_p is the kinetic energy measured from the Fermi surface; ϕ is a constant.

Eq.(1) gives the eigenvalues

$$E_p = \pm\sqrt{\epsilon_p^2 + \phi^2}. \tag{2}$$

So it takes an amount of energy $2|E_p| \geq 2\phi$ to excite such a quasi-electron from the lower to the upper state. The quantity ϕ is actually obtained as a self-consistent, self-energy (Hartree-Fock field) from the phonon-electron interaction,

$$\mid \phi \mid \approx \hbar\omega e^{-1/\rho} \tag{3}$$

where $\hbar\omega$ is the mean phonon frequency, and ρ the effective electron-electron interaction energy density on the Fermi surface.

Eqs.(1) and (2) bear a striking resemblance to the Dirac equation and its eigenvalues. In the Weyl representation, The Dirac equation reads

$$
\begin{aligned}
E\psi_1 &= \vec{\sigma} \cdot \vec{p}\psi_1 + m\psi_2 \\
E\psi_2 &= -\vec{\sigma} \cdot \vec{p}\psi_2 + m\psi_1 \\
E_p &= \pm\sqrt{p^2 + m^2}
\end{aligned} \tag{4}
$$

where ψ_1 and ψ_2 are the two eigenstates of the chirality operator γ_5.

This analogy may be a superficial one and devoid of physical significance. But it would also be interesting to see what would happen if we took the analogy seriously and pursued its consequences. The interpretation of Eq.(4) would be then first of all that the mass of a Dirac particle is a self-energy built up by some interaction, a statement which surprises nobody. Indeed we shall find that even though the starting point looks novel, there is nothing unconventional in our model. Nevertheless, we shall also see that the analogy casts a new light on old problems, and reveals some new things which have been overlooked in the usual discussion of the self-energy problem and the symmetry properties of elementary particles.

To give an idea about our program, we draw up a list of correspondences between superconductivity and the elementary particle theory.

Superconductivity	Elementary particles
free electrons	bare fermion (zero or small mass)
phonon interaction	some unknown interaction
energy gap	observed mass (nucleon)
collective excitation	meson , bound nucleon pair
charge	chirality
gauge invariance	γ_5−invariance (rigorous or approximate)

As we can see from the table, our problem will be to account for the nucleons (and hyperons) and the mesons in a unified way from some basic field. There is no strong reason why we should not also consider the leptons, but for the time being we would like to exclude them. The reason will become clear later on.

As for the exact nature of the basic interaction which would produce the baryons and mesons, our model does not say what it should be. Some other guiding principles are needed for this purpose, but we do not seem to possess any convincing ones yet. So looking around for some clues, we find two possibilities rather attractive for reasons of simplicity and elegance. One is the Heisenberg type theory[3] where we consider nonlinear spinor interactions. The other one is to use an analogy with the electromagnetic field. The electromagnetic field is inherently related to the conservation of charge, and the dynamics of interaction is uniquely determined by the gauge group. Attempts to generalize this idea to baryon problems have been made by Yang and Mills[4], Yang and Lee[5], Fujii[6], and recently by Sakurai[7].

Both types of theories have attractive points as well as difficulties. The most serious obstacle in any theory dealing with self-energies is the divergence problem, which is more pronounced in the Heisenberg type theory than in the other. The intermediate boson theory runs into trouble because gauge invariance requires such a field to be massless, yet massless boson fields other than electromagnetic and gravitational do not seem to exist. We do not know whether a finite observed mass can be compatible with the invariance assumption.

2

We will consider here the Heisenberg type theory because of its greater practical simplicity. The divergence will be disposed of by simple cut-off, as we do not claim to have found a way to resolve this difficulty.

Thus we adopt the following model Lagrangian for the nucleon. Isotopic spin is ignored.

$$L = -\bar{\psi}\gamma_\mu\partial_\mu\psi - g\left[\bar{\psi}\psi\bar{\psi}\psi - \bar{\psi}\gamma_5\psi\bar{\psi}\gamma_5\psi\right] \tag{5}$$

This Lagrangian is invariant under the transformations

$$\begin{aligned}&\text{(a)}\quad \psi \longrightarrow \exp[i\alpha]\psi \ ; \ \bar{\psi} \longrightarrow \bar{\psi}\exp[-i\alpha]\\ &\text{(b)}\quad \psi \longrightarrow \exp[i\alpha\gamma_5]\psi \ ; \ \bar{\psi} \longrightarrow \bar{\psi}\exp[+i\alpha\gamma_5]\end{aligned} \tag{6}$$

where α is a constant c number. (Local gauge transformation is not possible here.) a) implies the nucleon number conservation; b) will be called γ_5 invariance hereafter, which implies the conservation of chirality: the number of right-handed (bare)particles minus the number of left-handed particles is conserved. We get accordingly two conserved currents

$$\frac{\partial}{\partial x_\mu}\bar{\psi}\gamma_\mu\psi = 0 \ ; \ \frac{\partial}{\partial x_\mu}\bar{\psi}\gamma_5\gamma_\mu\psi = 0 \tag{7}$$

which can be directly verified.

Now we want to derive the observed nucleon mass in the Hartree-Fock approximation. Namely, we determine the mass by linearizing the interaction in which process the assumed mass is used in taking expectation values. We then have the relation

$$\begin{aligned} m &= 2g\left[\langle\bar{\psi}\psi\rangle - \gamma_5\langle\bar{\psi}\gamma_5\psi\rangle\right]\\ &= -2g\left[\mathrm{Tr}S^{(m)}(0) - \gamma_5\mathrm{Tr}\gamma_5 S^{(m)}(0)\right]\end{aligned} \tag{8}$$

where $S^{(m)}(x)$ is the nucleon Green's function having a mass m. In momentum space this becomes

$$m = -\frac{g}{(2\pi)^3} \int \frac{md^3p}{\sqrt{p^2+m^2}} \ . \tag{9}$$

A trivial solution is of course $m = 0$. But with a cut-off we find also a non-trivial one

$$\frac{\pi^2}{|g|K^2} = \sqrt{1+m^2/K^2} - (m^2/K^2)\sinh^{-1}|K/m| \tag{10}$$

provided that $g < 0$ and $\pi^2 < |g|K^2$. For $|m/K| \equiv x \ll 1$,

$$1 - \pi^2/|g|K^2 \approx x^2 \log(2/x), \tag{9'}$$

Eq.(9) is of the same form as the "energy gap equation" in the BCS theory. The non-analytic character of the solution with respect to the coupling constant is easily recognizable.

Our approximation scheme for the self-energy is illustrated by the following Feynman diagrams (Fig.1).

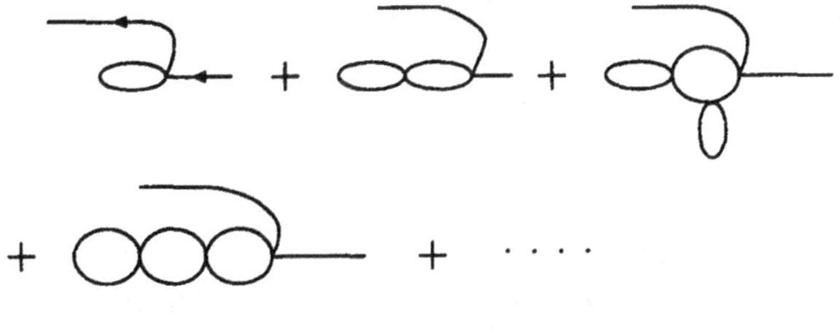

Fig. 1

Because of the attractive interaction ($g < 0$) between the virtual pair, the bubble diagrams give rise to a catastrophic change not obtained by perturbation expansion.

Thus we have created a mass out of nothing. But there are two solutions corresponding to $\neq m$, not to speak of the trivial solution $m = 0$. Presumably the vacuum corresponding to the latter solution is unstable like the normal state below the critical temperature for a superconductor. But would the two non-trivial solutions correspond to two different particles with equal mass? Heisenberg has found a similar situation in his theory. He wants to identify the two massive particles with the proton and the neutron.

Before discussing this problem, let us first worry about the conservation laws. Eq.(7) represents operator equations, so they should hold, among other things, for the matrix element between real one particle states. If we use the Dirac equation with a mass for ψ and $\bar{\psi}$, the ordinary current is all right, but the γ_5 current conservation breaks down:

$$\langle p_2|\frac{\partial}{\partial x_\mu}\bar{\psi}\gamma_5\gamma_\mu\psi|p_1\rangle = -2m\langle p_2|\bar{\psi}\gamma_5\psi|p_1\rangle \neq 0. \tag{11}$$

This means that $\gamma_5\gamma_\mu$ is not the correct vertex operator for the "dressed" particle where the mass is entirely due to the interaction. We would have to take into account the "radiative corrections" also for the vertex. The general form of the γ_5 current vertex operator $\Gamma_{5\mu}$ can be determined from Lorentz invariance and the continuity equation as follows:

$$\Gamma_{5\mu}(p_2, p_1) = \left(\gamma_5\gamma_\mu + \frac{2im\gamma_5 q_\mu}{q^2}\right) F(q^2), \quad q = p_2 - p_1 \tag{12}$$

where $F(q^2)$ is a form factor.

On the other hand, the local field theory requires that the coefficients of $\gamma_5\gamma_\mu$ and γ_5 obey dispersion relations of the type

$$f(q^2) = f(0) - \frac{q^2}{\pi}\int_{k_1^2}^{\infty}\frac{\mathrm{Im}f(-k^2)}{(q^2+k^2)k^2}dk^2. \tag{13}$$

If $F(0) \neq 0$, then Eq.(12) has a pole at $q^2 = 0$, which means, in terms of Eq.(13), the existence of a massless, pseudoscalar, boson contributing to the form factor. A Feynman diagram showing this situation is given in Fig.2.

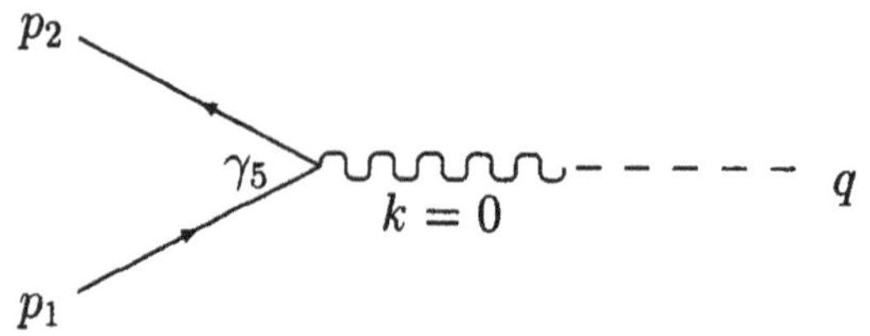

Fig.2

Since no such boson was assumed in our theory in the beginning, we have to manufacture it somehow out of the original fermion field. A natural way would be to interpret the boson as a bound state of a fermion pair—bound because of the attractive interaction. Thus we are forced to the conclusion that if a finite mass can arise from a γ_5 invariant theory there must also be zero-mass bound states of pairs. We would not have this situation if $F(0) = 0$, but then the total γ_5 charge

$$\langle p|\bar{\psi}\gamma_5\gamma_4\psi|p\rangle$$

would be zero, which leads to a contradiction (see the statement after Eq.(19)). We would not have the trouble either if the interaction were not γ_5 invariant, e.g. if the pseudoscalar term were missing in Eq.(5), which would not change Eq.(9). But we have invited the trouble deliberately because we need pseudoscalar bosons in the nucleon problem.

We can easily verify our conclusion within the present Hartree-Fock approximation. For this purpose let us set up the Bethe-Salpeter equation for the nucleon-antinucleon pair in the lowest order:

$$\begin{aligned}\Phi\left(p+\frac{q}{2},p-\frac{q}{2}\right) = ig\Big[\int \mathrm{Tr}S\left(p'+\frac{q}{2}\right)\Phi\left(p'+\frac{q}{2},p'-\frac{q}{2}\right)S\left(p'-\frac{q}{2}\right)d^4p' \\ -\gamma_5\int \mathrm{Tr}\gamma_5 S\left(p'+\frac{q}{2}\right)\Phi\left(p'+\frac{q}{2},p'-\frac{q}{2}\right)S\left(p'-\frac{q}{2}\right)d^4p'\Big] \quad (14)\end{aligned}$$

$\Phi\left(p+\frac{q}{2},p-\frac{q}{2}\right)$ is the wave function for a nucleon with momentum $p+\frac{q}{2}$ and an antinucleon with momentum $\frac{q}{2}-p$, the total momentum being q.

We also note that if we add an inhomogeneous term $\Gamma_0\left(p+\frac{q}{2},p-\frac{q}{2}\right)$ on the right-hand side of Eq.(14) and write Γ instead of Φ, then it would represent an integral equation for a vertex part Γ generated by Γ_0.

Now it is easy to see that the following functions satisfy the integral equation,

$$\begin{aligned}&\Gamma\left(p+\frac{q}{2},p-\frac{q}{2}\right) = L\left(p+\frac{q}{2}\right)\gamma_5+\gamma_5 L\left(p-\frac{q}{2}\right) = -i\gamma\cdot q\gamma_5 - 2m\gamma_5, \\ &\Gamma_0\left(p+\frac{q}{2},p-\frac{q}{2}\right) = L_0\left(p+\frac{q}{2}\right)\gamma_5+\gamma_5 L_0\left(p-\frac{q}{2}\right) = -i\gamma\cdot q\gamma_5, \\ &\left(L(p) = -i\gamma\cdot p - m = -i\left[S^{(m)}(p)\right]^{-1};\quad L_0(p) = -i\gamma\cdot p\right). \qquad (15)\end{aligned}$$

Taking $q = 0$, we get $\Gamma = -2m\gamma_5$, $\Gamma_0 = 0$. In other words $\Phi = 2m\gamma_5$ is a solution of the B-S equation for zero energy and momentum, which is a limiting case of a particle with zero rest mass. In fact we can construct bound state wave functions for q_0, $\vec{q} \neq 0$, $q^2 = 0$ starting from the above solution as the zeroth approximation.

For arbitrary q, Eq.(15) is a generalized Ward identity for the divergence of the γ_5 current $\Gamma_{5\mu}$ (Eq.(12 ($F(q^2) = 1$ in our approximation.)) A similar relation is known for the ordinary current[8], and can be derived in our case too. Since $L(p)|p\rangle = 0$ and $\langle p|L(p) = 0$, the continuity equation (11) is obviously satisfied.

Now the question of the mass degeneracy. By the γ_5 transformation $\psi \to e^{i\alpha\gamma_5}\psi$, the mass m changes into $m\exp[2i\alpha] = m(\cos 2\alpha + i\gamma_5 \sin 2\alpha)$. In fact we could have assigned this form to the mass operator from the beginning. There would be no change of physical content. This arbitrariness is due to the fact that we can add any number of the zero-mass, zero-energy "mesons" to the system. The vacuum itself is degenerate in this sense.

To see this, consider the vacuum state Ω defined by

$$\psi^{(+)}\Omega = \bar{\psi}^{(+)}\Omega = 0.$$

The infinitesimal γ_5 transformation is generated by

$$R = -\int \bar{\psi}\gamma_5\gamma_4\psi d^3x. \tag{16}$$

The zero-mass, zero-energy mesons are similar to the longitudinal photons encountered in quantum electrodynamics. The γ_5 and gauge transformations change the distribution of the respective quanta, but cause no physical effects. What is, then, the γ_5 quantum number for the vacuum or the one particle state? This is the eigenvalue of R, and may be written as

$$R = n_+ - n_- + \bar{n}_+ - \bar{n}_- \tag{17}$$

where $n_\pm(\bar{n}_\pm)$ is the number of bare ferminons (anti-fermions) with $\gamma_5 = \pm 1$. The nucleon number N is given by

$$N = n_+ + n_- - \bar{n}_+ - \bar{n}_- \tag{18}$$

so that $N \pm R = 2(n_\pm - \bar{n}_\mp)$ is an even number.

Real nucleons with finite mass certainly are not eigenstates of R if they transform in the conventional way under R. Such a situation would be possible only if we had a complete degeneracy with respect to R. This seems to be the present case. Our world is conveniently described as a superposition of states with different R's, but there are no realizable processes which change R (a superselection rule), and hence the degeneracy does not manifest itself as physical effects.

Such an interpretation may not be quite satisfactory, but we would like to point out that a similar situation appears in the BCS theory too with respect to charge conservation. At any rate we have not found pseudoscalar photons in nature,but rather we are inclined to identify them with the mesons, which have mass in reality. Thus we will have to admit that after all nature is not γ_5 invariant. We do not know whether there is any other way out. But if the violation of the invariance were small, the foregoing results would guarantee automatically the existence of meson states, this time with a finite mass.

There are two ways to achieve our goal. One is to assume a finite bare mass m_0, which should be small compared to the observed mass m. The other is to destroy the nice symmetry of the interaction a little bit. The former looks simpler, and esthetically less objectionable. In this case, we can confirm by calculation the following results. The nucleon self-energy has now lost the freedom of γ_5 rotation, but we still get two different masses of opposite sign, in addition to the trivial perturbation solution. As m_0 increases, the mass splitting grows larger, until at a certain point one of them merges with the trivial branch and disappears thereafter, leaving only the larger of the non-trivial solution.

On the other hand, the meson mass μ is proportional to $(m_0/m)^{1/2}$ so that only the solution with $m_0/m > 0$ (the largest of the three solutions), can give rise to a stable bound pair, while the other two give rise to "ghost" mesons.

Apart from the ghost trouble, this result raises the interesting question of the possibility of producing mass multiplets of nucleons and mesons since the superselection rule does not operate any more. But the question has to be left for future study.

In our model, of course, we have obtained only neutral mesons because isospin is neglected. But the generalization is easy. For example, the interaction

$$L_{int} = -g\left[(\bar{\psi}\psi) - \sum_{s=1}^{3}\bar{\psi}\gamma_5\tau_s\psi\bar{\psi}\gamma_5\tau_s\psi\right] \tag{19}$$

(where the τ's are the isotopic spin matrices) immediately leads to pseudoscalar mesons of isospin 1. The gauge group in this case consists of

$$\begin{aligned}\psi &\longrightarrow \exp[i\alpha]\psi\\ \psi &\longrightarrow \exp[i\vec{\alpha}\cdot\vec{\tau}]\psi\\ \psi &\longrightarrow \exp[i\vec{\alpha}\cdot\vec{\tau}\gamma_5]\psi\end{aligned} \tag{20}$$

which correspond respectively to nucleonic charge, isotopic spin, and the $\gamma_5\times$isotopic spin conservation[9]. The last two form a four-dimensional rotation group.

3

Because of some interesting features in their own right, we will discuss briefly the intermediate boson theory of primary interaction. As before, we would like to take a γ_5 invariant theory, which means that the boson is either vector or pseudovector[10]. We can immediately write down a self-energy equation in our Hartree approximation. Namely we equate the observed mass with the familiar lowest order self-energy. Actually the self-energy consists of two parts: $\Sigma(p) = i\gamma\cdot\vec{p}\Sigma_1(p) + \Sigma_2(p)$, Σ_1 being the wave function renormalization. Thus we get equations for Σ_1 and Σ_2 separately. These are direct analogs of the equations we encounter in superconductivity, where the boson means the phonon. It turns out that the vector interaction can give a non-trivial solution, whereas the pseudovector does not, because of the wrong sign of the self-energy. Physically speaking, the vector case causes an attractive interaction between virtual pairs, and hence a catastrophic change. Fig.3 shows the diagrams corresponding to our approximation and the above mentioned effect.

It should be interesting that this is the approximation considered by Landau[11] in his discussion of the Green's functions.

The qualitative feature of the solution is similar to the one obtained in the Heisenberg type theory, except for the nature of divergences. We also

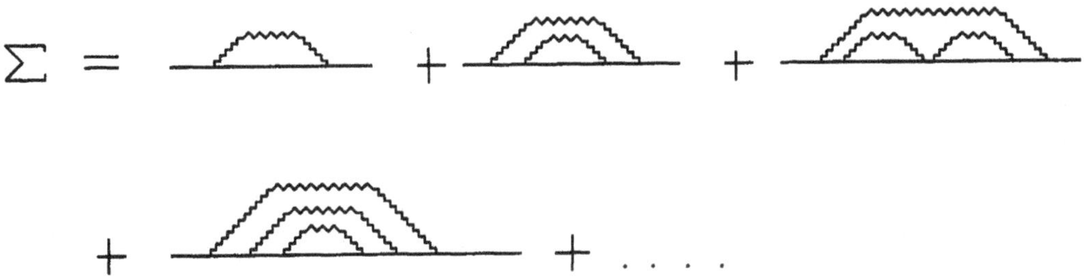

or

Fig. 3

obtain the zero mass bound states by solving the B-S equation in the ladder approximation. This solution was first discovered by Goldstein[12], but was considered as an abnormal object. Its raison d'etre has now become clear. According to Goldstein, however, the bound state wave function is not always normalizable. For small coupling constants, which is not really strong enough to cause such a binding, the norm of the wave function becomes negative (a ghost state).

Another intriguing question in this type of theory is whether one can produce a finite effective mass for the boson in a gauge invariant theory. Since we have done this for the fermion, it may not be impossible if some extra freedom is given to the field. Should the answer come out to be yes, the Yang-Mills-Yang-Lee-Sakurai theory of vector bosons would become very interesting indeed.

4

We finally come to the predictions or applications of our theory. The theory essentially boils down to the compound particle model of mesons, which has found some advocates in the past (Fermi-Yang, Sakata, Okun, Heisenberg,....). The new feature in our theory is that pseudoscalar mesons arise naturally and necessarily together with the nucleon mass as a consequence of the symmetry properties of the theory. Their type depends on the symmetry we assume. For example, it is possible to produce an isospin 1 meson but not an isospin 0 meson according to Eq.(20). There may, of course, be ordinary (or rather "accidental") bound states, but they will appear as excited or compound states of these basic mesons, like the now fashionable 2π and 3π resonances.

Being serious as we are about the compound particle model of the mesons, we should also try to calculate the meson-nucleon coupling constants from the basic constants. As is well known, there is no difference in the formal description of the particles whether they are elementary or compound. But if we know the wave function of a compound system, then the coupling constant is determined by it. In the case of a loosely bound system, the answer is simple. The pseudovector coupling constant is given by

$$\frac{f^2}{4\pi} = \frac{4a\mu}{m} \tag{21}$$

where m is the nucleon mass, μ the "meson" mass$= 2m - \epsilon$, and $a = \sqrt{m\epsilon}$ is the range of the wave function.

For strongly bound systems, however, the result should depend sensitively on the detailed dynamics since there are no "anomalous thresholds" in the dispersion relations for the form factors which reflect the structure of the wave function. Nevertheless let us extrapolate the above formula and see what happens. We find for the pion

$$\frac{f^2}{4\pi} = \frac{1}{n}\frac{4\mu}{m} \tag{22}$$

where we have also taken into account that different fermion pairs (proton, neutron, Λ, Σ, Ξ) may contribute to the pion state and n is the number of such pairs, assuming equal amount of contribution and neglecting mass fine structure. If we include all possible baryon combinations, then $n = 8$, and $f^2/4\pi = 0.075$!

We have to admit that we do not yet know what type of Lagrangian can possibly give rise to the observed baryon and meson spectrum. How much internal degree of freedom do we need to start with in order to account for the observed particles (and not to account for non-existing ones)? Why do the baryon masses split? Can we assume a high degree of symmetry properties in the beginning and yet come out with a smaller amount of apparent symmetries? The last question seems particularly relevant since we have found an example in the conflict between γ_5 invariance and finite mass. Here it should be helpful to seek analogies in solid state physics for better physical understanding.

For example, it is not hard to foresee how nature can manifest itself in an unsymmetric way while keeping the basic laws symmetric if we compare the situation with ferromagnetism. In an ordinary material, the ground state of a macroscopic body has spin zero practically, so that there is no preferred axis in space. In the ferromagnetic case, on the other hand, all the spins are parallel in the ground state, and they must point in some direction, thereby creating an asymmetry in reality. Such spontaneous polarizations may be happening in the world of elementary particles too[13].

We close this section with an application to weak interactions. So far this seems the most interesting and useful result coming out of our model.

The first question is , what is the renormalization of the vector and axial vector currents of the nucleon due to "strong" interactions, assuming there was a universal Fermi coupling in the beginning? Adopting our Heisenberg model, it can be shown that there will be no renormalization effect, and $g_V = g_A$ as long as we keep strict γ_5 invariance. The fact that g_A/g_V is only approximately unity implies then that there is a small violation of the invariance in agreement with the previous conclusion.

But this is not the whole story. We have already derived the nucleon axial vector vertex $\Gamma_5\mu$, which will also appear in the weak processes. The small violation of the invariance gives the meson mass μ, but will affect the form factor F of Eq.(12) relatively little. Thus we may be able to write

$$\Gamma_{5\mu}(p_2, p_1) = \gamma_5\gamma_\mu F_1(q^2) + \frac{2im\gamma_5 q_\mu}{q^2 + \mu^2} F_2(q^2)$$

$$\approx \left(\gamma_5\gamma_\mu + \frac{2im\gamma_5 q_\mu}{q^2+\mu^2}\right) F(q^2) \tag{23}$$

where now $F(0) = g_A/g_V = 1.25$. The second term of Eq.(23) is small compared to the first for the actual beta decay since $q^2 \ll \mu^2$. For large $q^2 \gg \mu^2$ we expect to recover the strict conservation: $F_1/F_2 \longrightarrow 1$ as $q^2 \longrightarrow \infty$.

Now we see that this second term enables one to determine the pion decay constant since, according to the dispersion theory, it represents the process going through the pion channel. Denoting the pion-nucleon (ps) coupling and the pion-lepton(pv) coupling as G_π and g_π respectively, we find

$$\sqrt{2}G_\pi g_\pi = 2mg_V F_2(-\mu^2) \approx 2mg_A. \tag{24}$$

Using $g_V = 10^{-5}/m^2$, $G^2/4\pi = 13.5$, we get $2.7 \times 10^{-8} sec$ for the pion life time, as compared to the observed $2.56 \times 10^{-8} sec$.

Eq.(24) is exactly the same as Goldberger-Treiman[14] formula derived by an entirely different approach and rather special assumptions. But we do not think that the agreement is a coincidence. There is a certain class of models which can more or less predict this relation[15]. The essential point seems to be that the pion is effectively treated as a bound state in the G-T theory. This manifests itself through the pion renormalization constant being zero or practically zero[16].

We can blindly generalize Eq.(24) to the strangeness changing axial vector current, where the pseudoscalar $K-$meson replaces the pion. Taking the ΛN vertex, for example, we again get a relation between the weak coupling g'_A, ΛNK coupling G_K and the $K-$lepton coupling g_K:

$$G_K g_K \approx (m_N + m_\Lambda) g'_A. \tag{25}$$

This relation does not contradict our present rather meager knowledge about these constants.

Since we are based on the compound particle model, all the considerations that have been made by various people in the past can be adopted in essence. The Gershtein-Zeldovich and Feynman-Gell-Mann idea of $\pi-$lepton vector coupling is, of course, a natural consequence of the model, though it has yet to be tested experimentally.

We feel that a systematic and quantitative calculation of the renormalization effects can be undertaken in our theory with more confidence than in the past because we have better understanding of the interrelation of different phenomena. So far we have tried to estimate the decay life times for most of the decay models of strange particles under very crude assumptions. The result is in general satisfactory, but it is not clear as to what it really means. We have not yet understood such basic questions as the $\Delta T = \frac{1}{2}$ rule and the smallness of the hyperon beta decay rate in any fundamental way.

References

[1] Bardeen, Cooper, and Schrieffer, Phys. Rev. 106, 162 (1957)

[2] N. N. Bogoliubov, J. Exptl. Theoret. Phys. (USSR) 34, 58, 73 (1958) (Soviet Phys. -JETP 34, 41, 51); Bogoliubov, Tolmachev, and Shirkov, "A New Method in the Theory of Superconductivity" (Academy of Sciences of USSR, Moscow, 1958). Also J. G. Valatin, Nuovo cimento 7, 843 (1958).

[3] W. Heisenberg, et al., Zeit.f. Naturf. 14, 441 (1959) Earlier papers are quoted there.

[4] C. N. Yang and R. L. Mills, Phys. Rev. 96, 191 (1954)

[5] T. D. Lee and C. N. Yang, Phys. Rev. 98, 1501 (1955)

[6] Y. Fujii, Progr. Theoret. Phys. (Kyoto) 21, 232 (1959)

[7] J. J. Sakurai, Annal. Phys., to be published.

[8] For example, Y. Takahashi, Nuovo cimento 6, 371 (1957)

[9] This type of theory has been discussed by F. Gursey, Nuovo cimento, to be published.

[10] Derivative interactions of scalar or pseudoscalar fields are also admissible.

[11] Landau, Abrikosov, and Khalatnikov, Dok. Akad. Nauk USSR 95, 497, 773 (1954); 96, 261 (1954): L. D. Landau, "Niels Bohr and the Development of Physics" (McGraw Hill Book Co., New York, 1955) p. 52

[12] J. Goldstein, Phys. Rev. 91, 1516 (1953).

[13] We find similar observations in Heisenberg's paper [3].

[14] M. L. Goldberger and S. B. Treiman, Phys. Rev. 110, 1178 (1958)

[15] Feynman, Gell-Mann, and Levy, to be published.

[16] K. Symanzik, Nuovo cimento 11, 269 (1958).
R. F. Sawyer, Phys. Rev. 116, 236 (1959)

DISCUSSION

Wightman: Is it true that the zero mass bound state moves up to become the $\pi-$meson when the γ_5 invariance is broken?

Jona-Lasinio: When you break the γ_5-invariance, for example by introducing a bare nucleon mass m_0, you dispose of an additional parameter which can be adjusted to give the desired mass.

Guth: How can you say that the model is a Heisenberg type theory if you do not specify the dynamics?

Jona-Lasinio: This theory is a Heisenberg type theory only in the sense that a nonlinear spinor equation is taken as starting point. The model is then linearized self-consistently and the cut-off which has to be introduced to eliminate divergences should be interpreted as a dynamical effect. The mechanism responsible for it is actually unspecified.

J. Sakurai: It is important to emphasize that much of what has been reported is independent of particular models. The only things that are relevant are: a) axial-vector conservation holds; and b) the force between a nucleon and an antinucleon is attractive.

Primakoff: When you break the γ_5 invariance to introduce the $\pi-$meson mass, how do you know that $\lim_{q^2 \to 0} F(q^2) \approx g_A/g_V$?

Jona-Lasinio: It is an assumption. In order to obtain both a finite $\pi-$meson mass and a renormalization of the axial vector coupling, we have to break the γ_5 invariance. So it is assumed the same γ_5 invariance violation is responsible for both effects.

PHYSICAL REVIEW VOLUME 122, NUMBER 1 APRIL 1, 1961

Dynamical Model of Elementary Particles Based on an Analogy with Superconductivity. I*

Y. NAMBU AND G. JONA-LASINIO†

The Enrico Fermi Institute for Nuclear Studies and the Department of Physics, The University of Chicago, Chicago, Illinois

(Received October 27, 1960)

It is suggested that the nucleon mass arises largely as a self-energy of some primary fermion field through the same mechanism as the appearance of energy gap in the theory of superconductivity. The idea can be put into a mathematical formulation utilizing a generalized Hartree-Fock approximation which regards real nucleons as quasi-particle excitations. We consider a simplified model of nonlinear four-fermion interaction which allows a γ_5-gauge group. An interesting consequence of the symmetry is that there arise automatically pseudoscalar zero-mass bound states of nucleon-antinucleon pair which may be regarded as an idealized pion. In addition, massive bound states of nucleon number zero and two are predicted in a simple approximation.

The theory contains two parameters which can be explicitly related to observed nucleon mass and the pion-nucleon coupling constant. Some paradoxical aspects of the theory in connection with the γ_5 transformation are discussed in detail.

I. INTRODUCTION

IN this paper we are going to develop a dynamical theory of elementary particles in which nucleons and mesons are derived in a unified way from a fundamental spinor field.[1] In basic physical ideas, it has thus the characteristic features of a compound-particle model, but unlike most of the existing theories, dynamical treatment of the interaction makes up an essential part of the theory. Strange particles are not yet considered.

The scheme is motivated by the observation of an interesting analogy between the properties of Dirac particles and the quasi-particle excitations that appear in the theory of superconductivity, which was originated with great success by Bardeen, Cooper, and Schrieffer,[2] and subsequently given an elegant mathematical formulation by Bogoliubov.[3] The characteristic feature of the BCS theory is that it produces an energy gap between the ground state and the excited states of a superconductor, a fact which has been confirmed experimentally. The gap is caused due to the fact that the attractive phonon-mediated interaction between electrons produces correlated pairs of electrons with opposite momenta and spin near the Fermi surface, and it takes a finite amount of energy to break this correlation.

Elementary excitations in a superconductor can be conveniently described by means of a coherent mixture of electrons and holes, which obeys the following equations[3,4]:

$$E\psi_{p+}=\epsilon_p\psi_{p+}+\phi\psi_{-p-}^*,\qquad E\psi_{-p-}^*=-\epsilon_p\psi_{-p-}^*+\phi\psi_{p+},\tag{1.1}$$

near the Fermi surface. ψ_{p+} is the component of the excitation corresponding to an electron state of momentum p and spin $+$ (up), and ψ_{-p-}^* corresponding to a hole state of momentum p and spin $+$, which means an absence of an electron of momentum $-p$ and spin $-$ (down). ϵ_p is the kinetic energy measured from the Fermi surface; ϕ is a constant. There will also be an equation complex conjugate to Eq. (1), describing another type of excitation.

Equation (1) gives the eigenvalues

$$E_p=\pm(\epsilon_p^2+\phi^2)^{\frac{1}{2}}.\tag{1.2}$$

The two states of this quasi-particle are separated in energy by $2|E_p|$. In the ground state of the system all the quasi-particles should be in the lower (negative) energy states of Eq. (2), and it would take a finite energy $2|E_p|\geqslant 2|\phi|$ to excite a particle to the upper state. The situation bears a remarkable resemblance to the case of a Dirac particle. The four-component Dirac equation can be split into two sets to read

$$E\psi_1=\sigma\cdot p\psi_1+m\psi_2,\qquad E\psi_2=-\sigma\cdot p\psi_2+m\psi_1,\qquad E_p=\pm(p^2+m^2)^{\frac{1}{2}},\tag{1.3}$$

where ψ_1 and ψ_2 are the two eigenstates of the chirality operator $\gamma_5=\gamma_1\gamma_2\gamma_3\gamma_4$.

According to Dirac's original interpretation, the ground state (vacuum) of the world has all the electrons in the negative energy states, and to create excited states (with zero particle number) we have to supply an energy $\geqslant 2m$.

In the BCS-Bogoliubov theory, the gap parameter ϕ, which is absent for free electrons, is determined essentially as a self-consistent (Hartree-Fock) representation of the electron-electron interaction effect.

* Supported by the U. S. Atomic Energy Commission.

† Fulbright Fellow, on leave of absence from Instituto di Fisica dell' Universita, Roma, Italy and Istituto Nazionale di Fisica Nucleare, Sezione di Roma, Italy.

[1] A preliminary version of the work was presented at the Midwestern Conference on Theoretical Physics, April, 1960 (unpublished). See also Y. Nambu, Phys. Rev. Letters **4**, 380 (1960); and Proceedings of the Tenth Annual Rochester Conference on High-Energy Nuclear Physics, 1960 (to be published).

[2] J. Bardeen, L. N. Cooper, and J. R. Schrieffer, Phys. Rev. **106**, 162 (1957).

[3] N. N. Bogoliubov, J. Exptl. Theoret. Phys. (U.S.S.R.) **34**, 58, 73 (1958) [translation: Soviet Phys.-JETP **34**, 41, 51 (1958)]; N. N. Bogoliubov, V. V. Tolmachev, and D. V. Shirkov, *A New Method in the Theory of Superconductivity* (Academy of Sciences of U.S.S.R., Moscow, 1958).

[4] J. G. Valatin, Nuovo cimento **7**, 843 (1958).

One finds that

$$\phi \approx \omega \exp[-1/\rho], \qquad (1.4)$$

where ω is the energy bandwidth ($\approx$ the Debye frequency) around the Fermi surface within which the interaction is important; ρ is the average interaction energy of an electron interacting with unit energy shell of electrons on the Fermi surface. It is significant that ϕ depends on the strength of the interaction (coupling constant) in a nonanalytic way.

We would like to pursue this analogy mathematically. As the energy gap ϕ in a superconductor is created by the interaction, let us assume that the mass of a Dirac particle is also due to some interaction between massless bare fermions. A quasi-particle in a superconductor is a mixture of bare electrons with opposite electric charges (a particle and a hole) but with the same spin; correspondingly a massive Dirac particle is a mixture of bare fermions with opposite chiralities, but with the same charge or fermion number. Without the gap ϕ or the mass m, the respective particle would become an eigenstate of electric charge or chirality.

Once we make this analogy, we immediately notice further consequences of special interest. It has been pointed out by several people[3,5–8] that in a refined theory of superconductivity there emerge, in addition to the individual quasi-particle excitations, collective excitations of quasi-particle pairs. (These can alternatively be interpreted as moving states of bare electron pairs which are originally precipitated into the ground state of the system.) In the absence of Coulomb interaction, these excitations are phonon-like, filling the gap of the quasi-particle spectrum.

In general, they are excited when a quasi-particle is accelerated in the medium, and play the role of a backflow around the particle, compensating the change of charge localized on the quasi-particle wave packet. Thus these excitations are necessary consequences of the fact that individual quasi-particles are not eigenstates of electric charge, and hence their equations are not gauge invariant; whereas a complete description of the system must be gauge invariant. The logical connection between gauge invariance and the existence of collective states has been particularly emphasized by one of the authors.[8]

This observation leads to the conclusion that if a Dirac particle is actually a quasi-particle, which is only an approximate description of an entire system where chirality is conserved, then there must also exist collective excitations of bound quasi-particle pairs. The chirality conservation implies the invariance of the theory under the so-called γ_5 gauge group, and from its nature one can show that the collective state must be a pseudoscalar quantity.

[5] D. Pines and J. R. Schrieffer, Nuovo cimento **10**, 496 (1958).
[6] P. W. Anderson, Phys. Rev. **110**, 827, 1900 (1958); **114**, 1002 (1959).
[7] G. Rickayzen, Phys. Rev. **115**, 795 (1959).
[8] Y. Nambu, Phys. Rev. **117**, 648 (1960).

It is perhaps not a coincidence that there exists such an entity in the form of the pion. For this reason, we would like to regard our theory as dealing with nucleons and mesons. The implication would be that the nucleon mass is a manifestation of some unknown primary interaction between originally massless fermions, the same interaction also being responsible for the binding of nucleon pairs into pions.

An additional support of the idea can be found in the weak decay processes of nucleons and pions which indicate that the γ_5 invariance is at least approximately conserved, as will be discussed in Part II. There are some difficulties, however, that naturally arise on further examination.

Comparison between a relativistic theory and a nonrelativistic, intuitive picture is often dangerous, because the former is severely restricted by the requirement of relativistic invariance. In our case, the energy-gap equation (4) depends on the energy density on the Fermi surface; for zero Fermi radius, the gap vanishes. The Fermi sphere, however, is not a relativistically invariant object, so that in the theory of nucleons it is not clear whether a formula like Eq. (4) could be obtained for the mass. This is not surprising, since there is a well known counterpart in classical electron theory that a finite electron radius is incompatible with relativistic invariance.

We avoid this difficulty by simply introducing a relativistic cutoff which takes the place of the Fermi sphere. Our framework does not yet resolve the divergence difficulty of self-energy, and the origin of such an effective cutoff has to be left as an open question.

The second difficulty concerns the mass of the pion. If pion is to be identified with the phonon-like excitations associated with a gauge group, its mass must necessarily be zero. It is true that in real superconductors the collective charge fluctuation is screened by Coulomb interaction to turn into the plasma mode, which has a finite "rest mass." A similar mechanism may be operating in the meson case too. It is possible, however, that the finite meson mass means that chirality conservation is only approximate in a real theory. From the evidence in weak interactions, we are inclined toward the second view.

The observation made so far does not yet give us a clue as to the exact mechanism of the primary interaction. Neither do we have a fundamental understanding of the isospin and strangeness quantum numbers, although it is easy to incorporate at least the isospin degree of freedom into the theory from the beginning. The best we can do here is to examine the various existing models for their logical simplicity and experimental support, if any. We will do this in Sec. 2, and settle for the moment on a nonlinear four-fermion interaction of the Heisenberg type. For reasons of simplicity in presentation, we adopt a model without isospin and strangeness degrees of freedom, and possessing complete γ_5 invariance. Once the choice is made, we can explore the whole idea mathematically, using essentially the formulation developed in reference 8. It is gratifying that the various field-theoretical techniques can be fully utilized. Section 3 will be devoted to introduction of the Hartree-Fock equation for nucleon self-energy, which will make the starting point of the theory. Then we go on to discuss in Sec. 4 the collective modes. In addition to the expected pseudoscalar "pion" states, we find other massive mesons of scalar and vector variety, as well as a scalar "deuteron." The coupling constants of these mesons can be easily determined. The relation of the pion to the γ_5 gauge group will be discussed in Secs. 5 and 6.

The theory promises many practical consequences. For this purpose, however, it is necessary to make our model more realistic by incorporating the isospin, and allowing for a violation of γ_5 invariance. But in doing so, there arise at the same time new problems concerning the mass splitting and instability. This refined model will be elaborated in Part II of this work, where we shall also find predictions about strong and weak interactions. Thus the general structure of the weak interaction currents modified by strong interactions can be treated to some degree, enabling one to derive the decay processes of various particles under simple assumptions. The calculation of the pion decay rate gives perhaps one of the most interesting supports of the theory. Results about strong interactions themselves are equally interesting. We shall find specific predictions about heavier mesons, which are in line with the recent theoretical expectations.

II. THE PRIMARY INTERACTION

We briefly discuss the possible nature of the primary interaction between fermions. Lacking any radically new concepts, the interaction could be either mediated by some fundamental Bose field or due to an inherent nonlinearity in the fermion field. According to our postulate, these interactions must allow chirality conservation in addition to the conservation of nucleon number. The chirality X here is defined as the eigenvalue of γ_5, or in terms of quantized fields,

$$X=\int\bar{\psi}\gamma_4\gamma_5\psi d^3x. \tag{2.1}$$

The nucleon number is, on the other hand

$$N=\int\bar{\psi}\gamma_4\psi d^3x. \tag{2.2}$$

These are, respectively, generators of the γ_5- and ordinary-gauge groups

$$\psi\rightarrow\exp[i\alpha\gamma_5]\psi,\quad \bar{\psi}\rightarrow\bar{\psi}\exp[i\alpha\gamma_5], \tag{2.3}$$

$$\psi\rightarrow\exp[i\alpha]\psi,\quad \bar{\psi}\rightarrow\bar{\psi}\exp[-i\alpha], \tag{2.4}$$

where α is an arbitrary constant phase.

Furthermore, the dynamics of our theory would require that the interaction be attractive between particle and antiparticle in order to make bound-state formation possible. Under the transformation (2.3), various tensors transform as follows:

$$\begin{aligned}
&\text{Vector:} && i\bar{\psi}\gamma_\mu\psi\rightarrow i\bar{\psi}\gamma_\mu\psi,\\
&\text{Axial vector:} && i\bar{\psi}\gamma_\mu\gamma_5\psi\rightarrow i\bar{\psi}\gamma_\mu\gamma_5\psi,\\
&\text{Scalar:} && \bar{\psi}\psi\rightarrow\bar{\psi}\psi\cos2\alpha+i\bar{\psi}\gamma_5\psi\sin2\alpha,\\
&\text{Pseudoscalar:} && i\bar{\psi}\gamma_5\psi\rightarrow i\bar{\psi}\gamma_5\psi\cos2\alpha-\bar{\psi}\psi\sin2\alpha,\\
&\text{Tensor:} && \bar{\psi}\sigma_{\mu\nu}\psi\rightarrow\bar{\psi}\sigma_{\mu\nu}\psi\cos2\alpha+i\bar{\psi}\gamma_5\sigma_{\mu\nu}\psi\sin2\alpha.
\end{aligned} \tag{2.5}$$

It is obvious that a vector or pseudovector Bose field coupled to the fermion field satisfies the invariance. The vector case would also satisfy the dynamical requirement since, as in the electromagnetic interaction, the forces would be attractive between opposite nucleon charges. The pseudovector field, on the other hand, does not meet the requirement as can be seen by studying the self-consistent mass equation discussed later.

The vector field looks particularly attractive since it can be associated with the nucleon number gauge group. This idea has been explored by Lee and Yang,[9] and recently by Sakurai.[10] But since we are dealing with strong interactions, such a field would have to have a finite observed mass in a realistic theory. Whether this is compatible with the invariance requirement is not yet clear. (Besides, if the bare mass of both spinor and vector field were zero, the theory would not contain any parameter with the dimensions of mass.)

The nonlinear fermion interaction seems to offer another possibility. Heisenberg and his co-workers[11] have been developing a comprehensive theory of elementary particles along this line. It is not easy, however, to gain a clear physical insight into their results obtained by means of highly complicated mathematical machinery.

We would like to choose the nonlinear interaction in this paper. Although this looks similar to Heisenberg's theory, the dynamical treatment will be quite different and more amenable to qualitative understanding.

The following Lagrangian density will be assumed ($\hbar=c=1$):

$$L=-\bar{\psi}\gamma_\mu\partial_\mu\psi+g_0[(\bar{\psi}\psi)^2-(\bar{\psi}\gamma_5\psi)^2]. \tag{2.6}$$

The coupling parameter g_0 is positive, and has dimensions $[\text{mass}]^{-2}$. The γ_5 invariance property of the interaction is evident from Eq. (2.5). According to the Fierz theorem, it is also equivalent to

$$-\tfrac{1}{2}g_0[(\bar{\psi}\gamma_\mu\psi)^2-(\bar{\psi}\gamma_\mu\gamma_5\psi)^2]. \tag{2.7}$$

This particular choice of γ_5-invariant form was taken without a compelling reason, but has the advantage

[9] T. D. Lee and C. N. Yang, Phys. Rev. **98**, 1501 (1955).
[10] J. J. Sakurai, Ann. Phys. **11**, 1 (1960).
[11] W. Heisenberg, Z. Naturforsch. **14**, 441 (1959). Earlier papers are quoted there.

that it can be naturally extended to incorporate isotopic spin.[12]

Unlike Heisenberg's case, we do not have any theory about the handling of the highly divergent singularities inherent in nonlinear interactions. So we will introduce, as an additional and independent assumption, an *ad hoc* relativistic cutoff or form factor in actual calculations. Thus the theory may also be regarded as an approximate treatment of the intermediate-boson model with a large effective mass.

As will be seen in subsequent sections, the nonlinear model makes mathematics particularly easy, at least in the lowest approximation, enabling one to derive many interesting quantitative results.

III. THE SELF-CONSISTENT EQUATION FOR NUCLEON MASS

We will assume that all quantities we calculate here are somewhow convergent, without asking the reason behind it. This will be done actually by introducing a suitable phenomenological cutoff.

Without specifying the interaction, let Σ be the unrenormalized proper self-energy part of the fermion, expressed in terms of observed mass m, coupling constant g, and cutoff Λ. A real Dirac particle will satisfy the equation

$$i\gamma\cdot p+m_0+\Sigma(p,m,g,\Lambda)=0 \qquad (3.1)$$

for $i\gamma\cdot p+m=0$. Namely

$$m-m_0=\Sigma(p,m,g,\Lambda)\,|_{i\gamma\cdot p+m=0}. \qquad (3.2)$$

The g will also be related to the bare coupling g_0 by an equation of the type

$$g/g_0=\Gamma(m,g,\Lambda). \qquad (3.3)$$

Equations (3.1) and (3.2) may be solved by successive approximation starting from m_0 and g_0. It is possible, however, that there are also solutions which cannot thus be obtained. In fact, there can be a solution $m\neq 0$ even in the case where $m_0=0$, and moreover the symmetry seems to forbid a finite m.

This kind of situation can be most easily examined by means of the generalized Hartree-Fock procedure[8,13] which was developed before in connection with the theory of superconductivity. The basic idea is not new in field theory, and in fact in its simplest form the method is identical with the renormalization procedure of Dyson, considered only in a somewhat different context.

Suppose a Lagrangian is composed of the free and interaction part: $L=L_0+L_i$. Instead of diagonalizing L_0 and treating L_i as perturbation, we introduce the self-energy Lagrangian L_s, and split L thus

$$L=(L_0+L_s)+(L_i-L_s)$$
$$=L_0'+L_i'.$$

For L_s we assume quite general form (quadratic or bilinear in the fields) such that L_0' leads to linear field equations. This will enable one to define a vacuum and a complete set of "quasi-particle" states, each particle being an eigenmode of L_0'. Now we treat L_i' as perturbation, and determine L_s from the requirement that L_i' shall not yield additional self-energy effects. This procedure then leads to Eq. (3.2). The self-consistent nature of such a procedure is evident since the self-energy is calculated by perturbation theory with fields which are already subject to the self-energy effect.

In order to apply the method to our problem, let us assume that $L_s=-m\bar{\psi}\psi$, and introduce the propagator $S_F^{(m)}(x)$ for the corresponding Dirac particle with mass m. In the lowest order, and using the two alternative forms Eqs. (2.6) and (2.7), we get for Eq. (3.2)

$$\Sigma=2g_0[\mathrm{Tr}S_F^{(m)}(0)-\gamma_5\,\mathrm{Tr}S_F^{(m)}(0)\gamma_5$$
$$-\tfrac{1}{2}\gamma_\mu\,\mathrm{Tr}\gamma_\mu S_F^{(m)}(0)+\tfrac{1}{2}\gamma_\mu\gamma_5\,\mathrm{Tr}\gamma_\mu\gamma_5 S_F^{(m)}(0)] \qquad (3.4)$$

in coordinate space.

This is quadratically divergent, but with a cutoff can be made finite. In momentum space we have

$$\Sigma=-\frac{8g_0 i}{(2\pi)^4}\int\frac{m}{p^2+m^2-i\epsilon}d^4p\,F(p,\Lambda), \qquad (3.5)$$

where $F(p,\Lambda)$ is a cutoff factor. In this case the self-energy operator is a constant. Substituting Σ from Eq. (3.5), Eq. (3.2) gives ($m_0=0$)

$$m=-\frac{g_0 mi}{2\pi^4}\int\frac{d^4p}{p^2+m^2-i\epsilon}F(p,\Lambda). \qquad (3.6)$$

This has two solutions: either $m=0$, or

$$1=-\frac{g_0 i}{2\pi^4}\int\frac{d^4p}{p^2+m^2-i\epsilon}F(p,\Lambda). \qquad (3.7)$$

The first trivial one corresponds to the ordinary perturbative result. The second, nontrivial solution will determine m in terms of g_0 and Λ.

If we evaluate Eq. (3.7) with a straight noninvariant cutoff at $|\mathbf{p}|=\Lambda$, we get

$$\frac{\pi^2}{g_0\Lambda^2}=\left(\frac{m^2}{\Lambda^2}+1\right)^{\frac{1}{2}}-\frac{m^2}{\Lambda^2}\ln\left[\left(\frac{\Lambda^2}{m^2}+1\right)^{\frac{1}{2}}+\frac{\Lambda}{m}\right]. \qquad (3.8)$$

If we use Eq. (3.5) with an invariant cutoff at $p^2=\Lambda^2$ after the change of path: $p_0\to ip_0$, we get

$$\frac{2\pi^2}{g_0\Lambda^2}=1-\frac{m^2}{\Lambda^2}\ln\left(\frac{\Lambda^2}{m^2}+1\right). \qquad (3.9)$$

[12] This will be done in Part II.

[13] N. N. Bogoliubov, Uspekhi Fiz. Nauk **67**, 549 (1959) [translation: Soviet Phys.-Uspekhi **67**, 236 (1959)].

Since the right-hand side of Eq. (3.8) or (3.9) is positive and $\leqslant 1$ for real Λ/m, the nontrivial solution exists only if

$$0<2\pi^2/g_0\Lambda^2<1. \tag{3.10}$$

Equation (3.9) is plotted in Fig. 1 as a function of m^2/Λ^2. As $g_0\Lambda^2$ increases over the critical value $2\pi^2$, m starts rising from 0. The nonanalytic nature of the solution is evident as m cannot be expanded in powers of g_0.

In the following we will assume that Eq. (3.10) is satisfied, so that the nontrivial solution exists. As we shall see later, physically this means that the nucleon-antinucleon interaction must be attractive ($g_0>0$) and strong enough to cause a bound pair of zero total mass. In the BCS theory, the nontrivial solution corresponds to a superconductive state, whereas the trivial one corresponds to a normal state, which is not the true ground state of the superconductor. We may expect a similar situation to hold in the present case.

In this connection, it must be kept in mind that our solutions are only approximate ones. We are operating under the assumption that the corrections to them are not catastrophic, and can be appropriately calculated when necessary. If this does not turn out to be so for some solution, such a solution must be discarded. Later we shall indeed find this possibility for the trivial solution, but for the moment we will ignore such considerations.

Let us define then the vacuum corresponding to the two solutions. Let $\psi^{(0)}$ and $\psi^{(m)}$ be quantized fields satisfying the equations

$$\gamma_\mu\partial_\mu\psi^{(0)}(x)=0, \tag{3.11a}$$

$$(\gamma_\mu\partial_\mu+m)\psi^{(m)}(x)=0, \tag{3.11b}$$

$$\psi^{(0)}(x)=\psi^{(m)}(x) \quad \text{for} \quad x_0=0. \tag{3.11c}$$

According to the standard procedure, we decompose the ψ's into Fourier components:

$$\begin{aligned}\psi_\alpha{}^{(i)}(x)&=\frac{1}{V^{\frac12}}\sum_{\substack{\mathbf{p},s\\ p_0=(p^2+m^2)^{\frac12}}}\left[u_\alpha{}^{(i)}(\mathbf{p},s)a^{(i)}(\mathbf{p},s)e^{ip\cdot x}\right.\\&\quad\left.+v_\alpha{}^{*(i)}(\mathbf{p},s)b^{(i)\dagger}(\mathbf{p},s)e^{-ip\cdot x}\right],\\ \psi_\alpha{}^{\dagger(i)}(x)&=\frac{1}{V^{\frac12}}\sum_{\substack{\mathbf{p},s\\ p_0=(p^2+m^2)^{\frac12}}}\left[u_\alpha{}^{(i)*}(\mathbf{p}\cdot s)a^{(i)\dagger}(\mathbf{p},s)\right.\\&\quad\left.\times e^{-ip\cdot x}+v_\alpha{}^{(i)}(\mathbf{p},s)b^{(i)}(\mathbf{p},s)e^{ip\cdot x}\right],\\ &i=0 \text{ or } m,\end{aligned} \tag{3.12}$$

where $u_\alpha{}^{(i)}(\mathbf{p},s)$, $v_\alpha{}^{(i)}(\mathbf{p},s)$ are the normalized spinor eigenfunctions for particles and antiparticles, with momentum p and helicity $s=\pm1$, and

$$\begin{aligned}&\{a^{(i)}(\mathbf{p},s),a^{(i)\dagger}(\mathbf{p}',s')\}\\&\quad=\{b^{(i)}(\mathbf{p},s),b^{(i)\dagger}(\mathbf{p}',s')\}=\delta_{\mathbf{p}\mathbf{p}'}\delta_{ss'}, \text{ etc.}\end{aligned} \tag{3.13}$$

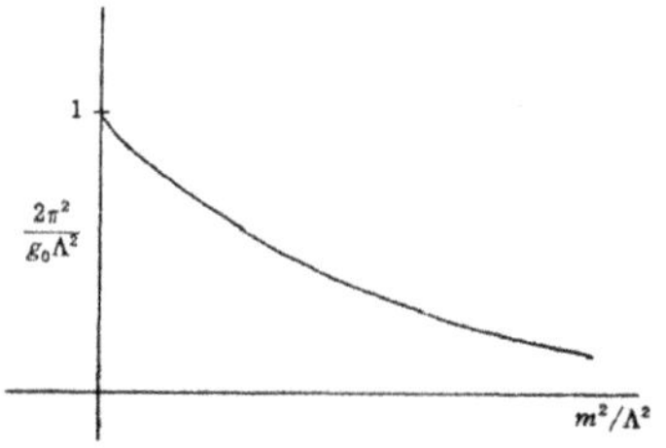

FIG. 1. Plot of the self-consistent mass equation (3.9).

The operator sets $(a^{(0)},b^{(0)})$ and $(a^{(m)},b^{(m)})$ are related by a canonical transformation because of Eq. (3.11c):

$$\begin{aligned}a^{(m)}(\mathbf{p},s)&=\sum_{\alpha,s'}\left[u_\alpha{}^{(m)*}(\mathbf{p},s)u_\alpha{}^{(0)}(\mathbf{p},s')a^{(0)}(\mathbf{p},s')\right.\\&\quad\left.+u_\alpha{}^{(m)*}(\mathbf{p},s)v_\alpha{}^{(0)*}(-\mathbf{p},s')b^{(0)\dagger}(-\mathbf{p},s')\right],\\ b^{(m)}(\mathbf{p},s)&=\sum_{\alpha,s'}\left[v_\alpha{}^{(m)*}(\mathbf{p},s)v_\alpha{}^{(0)}(\mathbf{p},s')b^{(0)}(\mathbf{p},s')\right.\\&\quad\left.+v_\alpha{}^{(m)*}(\mathbf{p},s)u_\alpha{}^{(0)*}(-\mathbf{p},s')a^{(0)\dagger}(-\mathbf{p},s')\right].\end{aligned} \tag{3.14}$$

Using Eq. (1.3), this is evaluated to give

$$\begin{aligned}a^{(m)}(\mathbf{p},s)&=[\tfrac12(1+\beta_p)]^{\frac12}a^{(0)}(\mathbf{p},s)\\&\quad+[\tfrac12(1-\beta_p)]^{\frac12}b^{(0)\dagger}(-\mathbf{p},s),\\ b^{(m)}(\mathbf{p},s)&=[\tfrac12(1+\beta_p)]^{\frac12}b^{(0)}(\mathbf{p},s)\\&\quad-[\tfrac12(1-\beta_p)]^{\frac12}a^{(0)\dagger}(-\mathbf{p},s),\\ \beta_p&=|\mathbf{p}|/(\mathbf{p}^2+m^2)^{\frac12}.\end{aligned} \tag{3.15}$$

The vacuum $\Omega^{(0)}$ or $\Omega^{(m)}$ with respect to the field $\psi^{(0)}$ or $\psi^{(m)}$ is now defined as

$$a^{(0)}(\mathbf{p},s)\Omega^{(0)}=b^{(0)}(\mathbf{p},s)\Omega^{(0)}=0, \tag{3.16}$$

$$a^{(m)}(\mathbf{p},s)\Omega^{(m)}=b^{(m)}(\mathbf{p},s)\Omega^{(m)}=0. \tag{3.16'}$$

Both $\psi^{(0)}$, $\bar\psi^{(0)}$ and $\psi^{(m)}$, $\bar\psi^{(m)}$ applied to $\Omega^{(0)}$ always create particles of mass zero, whereas the same applied to $\Omega^{(m)}$ create particles of mass m.

From Eqs. (3.15) and (3.16) we obtain

$$\begin{aligned}\Omega^{(m)}&=\prod_{\mathbf{p},s}\{[\tfrac12(1+\beta_p)]^{\frac12}\\&\quad-[\tfrac12(1-\beta_p)]^{\frac12}a^{(0)\dagger}(\mathbf{p},s)b^{(0)\dagger}(-\mathbf{p},s)\}\Omega^{(0)}.\end{aligned} \tag{3.17}$$

Thus $\Omega^{(m)}$ is, in terms of zero-mass particles, a superposition of pair states. Each pair has zero momentum, spin and nucleon number, and carries ±2 units of chirality, since chirality equals minus the helicity s for massless particles.

Let us calculate the scalar product $(\Omega^{(0)},\Omega^{(m)})$ from Eq. (3.15):

$$\begin{aligned}(\Omega^{(0)},\Omega^{(m)})&=\prod_{\mathbf{p},s}[\tfrac12(1+\beta_p)]^{\frac12}\\&=\exp\{\sum_{\mathbf{p},s}\tfrac12\ln[\tfrac12(1+\beta_p)]\}.\end{aligned} \tag{3.18}$$

For large p, $\beta_p\sim1-m^2/2\mathbf{p}^2$, so that the exponent diverges as $V\pi m^2\int d\mathbf{p}/(2\pi)^3$ (V=normalization volume). Hence

$$(\Omega^{(0)},\Omega^{(m)})=0. \qquad (3.19)$$

It is easy to see that any two states $\Psi^{(0)}$ and $\Psi^{(m)}$, obtained by applying a finite number of creation operators on $\Omega^{(0)}$ and $\Omega^{(m)}$ respectively, are also orthogonal.

Thus the two "worlds" based on $\Omega^{(0)}$ and $\Omega^{(m)}$ are physically distinct and outside of each other. No interaction or measurement, in the usual sense, can bridge them in finite steps.

What is the energy difference of the two vacua? Since both are Lorentz invariant states, the difference can only be either zero or infinity. Using the expression

$$\begin{aligned}H^{(m)}&=\sum_{\mathbf{p},s}(\mathbf{p}^2+m^2)^{\frac{1}{2}}\{a^{(m)\dagger}(\mathbf{p},s)a^{(m)}(\mathbf{p},s)\\&\qquad -b^{(m)}(\mathbf{p},s)b^{(m)\dagger}(\mathbf{p},s)\},\\H^{(0)}&=\sum_{\mathbf{p},s}|\mathbf{p}|\{a^{(0)\dagger}(\mathbf{p},s)a^{(0)}(\mathbf{p},s)\\&\qquad -b^{(0)}(\mathbf{p},s)b^{(0)\dagger}(\mathbf{p},s)\},\end{aligned} \qquad (3.20)$$

we get for the respective energies

$$E^{(m)}-E^{(0)}=-2\sum_{\mathbf{p}}[(\mathbf{p}^2+m^2)^{\frac{1}{2}}-|\mathbf{p}|], \qquad (3.21)$$

which is negative and quadratically divergent. So $\Omega^{(m)}$ may be called the "true" ground state, as was expected.

There remains finally the question of γ_5 invariance. The original Hamiltonian allowed two conservations X and N, Eqs. (2.1) and (2.2). Both $\Omega^{(0)}$ and $\Omega^{(m)}$ belong to $N=0$, and their elementary excitations carry $N=\pm1$. In the case of X, the same is true for the space $\Omega^{(0)}$, but $\Omega^{(m)}$ as well as its elementary excitations are not eigenstates of X, as is clear from the foregoing results. If the latter solution is to be a possibility, there must be an infinite degeneracy with respect to the quantum number X. A ground state will be in general a linear combination of degenerate states with different $X=0$, $\pm2,\cdots$:

$$\Omega^{(m)}=\sum_{n=-\infty}^{\infty}C_{2n}\Omega_{2n}^{(m)}. \qquad (3.22)$$

Equation (3.17) is in fact a particular case of this. The γ_5-gauge transformation Eq. (2.3) induces the change

$$\begin{aligned}a^{(0)}(\mathbf{p},\pm1)&\to e^{\mp i\alpha}a^{(0)}(\mathbf{p},\pm1),\\b^{(0)}(\mathbf{p},\pm1)&\to e^{\mp i\alpha}b^{(0)}(\mathbf{p},\pm1),\\a^{(0)\dagger}(\mathbf{p},\pm1)&\to e^{\pm i\alpha}a^{(0)\dagger}(\mathbf{p},\pm1),\\b^{(0)\dagger}(\mathbf{p},\pm1)&\to e^{\pm i\alpha}b^{(0)\dagger}(\mathbf{p},\pm1),\end{aligned} \qquad (3.23)$$

and the coefficients of Eq. (3.22) become

$$C_{2n}\to e^{-2ni\alpha}C_{2n}. \qquad (3.24)$$

In particular

$$\begin{aligned}\Omega^{(m)}&\to\Omega_\alpha^{(m)}\\&=\exp[-i\alpha X]\Omega^{(m)}\\&=\prod_{\mathbf{p},\pm}\{[\tfrac{1}{2}(1+\beta_p)]^{\frac{1}{2}}-[\tfrac{1}{2}(1-\beta_p)]^{\frac{1}{2}}\\&\qquad\times e^{\pm2i\alpha}a^{(0)\dagger}(p,\pm)b^{(0)\dagger}(-p,\pm)\}\Omega^{(0)}.\end{aligned} \qquad (3.25)$$

The Dirac equation (3.11b), at the same time, is transformed into

$$[\gamma_\mu\partial_\mu+m\cos2\alpha+im\gamma_5\sin2\alpha]\psi=0. \qquad (3.26)$$

The moral of this is that the self-consistent self-energy Σ is determined only up to a γ_5 transformation. This can be easily verified from Eq. (3.4), in which the second term on the right-hand side is nonvanishing when a propagator corresponding to Eq. (3.26) is used. Although Eq. (3.26) seems to violate parity conservation, it is only superficially so since $\Omega_\alpha^{(m)}$ is now not an eigenstate of parity. We could alternatively say that the parity operator undergoes transformation together with the mass operator. Despite the odd form of the equation (3.26), there is no change in the physical predictions of the theory. We shall see more of this later.

Let us calculate, as before, the scalar product of $\Omega_\alpha^{(m)}$ and $\Omega_{\alpha'}^{(m)}$. From Eqs. (3.17) and (3.25) we get

$$\begin{aligned}&(\Omega_\alpha^{(m)},\Omega_{\alpha'}^{(m)})\\&=\prod_{p,\pm}[\tfrac{1}{2}(1+\beta_p)-e^{\pm2i(\alpha'-\alpha)}\tfrac{1}{2}(1-\beta_p)]\\&=\prod_{p,\pm}[1+(e^{\pm2i(\alpha'-\alpha)}-1)\tfrac{1}{2}(1-\beta_p)]\\&=\exp\{\sum_{p,\pm}\ln[1+(e^{\pm2i(\alpha'-\alpha)}-1)\tfrac{1}{2}(1-\beta_p)]\}.\end{aligned} \qquad (3.27)$$

For large $|\mathbf{p}|$, the exponent goes like

$$\frac{V}{(2\pi)^3}\sum_{\pm}(e^{\pm2i(\alpha'-\alpha)}-1)\int\frac{m^2}{4\mathbf{p}^2}d^3p.$$

The integral is again divergent. Hence

$$\begin{aligned}(\Omega_\alpha^{(m)},\Omega_{\alpha'}^{(m)})&=(\Omega^{(m)},\exp[-i(\alpha'-\alpha)X]\Omega^{(m)})\\&=0,\quad \alpha'\neq\alpha(\mathrm{mod}\,2\pi),\end{aligned} \qquad (3.28)$$

and, of course

$$(\Omega^{(0)},\Omega_\alpha^{(m)})=0. \qquad (3.28')$$

We can evaluate $(\Omega_\alpha^{(m)},\Omega_{\alpha'}^{(m)})$ alternatively from Eqs. (3.22) and (3.24). Then

$$\sum_{m=-\infty}^{\infty}|C_{2n}|^2e^{2ni(\alpha-\alpha')}=0,\quad \alpha\neq\alpha'(\mathrm{mod}\,2\pi), \qquad (3.29)$$

implying that

$$|C_0|=|C_{\pm2}|=|C_{\pm4}|=\cdots=C. \qquad (3.30)$$

Thus there is an infinity of equivalent worlds described by $\Omega_\alpha^{(m)}$, $0\leqslant\alpha<2\pi$. The states Ω_{2n} of Eq. (3.22) are then expressed in terms of $\Omega_\alpha^{(m)}$ as

$$C_{2n}\Omega_{2n}^{(m)}=\frac{1}{2\pi}\int_0^{2\pi}e^{2ni\alpha}\Omega_\alpha^{(m)}d\alpha, \qquad (3.31)$$

which form another orthogonal set. Since the original total H commutes with X, it will have no matrix elements connecting different "worlds." Moreover, as was the case with $\Omega^{(m)}$ and $\Omega^{(0)}$, no finite measurement can induce similar transitions. This is a kind of superselection rule, which effectively avoids the apparent degeneracy to show up as physical effects.[14] The usual description of the world by means of $\Omega^{(m)}$ and ordinary Dirac particles must be regarded as only the most convenient one.

We still are left with some paradoxes. The X conservation implies the existence of a conserved X current:

$$j_{\mu5}=i\bar{\psi}\gamma_\mu\gamma_5\psi, \tag{3.32}$$

$$\partial_\mu j_{\mu5}=0, \tag{3.32'}$$

which can readily be verified from Eq. (2.6). On the other hand, for a massive Dirac particle the continuity equation is not satisfied:

$$\partial_\mu\bar{\psi}^{(m)}\gamma_\mu\gamma_5\psi^{(m)}=2m\bar{\psi}^{(m)}\gamma_5\psi^{(m)}. \tag{3.33}$$

If a massive Dirac particle has to be a real eigenstate of the system, how can this be reconciled? The answer would be that the X-current operator taken between real one-nucleon states should not be given simply by $i\gamma_\mu\gamma_5$ because of the "radiative corrections." We expect instead

$$\langle p'|j_{\mu5}|p\rangle=\bar{u}(p')X_\mu(p',p)u(p), \tag{3.34}$$

where the renormalized quantity $X_{\mu5}$ should be, from relativistic invariance grounds, of the form

$$X_\mu(p',p)=F_1(q^2)i\gamma_\mu\gamma_5+F_2(q^2)\gamma_5q_\mu,\quad q=p'-p,\quad p^2=p'^2=-m^2. \tag{3.35}$$

The continuity equation (3.32′), together with Eq. (3.33), further reduces this to

$$F_1=F_2q^2/2m\equiv F,\quad X_\mu(p',p)=F(q^2)\left(i\gamma_\mu\gamma_5+\frac{2m\gamma_5q_\mu}{q^2}\right). \tag{3.36}$$

The real nucleon is not a point particle. Its X-current (3.36) is provided with the dramatic "anomalous" term.

To understand the physical meaning of the anomalous term, we have to make use of the dispersion relations. The form factors F_1 and F_2 will, in general, satisfy dispersion relations of the form

$$F_i(q^2)=F_i(0)-\frac{q^2}{\pi}\int\frac{\mathrm{Im}F_i(-\kappa^2)}{(q^2+\kappa^2-i\epsilon)\kappa^2}d\kappa^2, \tag{3.37}$$

assuming one subtraction. Each singularity at κ^2 corresponds to some physical intermediate state. Thus if $F(0)\neq0$, Eq. (3.36) indicates that there is a pole at $q^2=0$ for F_2 (and no subtraction), which means in turn that there is an isolated intermediate state of zero mass.

[14] This was discussed by R. Haag, Kgl. Danske Videnskab. Selskab, Mat.-fys. Medd. **29**, No. 12 (1955). See also L. van Hove, Physica **18**, 145 (1952).

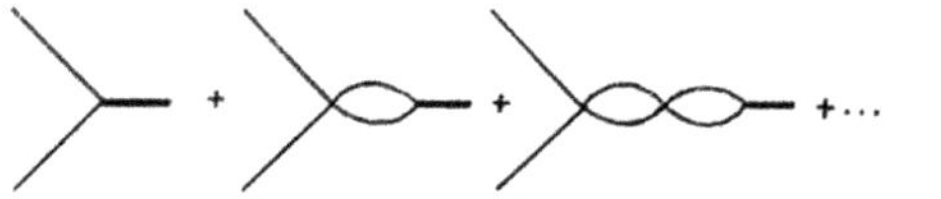

FIG. 2. Graphs corresponding to the Bethe-Salpeter equation in "ladder" approximation. The thick line is a bound state.

To see its nature, we take a time-like q in its own rest frame and go to the limit $q^2\to0$. The anomalous term has then only the time component, and is proportional to the amplitude for creation of a nucleon pair in a $J=0^-$ state. Hence the zero mass state must have the same property as this pair. It belongs to nucleon number zero, so that we may call it a zero-mass pseudoscalar meson. *In order for a γ_5-invariant Hamiltonian such as Eq. (2.6) to allow massive nucleon states and a nonvanishing X current for $q=0$, it is therefore necessary to have at the same time pseudoscalar zero-mass mesons coupled with the nucleons.* Since we did not have such mesons in the theory, they must be regarded as secondary products, i.e., bound states of nucleon pairs. This conclusion would not hold if in Eq. (3.36) $F(q^2)=O(q^2)$ near $q^2=0$. A nucleon then would have always $X=0$. Such a possibility cannot be excluded. We will show, however, that the pseudoscalar zero-mass bound states do follow explicitly, once we assume the nontrivial solution of the self-energy equation.

IV. THE COLLECTIVE STATES

From the general discussion of Secs. 2 and 3, we may expect the existence of collective states of the fundamental field which would manifest themselves as stable or unstable particles. In particular we have argued that, as a consequence of the γ_5 invariance, a pseudoscalar zero-mass state must exist. We want now to discuss the problem in detail, trying to determine the mass spectrum of the collective excitations (at least its general features) and the strength of their coupling with the nucleons. These states must be considered as a direct effect of the same primary interaction which produces the mass of the nucleon, which itself is a collective effect. We will study the bound-state problem through the use of the Bethe-Salpeter equation, taking into account explicitly the self-consistency conditions. We first verify in the following the existence of the zero-mass pseudoscalar state.

The Bethe-Salpeter equation for a bound pair B deals with the amplitude

$$\Phi(x,y)=\langle0|T(\psi(x)\bar{\psi}(y))|B\rangle. \tag{4.1}$$

As is well known, the equation is relatively easy to handle in the ladder approximation. In our case we have a four-spinor point interaction and the analog of the "ladder" approximation would be the iteration of the simplest closed loop (see Fig. 2) in which all lines represent dressed particles. We introduce the vertex function

Γ related to Φ by

$$\Phi(p)=S_F^{(m)}(p+\tfrac{1}{2}q)\Gamma(p+\tfrac{1}{2}q,\,p-\tfrac{1}{2}q)S_F^{(m)}(p-\tfrac{1}{2}q). \quad (4.2)$$

All we have to do then is to set up the integral equation generated by the chain of diagrams, looking for solutions having the symmetry properties of a pseudoscalar state. This means that our solutions must be proportional to γ_5. This requirement makes only the pseudoscalar and axial vector part of the interaction contribute to the integral equation. We have

$$\begin{aligned}\Gamma(p+\tfrac{1}{2}q,\,p-\tfrac{1}{2}q)&\\ =\frac{2ig_0}{(2\pi)^4}\gamma_5\int &\mathrm{Tr}[\gamma_5 S_F^{(m)}(p'+\tfrac{1}{2}q)\\ &\times\Gamma(p'+\tfrac{1}{2}q,\,p'-\tfrac{1}{2}q)S_F^{(m)}(p'-\tfrac{1}{2}q)]d^4p'\\ -\frac{ig_0}{(2\pi)^4}\gamma_5\gamma_\mu\int &\mathrm{Tr}[\gamma_5\gamma_\mu S^{(m)}(p'+\tfrac{1}{2}q)\\ &\times\Gamma(p'+\tfrac{1}{2}q,\,p'-\tfrac{1}{2}q)S_F^{(m)}(p'-\tfrac{1}{2}q)]d^4p'. \quad (4.3)\end{aligned}$$

For the moment let us ignore the pseudovector term on the right-hand side. It then follows that the equation has a constant solution $\Gamma=C\gamma_5$ if $q^2=0$. To see this, first observe that for the special case $q=0$, Eq. (4.3) reduces to

$$1=-\frac{8ig_0}{(2\pi)^4}\int\frac{d^4p}{p^2+m^2-i\epsilon}, \quad (4.4)$$

which is nothing but the self-consistency condition (3.7), provided that the same cutoff is applied. Since the pseudoscalar term of Eq. (4.3) gives a function of q^2 only, the same condition remains true as long as $q^2=0$.

When the pseudovector term is included, we have still the same eigenvalue $q^2=0$ with a solution of the form $\Gamma=C\gamma_5+iD\gamma_5\gamma\cdot q$, which is not difficult to verify (see Appendix).

We now add some remarks. First, the bound state amplitude for this solution spreads in space over a region of the order of the fermion Compton wavelength $1/m$ because of Eq. (4.2), making the zero-mass particle only partially localizable. We want also to stress the role played by the γ_5 invariance in the argument. We had in fact already inferred the existence of the pseudoscalar particle from relativistic and γ_5 invariance alone, and at first sight the same result seems to follow now essentially from the self-consistency equation. However, we must notice that only the scalar term of the Lagrangian appears in this equation while only the pseudoscalar part contributes in the Bethe-Salpeter equation. It is because of the γ_5-invariant Lagrangian that the Bethe-Salpeter equation can be reduced to the self-consistency condition.

Along the same line we could try to see whether other bound states exist in the "ladder" approximation. However, besides calculating the spectrum, it is also important to determine the interaction properties of these collective states with the fermions. For this purpose the study of the two-"nucleon" scattering amplitude appears much more suitable, as we shall realize after the following remark. Once we have recognized that in the ladder approximation the collective states would appear as real stable particles, we must expect to the same degree of approximation poles in the scattering matrix of two nucleons corresponding to the possibility of the virtual exchange of these particles. For definiteness we shall refer again as an example to the pseudoscalar zero-mass particle. Let us indicate by $J_p(q)$ the analytical expression corresponding to the graph whose iteration produces the bound state [Fig. 3(a)]. We construct next the scattering matrix generated by the exchange of all possible simple chains built with this element. This means that we consider the set of diagrams in Fig. 3(b). The series is easily evaluated and we obtain

$$2g_0 i\gamma_5\frac{1}{1-J_p(q)}i\gamma_5, \quad (4.5)$$

where the γ_5's refer to the pairs (1,1′) and (2,2′), respectively. The meaning of this result is clear: because of the self-consistent equation $J_p(0)=1$, Eq. (4.5) is equivalent to a phenomenological exchange term where the intermediate particle is our pseudoscalar massless boson (Fig. 4). The coupling constant G can now be evaluated by straightforward comparison. Before doing this calculation we need the explicit expression of $J_P(q)$. Using the ordinary rules for diagrams, we have

$$J_P(q)=-\frac{2ig_0}{(2\pi)^4}\times\int\frac{4(m^2+p^2)-q^2}{[(p+\frac{1}{2}q)^2+m^2][(p-\frac{1}{2}q)^2+m^2]}d^4p. \quad (4.6)$$

It is however more convenient to rewrite J_P in the form of a dispersive integral, and if we forget for a moment that it is a divergent expression, a simple manipulation gives

$$J_P(q)=\frac{g_0}{4\pi^2}\int_{4m^2}^{\Lambda^2}\frac{\kappa^2(1-4m^2/\kappa^2)^{\frac{1}{2}}}{q^2+\kappa^2}d\kappa^2. \quad (4.6')$$

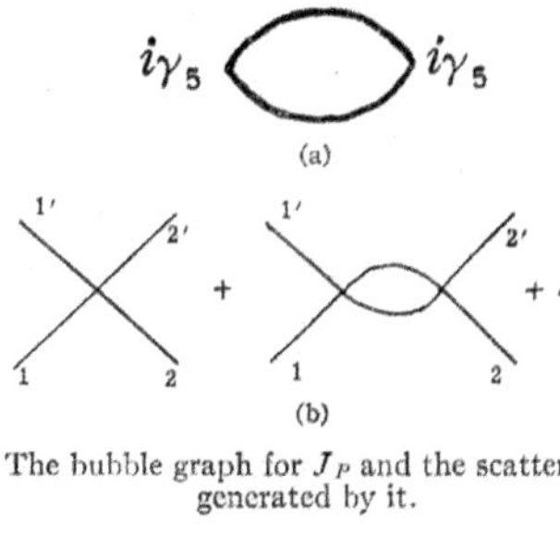

FIG. 3. The bubble graph for J_P and the scattering matrix generated by it.

In order for this expression to be meaningful, a new cutoff Λ must be introduced. There is no simple relation between this and the previous cutoffs. The dispersive form is more comfortable to handle and accordingly we shall reformulate the self-consistent condition $J_P(0)=1$, or

$$1=\frac{g_0}{4\pi^2}\int_{4m^2}^{\Lambda^2}(1-4m^2/\kappa^2)^{\frac{1}{2}}d\kappa^2. \tag{4.7}$$

It may be of interest to remark at this point that Eq. (4.7) can be obtained also if we think of our theory as a theory with intermediate pseudoscalar boson in the limit of infinite boson mass. We are now in a position to evaluate the phenomenological coupling constant G. From Eqs. (4.6′) and (4.7) we have

$$J_P(q^2)=1-q^2\frac{g_0}{4\pi^2}\int_{4m^2}^{\Lambda^2}\frac{(1-4m^2/\kappa^2)^{\frac{1}{2}}}{q^2+\kappa^2}d\kappa^2, \tag{4.8}$$

which leads immediately to the result

$$\frac{G_P^2}{4\pi}=2\pi\left[\int_{4m^2}^{\Lambda^2}\frac{(1-4m^2/\kappa^2)^{\frac{1}{2}}}{\kappa^2}d\kappa^2\right]^{-1}. \tag{4.9}$$

This equation is interesting since it establishes a connection between the phenomenological constant G_P and the cutoff independently of the value of the fundamental coupling g_0. This fact exhibits the purely dynamical origin of the phenomenological coupling G_P. Actually g_0 is buried in the value of the mass m.

So far we have exploited only the γ_5 vertex. What happens then if the scalar part is iterated to form chains of bubbles similar to those we have already discussed? The procedure just explained can be followed again, and a quantity $J_S(q)$ can be defined similarly with the result

$$J_S(q)=\frac{g_0}{4\pi^2}\int_{4m^2}^{\Lambda^2}\frac{(\kappa^2-4m^2)(1-4m^2/\kappa^2)^{\frac{1}{2}}}{q^2+\kappa^2}d\kappa^2. \tag{4.10}$$

It is immediately seen that because of Eq. (4.7)

$$J_S(-4m^2)=1, \tag{4.11}$$

which causes a new pole to appear in the S matrix for $q^2=-4m^2$. This means that we have another collective state of mass $2m$, parity $+$ and spin 0! We observe that it is necessary to assume the same cutoff as in the pseudoscalar case in order that this result may be obtained. The choice of the same cutoff in both cases seems to be suggested by the γ_5 invariance as will be seen later. We also notice the peculiar symmetry existing between the pseudoscalar and the scalar state: the first has zero mass and binding energy $2m$, while the opposite is true for the scalar particle. So in the bound-state picture the scalar particle would not be a true bound state and should be, rather, interpreted as a

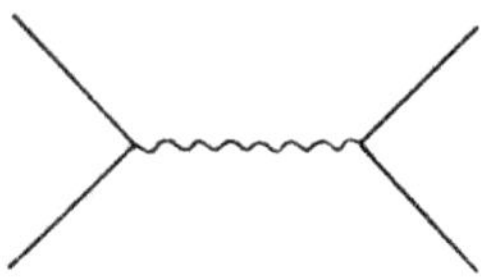

FIG. 4. The equivalent phenomenological one-meson exchange graph.

correlated exchange of pairs in the scattering process.[15] The "nucleon-nucleon" forces induced by the exchange of the scalar particle are, of course, of rather short range. The general physical implications of these results will be discussed more thoroughly later.

The phenomenological coupling constant G_S for the scalar meson is given by

$$\frac{G_S^2}{4\pi}=2\pi\left[\int_{4m^2}^{\Lambda^2}\frac{(1-4m^2/\kappa^2)^{\frac{1}{2}}}{(\kappa^2-4m^2)}d\kappa^2\right]^{-1}. \tag{4.12}$$

Let us next turn to the vector state generated by iteration of the vector interaction. In this case we obtain for each "bubble" a tensor

$$\begin{aligned} J_{V\mu\nu}&=(\delta_{\mu\nu}-q_\mu q_\nu/q^2)J_V,\\ J_V&=-\frac{g_0}{4\pi^2}\frac{q^2}{3}\int_{4m^2}^{\Lambda^2}\frac{d\kappa^2}{q^2+\kappa^2}\\ &\quad\times\left(1+\frac{2m^2}{\kappa^2}\right)(1-4m^2/\kappa^2)^{\frac{1}{2}}. \end{aligned} \tag{4.13}$$

Perhaps a remark is in order here regarding the evaluation of J_V. It suffers from an ambiguity of subtraction well known in connection with the photon self-energy problem. The above result is of the conventional gauge invariant form, which we take to be the proper choice.

Equation (4.13) leads to the scattering matrix

$$g_0\left[\gamma_\mu\frac{1}{1-J_V}\gamma_\mu-\gamma\cdot q\frac{J_V}{(1-J_V)q^2}\gamma\cdot q\right], \tag{4.14}$$

where the second term is, of course, effectively zero. It can be easily seen that the denominator can produce a pole below $4m^2$ for sufficiently small Λ^2. In fact, from Eqs. (4.7) and (4.13), we find

$$(8/3)m^2<\mu_V^2. \tag{4.15}$$

The coupling constant is given by

$$\frac{G_V^2}{4\pi}=3\pi\left[\int_{4m^2}^{\Lambda^2}d\kappa^2\frac{\kappa^2+2m^2}{(\kappa^2-\mu_V^2)^2}(1-4m^2/\kappa^2)^{\frac{1}{2}}\right]^{-1}. \tag{4.16}$$

It must be noted that the mass of the vector meson now depends on the cutoff, unlike the previous two cases.

Finally we are left with the pseudovector state. We

[15] Of course this and other heavy mesons will in general become unstable in higher order approximation, which is beyond the scope of the present paper.

find for the bubble[16]

$$J_{A\mu\nu}=-J_{V\mu\nu}+J_A'\delta_{\mu\nu},$$
$$J_A'=\frac{g_0m^2}{2\pi^2}\int_{4m^2}^{\Lambda^2}\frac{d\kappa^2}{q^2+\kappa^2}(1-4m^2/\kappa^2)^{\frac{1}{2}}. \tag{4.17}$$

In view of the self-consistency condition (4.7), it can be seen that this does not produce a pole of the scattering matrix for $-q^2<4m^2$, corresponding to a pseudovector meson.

So far we have considered only iterations of the same kind of interactions. In the ladder approximation there is actually a coupling between pseudoscalar and pseudovector interactions as was explicitly considered in Eq. (4.3). However, the coupling between scalar and vector interactions vanish because of the Furry's theorem.

This coupling of pseudoscalar and pseudovector interactions does not change the pion pole of the scattering matrix, but it affects the coupling of the pion to the nucleon since a chain of the pseudoscalar can join the external nucleon with an axial vector interaction. In other words, the pion-nucleon coupling is in general a mixture of pseudoscalar and derivative pseudovector types (Appendix).

We would like to inject here a remark concerning the trivial solution of the self-energy equation, against which we had no decisive argument. So let us also try to apply our scattering formula to this solution. For the pseudoscalar state we now find $J_P(q=0)>1$, provided that the cutoff Λ is kept fixed and m is set equal to zero in Eq. (4.6′). (The pseudovector interference vanishes.) In other words, there will be a pole for some $q^2>0$ ($\mu^2<0$). This is again a supporting evidence that the trivial solution could be unstable, capable of decaying by emitting such mesons. The final answer, however, depends on the exact nature of the cutoff.

Finally we would like to discuss the nucleon-nucleon scattering in the same spirit and approximation as for the nucleon-antinucleon scattering. In order to make a correspondence with the previous cases, it is convenient to rewrite the Hamiltonian in the following way:

$$\begin{aligned}H_1&=-g_0[\bar\psi\psi\bar\psi^c\psi^c-\bar\psi\gamma_5\psi\bar\psi^c\gamma_5\psi^c]\\&=\tfrac{1}{2}g_0[\bar\psi\gamma_\mu\psi^c\bar\psi^c\gamma_\mu\psi-\bar\psi\gamma_\mu\gamma_5\psi^c\bar\psi^c\gamma_\mu\gamma_5\psi]\\&=-\tfrac{1}{2}g_0[\bar\psi\gamma_\mu C\bar\psi\psi C^{-1}\gamma_\mu\psi-\bar\psi\gamma_\mu\gamma_5C\bar\psi\psi C^{-1}\gamma_\mu\gamma_5\psi],\end{aligned} \tag{4.18}$$

where ψ^c, $\bar\psi^c$ are the charge-conjugate fields.

The last form of Eq. (4.18) is suitable for our purpose. We note first that the vector part of the interaction is identically zero because of the anticommutativity of ψ. Thus only the pseudovector part survives. A "bubble" made of this interaction then is seen to give rise to the same integral J_A, Eq. (4.17). Since the interfering pseudoscalar interaction is missing in the present case,

[16] We meet here again the problem of subtraction. Our choice follows naturally from comparison with the vector case, and is consistent with Eq. (3.33).

TABLE I. Mass spectrum.

Nucleon number	Mass μ	Spin-parity	Spectroscopic notation
0	0	0^-	$^1S_0$
0	$2m$	0^+	$^3P_0$
0	$(8/3)m^2<\mu^2$	1^-	$^3P_1$
±2	$2m^2<\mu^2$	0^+	$^1S_0$

we get the complete scattering matrix by iterating J_A:

$$\begin{aligned}&-\gamma_\mu\gamma_5C\left[\frac{\delta_{\mu\nu}-q_\mu q_\nu/q^2}{1-J_A}+\frac{q_\mu q_\nu/q^2}{1-J_A'}\right]C^{-1}\gamma\gamma_5\\&=\gamma_\mu\gamma_5C\frac{1}{1-J_A}C^{-1}\gamma_\mu\gamma_5\\&\qquad+\gamma\cdot q\gamma_5C\frac{J_V/q^2}{(1-J_A')(1-J_A)}C^{-1}\gamma\cdot q\gamma_5,\end{aligned} \tag{4.19}$$

$J_A\equiv J_A'-J_V.$

The first term, corresponding to a scattering in the $J=1^-$ state, does not have a pole. The second term can have one below $4m^2$ for $1=J_A'$. With Eqs. (4.7) and (4.17), this determines the mass μ_D:

$$2m^2<\mu_D^2. \tag{4.20}$$

In this second term of the scattering matrix, the wave function is proportional to $C\gamma\cdot q\gamma_5$, so that the bound state behaves like a scalar "deuteron" (a singlet S state). The residue of the pole determines the nucleon-"deuteron" coupling constant (derivative) G_D^2, which is positive as it should be.

Table I summarizes the main results of this section. Although our approximation is a very crude one, we believe that it reflects the real situation at least qualitatively, because all the results are understandable in simple physical terms. Thus in the nonrelativistic sense, our Hamiltonian contains spin-independent attractive scalar and vector interactions plus a spin-dependent axial vector interaction between a particle and an antiparticle. Between particles, the vector part turns into a repulsion. Table I is just what we expect for the level ordering from this consideration.

V. PHENOMENOLOGICAL THEORY AND γ_5 INVARIANCE

In the previous section special subsets of diagrams were taken into account, and the existence of various boson states was established, together with their couplings with the nucleons. As was discussed there, we can reasonably expect that these results are essentially correct in spite of the very simple approximations. Because the bosons have in general small masses (compared to the unbound nucleon states), they will play important roles in the dynamics of strong interactions at least at energies comparable to these masses.

Thus if we are willing to accept the conclusions of our lowest order approximation, what we should do then is to study the dynamics of systems consisting of nucleons and the different kinds of bosons which all together represent the primary manifestation of the fundamental interaction. These particles will be now assumed to interact via their phenomenological couplings. So we may describe our purpose as an attempt to construct a theory in the conventional sense in which a separate field is introduced for each kind of particle. However, this is not a simple and unambiguous problem because our fundamental theory is completely γ_5 invariant and we must make sure that this invariance is preserved at any stage of our calculations in order that the results be meaningful. For a better understanding of the problem, let us consider our Lagrangian in the lowest self-consistent approximation. We have

$$L'=L_0'+L_I',$$

where

$$L_0'=-(\bar{\psi}\gamma_\mu\partial_\mu\psi+m\bar{\psi}\psi),$$
$$L_I'=g_0[(\bar{\psi}\psi)^2-(\bar{\psi}\gamma_5\psi)^2]+m\bar{\psi}\psi. \tag{5.1}$$

L' is obviously γ_5 invariant. In order to preserve this invariance we must study the S matrix generated by L_I'. Some subsets of diagrams have been considered in the previous section and it will be shown now how those calculations comply with γ_5 invariance. This point must be understood clearly so that we shall discuss it in a rather systematic way. Let us recall first how we constructed the scattering matrix in the "ladder" approximation. The lowest-order contribution is certainly invariant as no internal massive line appears. But what will happen to the next-order terms [Fig. 3(b)]? To these diagrams corresponds the expression

$$J_S(q^2)-\gamma_5 J_P(q^2)\gamma_5+iJ_{SP}(q^2)\gamma_5+i\gamma_5 J_{PS}(q^2). \tag{5.2}$$

In the gauge in which our calculations were performed, the last two terms happened to be zero. We write down next the transformation properties of the quantities appearing above. By straightforward calculation we find

$$\begin{aligned}
\gamma_5 &\to \gamma_5\cos 2\alpha+i\sin 2\alpha,\\
1 &\to \cos 2\alpha+i\gamma_5\sin 2\alpha,\\
J_P &\to J_P\cos^2 2\alpha+J_S\sin^2 2\alpha,\\
J_S &\to J_S\cos^2 2\alpha+J_P\sin^2 2\alpha,\\
J_{SP} &\to (J_P-J_S)\sin 2\alpha\cos 2\alpha,\\
J_{PS} &\to (J_P-J_S)\sin 2\alpha\cos 2\alpha.
\end{aligned} \tag{5.3}$$

By simple substitution the invariance follows easily. The argument can now be extended to all orders, provided at each order all the possible combinations of S and P are included. The invariance of the scattering in the "ladder" approximation is thus established. It may look surprising that the SP and PS contributions do not vanish identically. This can be understood by considering the fact that the γ_5 transformation changes the parity of the vacuum which will be in general a superposition of states of opposite parities. In this way products of fields of different parities (as the SP propagator) may have a nonvanishing average value in the vacuum state.

We may now attempt the construction of the phenomenological coupling by introducing two local fields Φ_P and Φ_S describing the pseudoscalar and the scalar particles, respectively. We start by observing that, in the same gauge in which the previous calculations were made, we can write the meson-nucleon interaction as

$$L_I=G_P i\bar{\psi}\gamma_5\psi\Phi_P+G_S\bar{\psi}\psi\Phi_S. \tag{5.4}$$

In order to find the general expression valid in any gauge, it is convenient to introduce the following two-dimensional notation

$$\varphi\equiv\begin{pmatrix} i\bar{\psi}\gamma_5\psi \\ \bar{\psi}\psi\end{pmatrix},\quad \Phi\equiv\begin{pmatrix}\Phi_P\\ \Phi_S\end{pmatrix},\quad G\equiv\begin{pmatrix}G_P & 0\\ 0 & G_S\end{pmatrix}. \tag{5.5}$$

The interaction Lagrangian Eq. (5.4) can be written in this notation in a compact form,

$$L_I=\varphi G\Phi. \tag{5.6}$$

The effect of the γ_5 transformation on φ is described with the aid of the matrix

$$U\equiv\begin{pmatrix}\cos 2\alpha & -\sin 2\alpha\\ \sin 2\alpha & \cos 2\alpha\end{pmatrix}, \tag{5.7}$$

which satisfies $UU^+=UU^{-1}=UU^T=1$. In other words, the γ_5 transformation induces a unitary transformation in the two-dimensional space, and Eq. (5.6) remains invariant if

$$G\to UGU^{-1},\quad \Phi\to U\Phi. \tag{5.8}$$

To complete the construction of the theory, the free Lagrangian for the fields Φ_P and Φ_S must be added. If we work again in the special gauge $\alpha=0$, we may write

$$L_0=-\tfrac{1}{2}\partial_\mu\Phi_P\partial_\mu\Phi_P-\tfrac{1}{2}\partial_\mu\Phi_S\partial_\mu\Phi_S-\tfrac{1}{2}\mu^2\Phi_S^2, \tag{5.9}$$

where $\mu^2=4m^2$. We use again the two-dimensional notation, and defining the mass operator

$$M^2\equiv\begin{pmatrix}0 & 0\\ 0 & \mu^2\end{pmatrix}, \tag{5.10}$$

we write Eq. (5.9) in the invariant form

$$L_0=-\tfrac{1}{2}\partial_\mu\Phi\partial_\mu\Phi-\tfrac{1}{2}\Phi M^2\Phi. \tag{5.11}$$

In this way we have given a formal prescription for the γ_5 transformation in the phenomenological treatment. We have to emphasize here that the Lagrangians (5.9) and (5.11) are *not* γ_5 invariant in the ordinary sense of the word. In our theory, where the mesons are only phenomenological substitutes which partially represent the dynamical contents of the theory, they may be, however, called γ_5 covariant. In other words, *the masses and the coupling constants are not fixed parameters, but rather dynamical quantities which are subject to transformations when the representation is changed.* It will be legitimate to ask whether this situation corresponds to the one obtained in the framework of the fundamental theory and discussed in the "ladder" approximation in the previous section. We shall examine the transformation rule for the mass operator M^2, since this illustrates the case in point. Let us calculate explicitly M^2 in an arbitrary gauge α. We have

$$M^2 \to UM^2U^{-1}$$

$$=\mu^2\begin{pmatrix} \sin^2 2\alpha & -\sin 2\alpha\cos 2\alpha \\ -\sin 2\alpha\cos 2\alpha & \cos^2 2\alpha \end{pmatrix}. \quad (5.12)$$

The meaning of this equation is that the pseudoscalar and the scalar particle will have generally different masses in different gauges. In particular we see that the pseudoscalar particle has in the gauge α a mass $\sin 2\alpha\mu$. If this is the case we must expect that after the transformation the pole in the corresponding propagator will move from $q^2=0$ to $q^2=-(\sin^2 2\alpha)\mu^2$. This actually may be verified directly in the "ladder" approximation which shows that the pion propagator changes according to

$$iG_P{}^2\Delta_{FP}=\frac{2g_0}{1-J_P}\to\frac{2g_0}{1-J_P\cos^2 2\alpha-J_S\sin^2 2\alpha}. \quad (5.13)$$

Using the results of the previous section, it is seen that the denominator of the right-hand side vanishes for $q^2=-(\sin^2 2\alpha)4m^2$. In this way we have seen how our γ_5-invariant theory can be approximated by a phenomenological description in terms of pseudoscalar and scalar mesons. Of course one may add the vector meson as well. Such a description does not look γ_5 invariant. It is only γ_5 covariant, and the masses and coupling constants must be understood to be matrices which, however, can be simultaneously diagonalized.

The reason for this situation is the degeneracy of the vacuum and the world built upon it. Only after combining all the equivalent but nonintersecting worlds labeled with different α do we recover complete γ_5 invariance. Nevertheless, even in a particular world we can find manifestations of the invariance, such as the zero-mass pseudoscalar meson and the conserved γ_5 current.

VI. THE CONSERVATION OF AXIAL VECTOR CURRENT

In this section we will discuss another paradoxical aspect of the theory regarding the γ_5 invariance. In Sec. 3 we argued that the X current should really be conserved, and that this is possible if a nucleon X current possesses a peculiar anomalous term. We now verify the statement explicitly in our approximation.

First we have to realize that the problem is again how to keep the γ_5-invariant nature of the theory at every stage of approximation. It is well known in quantum electrodynamics that, in order to observe the ordinary gauge invariance, a certain set of graphs have to be combined together in a given approximation. The necessity for this is based on a general proof which makes use of the so-called Ward identity. In our present case there also exists an analog of the Ward identity. In order to derive it, let us first consider the proper self-energy part of our fermion in the presence of an external axial vector field B_μ with the interaction $L_B=-j_{\mu5}B_\mu$. The self-energy operator is now a matrix $\Sigma^{(B)}(p',p)$ depending on initial and final momenta. Expanding Σ in powers of B, we have

$$\Sigma^{(B)}(p',p)=\Sigma(p)+\Lambda_{\mu5}(p',p)B_\mu(p'-p)+\cdots. \quad (6.1)$$

We readily realize that the coefficient of the second term gives the desired X-current vertex correction.

On the other hand, the entire Lagrangian remains invariant under a *local* γ_5 transformation if Eq. (2.3) is accompanied by

$$B_\mu\to B_\mu-\partial_\mu\alpha, \quad (6.2)$$

where α is now an arbitrary function. In other words,

$$e^{i\alpha\gamma_5}\Sigma^{(B-\partial\alpha)}e^{i\alpha\gamma_5}=\Sigma^{(B)} \quad (6.3)$$

in a symbolical way of writing.[17]

Expanding (6.3) after putting $B=0$, we get

$$i\alpha(p'-p)[\gamma_5\Sigma(p)+\Sigma(p')\gamma_5] =i\alpha(p'-p)(p'-p)_\mu\Lambda_{\mu5}(p',p),$$

or

$$\gamma_5\Sigma(p)+\Sigma(p')\gamma_5=(p'-p)_\mu\Lambda_{\mu5}(p',p). \quad (6.4)$$

The entire vertex $\Gamma_{\mu5}=i\gamma_\mu\gamma_5+\Lambda_{\mu5}$ then satisfies

$$\begin{aligned}\gamma_5L'(p)+L'(p')\gamma_5&=-(p'-p)_\mu\Gamma_{\mu5}(p',p),\\ L'(p)&\equiv-i\gamma\cdot p-\Sigma(p),\end{aligned} \quad (6.5)$$

which is the desired generalized Ward identity.[18] The right-hand side of Eq. (6.5) is the divergence of the X current, while the left-hand side vanishes when p and p' are on the mass shell of the actual particle. The X-current conservation is thus established. Moreover, the way the anomalous term arises is now clear. For if we assume $\Sigma(p)=m$, Eq. (6.4) gives

$$2m\gamma_5=(p'-p)_\mu\Lambda_{\mu5}(p'-p), \quad (6.6)$$

so that we may write the longitudinal part of Λ as

$$\Lambda_{\mu5}{}^{(l)}(p',p)=2m\gamma_5q_\mu/q^2,\quad q=p'-p, \quad (6.7)$$

which is of the desired form.

Next we have to determine what types of graphs

[17] We assume here that $\alpha(x)$ is different from zero only over a finite space-time region, so that the gauge of the nontrivial vacuum, which we may fix at remote past, is not affected by the transformation. The limiting process of going over to constant α is then ill-defined as we can see from the fact that the anomalous term in $\Gamma_{\mu5}$ has no limit as $q\to0$.

[18] See also J. Bernstein, M. Gell-Mann, and L. Michel, Nuovo cimento **16**, 560 (1960).

should be considered for Γ_μ in our particular approximation of the self-energy. Examining the way in which the relation (6.3) is maintained in a perturbation expansion, we are led to the conclusion that our self-energy represented by Fig. 5(a) gives rise to the series of vertex graphs [Fig. 5(b)]. The summation of the graphs is easily carried out to give

$$\Lambda_{\mu 5}=i\gamma_5\frac{1}{1-J_P}J_{PA}, \tag{6.8}$$

where J_P was obtained before [Eq. (4.8)], and

$$\begin{aligned} J_{PA}&=\frac{2ig_0}{(2\pi)^4}\int \mathrm{Tr}\gamma_5 S(p+q/2)\gamma_\mu\gamma_5 S(p-q/2)d^4p \\ &=-\frac{g_0}{2\pi^2}imq_\mu\int_{4m^2}^{\Lambda^2}\frac{d\kappa^2}{q^2+\kappa^2}(1-4m^2/\kappa^2)^{\frac{1}{2}}. \end{aligned} \tag{6.9}$$

Thus

$$\begin{aligned} \Gamma_{\mu 5}&=i\gamma_\mu\gamma_5+\Lambda_{\mu 5} \\ &=i\gamma_\mu\gamma_5+2m\gamma^5 q_\mu/q^2, \end{aligned} \tag{6.10}$$

in agreement with the general formula. We see also that there is no form factor in this approximation.

This example will suffice to show the general procedure necessary for keeping γ_5 invariance. When we consider further corrections, the procedure becomes more involved, but we can always find a set of graphs which are sufficient to maintain the X-current conservation. We shall come across this problem in connection with the axial vector weak interactions.

VII. SUMMARY AND DISCUSSION

We briefly summarize the results so far obtained. Our model Hamiltonian, though very simple, has been found to produce results which strongly simulate the general characteristics of real nucleons and mesons. It is quite appealing that both the nucleon mass and the pseudoscalar "pion" are of the same dynamical origin, and the reason behind this can be easily understood in terms of (1) classical concepts such as attraction or repulsion between particles, and (2) the γ_5 symmetry.

According to our model, the pion is not the primary agent of strong interactions, but only a secondary effect. The primary interaction is unknown. At the present stage of the model the latter is only required to have appropriate dynamical and symmetry properties, although the nonlinear four-fermion interaction, which we actually adopted, has certain practical advantages.

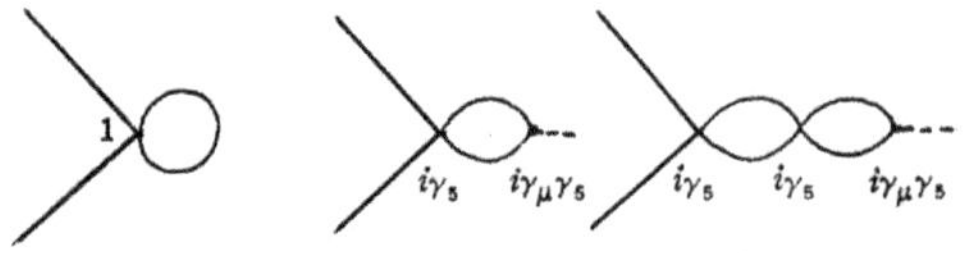

FIG. 5. Graphs for self-energy and matching radiative corrections to an axial vector vertex.

FIG. 6. A class of higher order self-energy graphs.

In our model the idealized "pion" occupies a special position in connection with the γ_5-gauge transformation. But there are also other massive bound states which may be called heavy mesons and deuterons. The conventional meson field theory must be regarded, from our point of view, as only a phenomenological description of events which are actually dynamic processes on a higher level of understanding, in the same sense that the phonon field is a phenomenological description of interatomic dynamics.

Our theory contains two parameters, the primary coupling constant and the cutoff, which can be translated into observed quantities: nucleon mass and the pion-nucleon coupling constant. It is interesting that the pion coupling depends only on the cutoff in our approximation. In order to make the pion coupling as big as the observed one (≈ 15) the cutoff has to be rather small, being of the same order as the nucleon mass.

We would like to make some remarks about the higher order approximations. If the higher order corrections are small, the usual perturbation calculation will be sufficient. If they are large compared to the lowest order estimation, the self-consistent procedure must be set up, including these effects from the beginning. This is complicated by the fact that the pions and other mesons have to be properly taken into account.

To get an idea about the importance of the corrections, let us take the next order self-energy graph (Fig. 6). This is only the first term of a class of corrections shown in Fig. 6, the sum of which we know already to give rise to an important collective effect, i.e., the mesons. It would be proper, therefore, to consider the entire class put together. The correction is then equivalent to the ordinary second order self-energy due to mesons, plus modifications arising at high momenta. Thus strict perturbation with respect to the bare coupling g_0 will not be an adequate procedure. Evaluating, for example, the pion contribution in a phenomenological way, we get

$$\frac{\delta m}{m}=\frac{G_P^2}{32\pi^2}\int_{m^2}^{\Lambda'^2}\frac{d\kappa^2}{\kappa^2}\left(1-\frac{m^2}{\kappa^2}\right), \tag{7.1}$$

where Λ' is an effective cutoff. Substituting G_P^2 from Eq. (4.9), this becomes

$$\frac{\delta m}{m}=\frac{1}{4}\int_{4m^2}^{4\Lambda'^2}\frac{d\kappa^2}{\kappa^2}\left(1-\frac{4m^2}{\kappa^2}\right)\bigg/\int_{4m^2}^{\Lambda^2}\frac{d\kappa^2}{\kappa^2}\left(1-\frac{4m^2}{\kappa^2}\right)^{\frac{1}{2}}. \tag{7.2}$$

As Λ and Λ' should be of the same order of magnitude, the higher order corrections are in general not negligible. We may point out, on the other hand, that there is a tendency for partial cancellation between contributions from different mesons or nucleon pairs.

We already remarked before that the model treated here is not realistic enough to be compared with the actual nucleon problem. Our purpose was to show that a new possibility exists for field theory to be richer and more complex than has been hitherto envisaged, even though the mathematics is marred by the unresolved divergence problem.

In the subsequent paper we will attempt to generalize the model to allow for isospin and finite pion mass, and draw various consequences regarding strong as well as weak interactions.

APPENDIX

We treat here, for completeness, the problem created by the coupling of pseudoscalar and pseudovector terms encountered in the text. As we have seen, such an effect is not essential for the discussion of γ_5 invariance, but rather adds to complication, which however naturally appears in the ladder approximation.

First let us write down the integral equation for a vertex part Γ:

$$\begin{aligned}\Gamma(p+\tfrac{1}{2}q,\, p-\tfrac{1}{2}q) &= \gamma(p+\tfrac{1}{2}q,\, p-\tfrac{1}{2}q)+\frac{2ig_0}{(2\pi)^4}\gamma_5\int \mathrm{Tr}[\gamma_5 S(p'+\tfrac{1}{2}q)\\ &\quad\times\Gamma(p'+\tfrac{1}{2}q,\, p-\tfrac{1}{2}q)S_F(p-\tfrac{1}{2}q)]d^4p'\\ &\quad-\frac{ig_0}{(2\pi)^4}\gamma_5\gamma_\mu\int \mathrm{Tr}[\gamma_5\gamma_\mu S(p'+\tfrac{1}{2}q)\\ &\quad\times\Gamma(p'+\tfrac{1}{2}q,\, p-\tfrac{1}{2}q)S_F(p-\tfrac{1}{2}q)]d^4p'. \qquad \text{(A.1)}\end{aligned}$$

This embraces three special cases depending on the inhomogeneous term γ:

(a) $\gamma=0$ for the Bethe-Salpeter equation for the pseudoscalar meson;
(b) $\gamma=i\gamma_\mu\gamma_5$ for the pseudovector vertex function $\Gamma_{\mu5}$;
(c) $\gamma=2g_0(\gamma_5)_f(\gamma_5)_i-g_0(\gamma_\mu\gamma_5)_f(\gamma_\mu\gamma_5)_i$ for the nucleon-antinucleon scattering through these interactions.

Here i and f refer to initial and final states, and the integral kernel of Eq. (A.1) operates on the f part.

We will consider them successively.

(a) We make the ansatz $\Gamma=C\gamma_5+iD\gamma_5\gamma\cdot q$. The integrals in Eq. (A.1) then reduce to the standard forms considered in the text. Making use of Eqs. (4.9), (4.17), and (6.9), we get[16]

$$\begin{aligned}C&=C-(C+2mD)q^2I,\\ D&=(C+2mD)mI,\\ I(q^2)&=\frac{g_0}{4\pi^2}\int\frac{d\kappa^2}{q^2+\kappa^2}\left(1-\frac{4m^2}{\kappa^2}\right)^{\frac{1}{2}},\end{aligned} \qquad \text{(A.2)}$$

which lead to $q^2=0$, and $C:D=1-2m^2I(0):mI(0)$. From Eq. (4.8), we have $0<2m^2I(0)<\frac{1}{2}$.

(b) Put
$$\begin{aligned}\Gamma_{\mu5}&=(i\gamma_\mu\gamma_5+2m\gamma_5q_\mu/q^2)F_1(q^2)\\ &\quad+(i\gamma_\mu\gamma_5-i\gamma\cdot q\gamma_5q_\mu/q^2)F_2(q^2). \qquad \text{(A.3)}\end{aligned}$$

This is seen to satisfy the integral equation if

$$\begin{aligned}F_1&=1,\\ F_2&=J_A(q^2)/[1-J_A(q^2)],\\ J_A(q^2)&=2m^2I(q^2)-J_V(q^2),\end{aligned} \qquad \text{(A.4)}$$

where $J(q^2)$ was defined in Eq. (4.13).

On the mass shell, $\Gamma_{\mu5}$ reduces to

$$\begin{aligned}&(i\gamma_\mu\gamma_5+2m\gamma_5q_\mu/q^2)F(q^2),\\ &F(q^2)=1+F_2(q^2)=1/[1-J_A(q^2)].\end{aligned} \qquad \text{(A.5)}$$

For $q^2=0$, we have $J(q^2)=0$ so that $1<F(0)=1/[1-2m^2I(0)]<2$.

(c) From the structure of the inhomogeneous term, it is clear that the scattering matrix is given by

$$M=2g_0(\Gamma_5)_f(\gamma_5)_i+g_0(\Gamma_{\mu5})_f(i\gamma_\mu\gamma_5)_i,$$

where Γ_5 is the pseudoscalar vertex function.

Again, from Eq. (A.1), Γ_5 is determined as

$$\Gamma_5=\gamma_5[1-2m^2I(q^2)]/q^2I(q^2)-mi\gamma\cdot q\gamma_5/q^2, \qquad \text{(A.6)}$$

which has an entirely different behavior from the bare γ_5 for small q^2. The scattering matrix is then

$$\begin{aligned}M&=(\gamma_5)_f(\gamma_5)_i2g_0[1-2m^2I(q^2)]/q^2I(q^2)\\ &\quad-[(i\gamma\cdot q\gamma_5)_f(\gamma_5)_i-(\gamma_5)_f(i\gamma\cdot q\gamma_5)_i]2mg_0/q^2\\ &\quad-(i\gamma\cdot q\gamma_5)_f(i\gamma\cdot q\gamma_5)_ig_0J_A(q^2)/q^2[1-J_A(q^2)]\\ &\quad+(i\gamma_\mu\gamma_5)_f(i\gamma_\mu\gamma_5)_ig_0/[1-J_A(q^2)]. \qquad \text{(A.7)}\end{aligned}$$

The first three terms have a pole at $q^2=0$. The coupling constants of the pseudoscalar meson are then

pseudoscalar coupling:

$$G_p^2=2g_0[1-2m^2I(0)]/I(0),$$

pseudovector coupling:

$$\begin{aligned}G_{pv}^2&=g_0J_A(0)/[1-J_A(0)]\\ &=g_02m^2I(0)/[1-2m^2I(0)]. \qquad \text{(A.8)}\end{aligned}$$

Their relative sign is such that the equivalent pseudoscalar coupling on the mass shell is

$$G_p'^2=4m^2g_0\left\{\left[\frac{1-2m^2I(0)}{2m^2I(0)}\right]^{\frac{1}{2}}+\left[\frac{2m^2I(0)}{1-2m^2I(0)}\right]^{\frac{1}{2}}\right\}^2. \qquad \text{(A.9)}$$

PHYSICAL REVIEW VOLUME 124, NUMBER 1 OCTOBER 1, 1961

Dynamical Model of Elementary Particles Based on an Analogy with Superconductivity. II*

Y. NAMBU AND G. JONA-LASINIO†
Enrico Fermi Institute for Nuclear Studies and Department of Physics, University of Chicago, Chicago, Illinois

(Received May 10, 1961)

Continuing the program developed in a previous paper, a "superconductive" solution describing the proton-neutron doublet is obtained from a nonlinear spinor field Lagrangian. We find the pions of finite mass as nucleon-antinucleon bound states by introducing a small bare mass into the Lagrangian which otherwise possesses a certain type of the γ_5 invariance. In addition, heavier mesons and two-nucleon bound states are obtained in the same approximation. On the basis of numerical mass relations, it is suggested that the bare nucleon field is similar to the electron-neutrino field, and further speculations are made concerning the complete description of the baryons and leptons.

I. INTRODUCTION

IN Part I of this paper[1] we have proposed a model of strong interactions based on an analogy with the BCS-Bogoliubov theory of superconductivity. It is characterized by a nonlinear spinor field possessing γ_5 invariance, and simulates some important features of the meson-nucleon system. The basic principle underlying the model is the idea that field theory may admit, as a result of dynamical instability, extraordinary (nontrivial) solutions that have less symmetries than are built into the Lagrangian.[2] In fact we have obtained as an extraordinary solution a massive fermion and a massless pseudoscalar boson as idealized proton and pion, together with other heavy mesons.

If we now try to make our model more realistic, a number of problems spring up naturally. First of all, we would have to account for the isospin and strangeness quantum numbers. It seems rather obvious that these degrees of freedom have to be built into the theory from the beginning, although there may be some possibility of utilizing both the ordinary and extraordinary solutions to enlarge the Hilbert space, as will be discussed later.

These quantum numbers will not yet be enough to determine our theory satisfactorily, as we expect to have more additional symmetries which are at least approximately satisfied. Among other things, we have postulated the γ_5 invariance as a cornerstone of our previous model. What would be the proper generalization of the γ_5 invariance? Then there also arises the inevitable question of any possible symmetry among baryons of different strangenesses. Since such a symmetry is at any rate only approximate, the test of the theory will depend on its ability to account for the violation of the symmetry as well.

Finally, we face the problem of the baryon versus the lepton, the electromagnetism, and the weak processes. Here our theory creates a particular incentive for speculation concerning the baryon-lepton problem, since the ordinary and extraordinary solutions immediately remind us of these two families of particles.

We do not profess to have any clear-cut answers to these problems. In the present paper we shall again content ourselves with a rather modest task. We will first discuss a generalization of our model which incorporates the isospin for the nucleon and guarantees the existence of the pion. This can be done by demanding a $\gamma_5\times$isospin gauge group with a slight violation so as to give the pion its finite mass. We find that the bare mass necessary to achieve the latter end is at most several Mev. On this basis a suggestion is made that the bare nucleon field is essentially the same as the electron-neutrino field.

The complete picture of the baryon symmetries and the baryon-lepton problem is largely beyond the scope of the present paper, but some relevant discussions on this subject will also be presented, especially those concerned with the Sakata model and the general γ_5 symmetry.

II. MODEL LAGRANGIAN FOR THE NUCLEON

First we would like to observe that the nonlinear spinor field adopted in I is not an essential element of our theory, as is the case with the Heisenberg theory[3] but is rather a model adopted to study our dynamical principles. At least in the present stage of the game, the controlling factors are the symmetry properties and qualitative dynamical characteristics of the basic fermion-fermion interaction, and whether the interaction is due to some fundamental boson, or fundamental nonlinearity (or something entirely new) is of secondary importance. Nevertheless, we have to choose

* This work was supported by the U. S. Atomic Energy Commission.

† Present address: Istituto di Fisica dell'Universita, Roma, and Istituto Nazionale di Fisica Nucleare, Sezione di Roma, Italy.

[1] Y. Nambu and G. Jona-Lasinio, Phys. Rev. **122**, 345 (1961); referred to hereafter as I. Y. Nambu, *Proceedings of the 1960 Annual International Conference on High-Energy Physics at Rochester* (Interscience Publishers, Inc., New York, 1960), p. 858.

[2] See also J. Goldstone, Nuovo cimento **19**, 154 (1961), N. N. Bogoliubov (to be published), V. G. Vaks and A. I. Larkin, *Proceedings of the 1960 Annual International Conference on High-Energy Physics at Rochester* (Interscience Publishers, Inc., New York, 1960), p. 871.

[3] H. P. Duerr, W. Heisenberg, H. Mitter, S. Schlieder, and K. Yamazaki, Z. Naturforsch. **14**, 441 (1959); W. Heisenberg, *Proceedings of the 1960 Annual International Conference on High-Energy Physics at Rochester* (Interscience Publishers, Inc., 1960), p. 851.

some model, and naturally there will arise certain predictions specific to the particular model. We take notice of the fact that the pion, the lightest of the meson family, is pseudoscalar and isovector, whereas its isoscalar counterpart of comparable mass does not seem to exist.[4] If the pion is to be intimately related to a symmetry property as in our previous model, this would imply that the model of nucleons should allow an (approximate) invariance under the $\gamma_5 \times$isospin gauge group of Gürsey,[5] but not under the simple (Touschek) γ_5 gauge group, at least not so well as in the former case. For this reason, we would altogether consider the following gauge groups:

$$\psi \to e^{i\alpha}\psi, \qquad \bar{\psi} \to \bar{\psi}e^{-i\alpha}, \tag{2.1a}$$

$$\psi \to \exp(i\boldsymbol{\tau}\cdot\boldsymbol{\alpha}')\psi, \qquad \bar{\psi} \to \bar{\psi}\exp(-i\boldsymbol{\tau}\cdot\boldsymbol{\alpha}'), \tag{2.1b}$$

$$\psi \to \exp(i\gamma_5\boldsymbol{\tau}\cdot\boldsymbol{\alpha}'')\psi, \qquad \bar{\psi} \to \bar{\psi}\exp(i\gamma_5\boldsymbol{\tau}\cdot\boldsymbol{\alpha}''), \tag{2.1c}$$

where τ denotes the nucleon isospin matrices.

Obviously, the first two are generators of the nucleon number gauge and the isospin transformation, respectively. The second and third transformations combined form a four-dimensional rotation group on the four components composed by the proton and neutron of both handednesses.[5] Thus we may also replace Eqs. (2.1a) and (2.1b) by the following transformations

$$\begin{aligned} &\psi_R \to \exp(i\boldsymbol{\tau}\cdot\boldsymbol{\alpha}_R)\psi_R, \quad \psi_R^\dagger \to \psi_R^\dagger\exp(-i\boldsymbol{\tau}\cdot\boldsymbol{\alpha}_R),\\ &\psi_L \to \exp(i\boldsymbol{\tau}1\alpha_L)\psi_L, \quad \psi_L^\dagger \to \psi_L^\dagger\exp(-i\boldsymbol{\tau}\cdot\boldsymbol{\alpha}_L), \end{aligned} \tag{2.2}$$

where ψ_R and ψ_L are the right- and left-handed components.

As the simplest Lagrangian that meets our requirements, we adopt the form

$$\begin{aligned} L = &-\bar{\psi}\gamma_\mu\partial_\mu\psi - \bar{\psi}M^0\psi\\ &+g_0[\bar{\psi}\psi\bar{\psi}\psi - \sum_{i=1}^{3}\bar{\psi}\gamma_5\tau_i\psi\bar{\psi}\gamma_5\tau_i\psi]. \end{aligned} \tag{2.3}$$

If the bare mass operator $M^0=0$, this Lagrangian possesses, in addition to Eq. (2.1), an invariance under the discrete "mass reversal" group:

$$\psi \to \gamma_5\psi, \quad \bar{\psi} \to -\bar{\psi}\gamma_5. \tag{2.4}$$

The bare mass operator M^0 is a possible agent for the breakdown of the Gürsey group, and will be related to the finite pion mass.[6] For the moment, we will assume $M^0=0$. Before going to solve the self-consistent equation for the mass, we give the result of the Fierz transformation on Eq. (2.3): The interaction becomes

$$\begin{aligned} L_{\text{int}} = &\tfrac{1}{4}g_0[\bar{\psi}\psi\bar{\psi}\psi - \bar{\psi}\gamma_5\tau_i\psi\bar{\psi}\gamma_5\tau_i\psi]\\ &+\tfrac{1}{4}g_0[\bar{\psi}\gamma_5\psi\bar{\psi}\gamma_5\psi - \bar{\psi}\tau_i\psi\bar{\psi}\tau_i\psi]\\ &-\tfrac{1}{2}g_0[\bar{\psi}\gamma_\mu\psi\bar{\psi}\gamma_\mu\psi - \bar{\psi}\gamma_\mu\gamma_5\psi\bar{\psi}\gamma_\mu\gamma_5\psi]\\ &+\tfrac{1}{8}g_0[\bar{\psi}\sigma_{\mu\nu}\psi\bar{\psi}\sigma_{\mu\nu}\psi - \bar{\psi}\sigma_{\mu\nu}\tau_i\psi\bar{\psi}\sigma_{\mu\nu}\tau_i\psi], \end{aligned} \tag{2.5}$$

which is a rather complicated combination of all kinds of terms.

We now apply the linearization procedure of I to Eqs. (2.3) and (2.4), and obtain the self-energy

$$\begin{aligned} m &= (1+\tfrac{1}{4})g_0\,\mathrm{Tr}S_F^{(m)}(0)\\ &= -i\frac{10g_0}{(2\pi)^4}\int\frac{d^4pm}{p^2+m^2}F(p,\Lambda). \end{aligned} \tag{2.6}$$

Note that the trace refers to both spin and isospin variables. This differs from Eq. (3.6) of I only by the change of the effective coupling $g_0 \to 5g_0/2 \equiv g_0'$. So we can simply take over the previous formulas, namely,

$$1 = \frac{g_0'}{4\pi^2}\int_{4m^2}^{\Lambda 2} d\kappa^2\left(1-\frac{4m^2}{\kappa^2}\right)^{\frac{1}{2}}, \tag{2.7}$$

for the nontrivial solution if the dispersion integral (4.7) of I is used.

III. DETERMINATION OF MESON STATES

Since the interaction Lagrangian in Eqs. (1.1) and (1.3) contains a number of different couplings, we expect to get various kinds of "mesons" as bound nucleon-antinucleon pairs in our simple ladder approximation. As was explained in I, this is the proper approximation to match our self-energy equation at least for the pseudoscalar meson which is expected to have zero mass; moreover, even for other types of bound states we may reasonably trust its qualitative validity in predicting the existence and level ordering of possible bound states to the extent that our interaction is regarded basically as a short-range potential between spinor particles.

For general discussion, it is convenient to follow the procedure given in the Appendix of I. The basic equation to be considered is of the type

$$\begin{aligned} \Gamma(p+\tfrac{1}{2}q, p-\tfrac{1}{2}q) = \gamma + i\textstyle\sum_n g_nO_n\\ \times\int \mathrm{Tr}[O_nS_F(p'+\tfrac{1}{2}q)\Gamma(p'+\tfrac{1}{2}q, p'-\tfrac{1}{2}q)\\ \times S_F(p'-\tfrac{1}{2}q)]d^4p', \end{aligned} \tag{3.1}$$

where the summation on the right-hand side is over the various tensor forms in the interaction Lagrangian. The "vertex function" $\Gamma(p+\frac{1}{2}q, p-\frac{1}{2}q)$ reduces to a bound state wave function when it becomes a homogeneous solution $(\gamma=0)$ for a particular value of $q^2=-\mu^2$. We will briefly discuss those two-nucleon states for which there is a possibility of binding.

A. Pseudoscalar, Isovector Meson

Unlike the case in I, only the pseudoscalar interaction $\sim\bar{\psi}\gamma_5\tau_i\psi\bar{\psi}\gamma_5\tau_i\psi$ contributes to this state. Assuming

[4] It may not be impossible that the ordinary γ_5 invariance is violated more strongly than the Gürsey γ_5 invariance so that the mass of the $\pi_0{}^0$ meson may come sufficiently high. But to achieve this end by means of a bare mass does not seem to be feasible.

[5] F. Gürsey, Nuovo cimento **16**, 230 (1960).

[6] For its possible origin, see Sec. V.

$\Gamma_i{}^P=\gamma_5\tau_i\Gamma^P$, we obtain

$$\Gamma^P=\gamma^P+\Gamma^P[1-q^2I^P(q^2)],$$

$$I^P(q^2)=\frac{g_0'}{4\pi^2}\int_{4m^2}^{\Lambda^2}\frac{d\kappa^2}{q^2+\kappa^2}\left(1-\frac{4m^2}{\kappa^2}\right)^{\frac{1}{2}}, \tag{3.2}$$

where, of course, Eq. (2.7) was utilized. This has a homogeneous solution for $q^2=0$, corresponding to the zero-mass "pion." This pion-nucleon coupling is of pure pseudoscalar type, which can be calculated from the inhomogeneous equation with $\gamma^P=g_0'\gamma_5\tau_i$, as was done in the Appendix of I. We get, namely,[7]

$$G_P^2/4\pi=g_0'[4\pi I^P(0)]^{-1}$$

$$=\pi\left[\int_{4m^2}^{\Lambda^2}\frac{d\kappa^2}{\kappa^2}\left(1-\frac{4m^2}{\kappa^2}\right)^{\frac{1}{2}}\right]^{-1}. \tag{3.3}$$

B. Scalar, Isoscalar Meson

With the ansatz $\Gamma=\Gamma^S$ we have

$$\Gamma^S=\gamma^S+\Gamma^SI^S(q^2),$$

$$I^S(q^2)=\frac{g_0'}{4\pi^2}\int_{4m^2}^{\Lambda^2}\frac{\kappa^2-4m^2}{q^2+\kappa^2}d\kappa^2\left(1-\frac{4m^2}{\kappa^2}\right)^{\frac{1}{2}} \tag{3.4}$$

$$=1-(q^2+4m^2)I^P(q^2).$$

This leads to a zero-binding state: $q^2=-4m^2$ with the scalar nucleon coupling constant

$$G_S^2/4\pi=g_0'[4\pi I^P(-4m^2)]^{-1}. \tag{3.5}$$

C. Vector Mesons

There are two vector mesons, with isospin 1 and 0. The isovector meson arises from the tensor interaction $\sim\bar\psi\sigma_{\mu\nu}\tau_i\psi\bar\psi\sigma_{\mu\nu}\tau_i\psi$ with the wave function of the type

$$\Gamma_{\mu i}{}^V=\sigma_{\mu\nu}q_\nu\tau_i\Gamma^V.$$

The mass is determined from[8]

$$1=-\frac{g_0'}{60\pi^2}\int_{4m^2}^{\Lambda^2}\frac{d\kappa^2}{\kappa^2-\mu^2}\left(1-\frac{4m^2}{\kappa^2}\right)^{\frac{1}{2}}$$

$$\times\left[\kappa^2-4m^2-\mu^2\left(2+\frac{4m^2}{\kappa^2}\right)\right],$$

which has a solution (for sufficiently small Λ^2)

$$\mu^2\gtrsim 10m^2/3.$$

The coupling of this meson to the nucleon is necessarily of the derivative type.

[7] Note that this is half the value of I because a pion (e.g., π_0) consists of two substates $\bar{p}p$ and $\bar{n}n$, which changes the normalization of the pion wave function.

[8] The ambiguity about the subtraction of the most divergent part was discussed in I, section 4. The gross qualitative feature is not altered even if we do not make a subtraction.

To the isoscalar meson, both vector and tensor interactions contribute, the former being attractive and the latter repulsive. The wave function will have the form

$$\Gamma_\mu{}^V=\gamma_\mu\Gamma_1{}^V+\sigma_{\mu\nu}q_\nu\Gamma_2{}^V,$$

which yields a coupled equation for Γ_1 and Γ_2. This coupling, however, is rather small, so that we get a solution by neglecting Γ_2:

$$1=\frac{g_0'}{15\pi^2}\mu^2\int_{4m^2}^{\Lambda^2}\frac{d\kappa^2}{\kappa^2-\mu^2}\left(1-\frac{4m^2}{\kappa^2}\right)^{\frac{1}{2}}\left(1+\frac{2m^2}{\kappa^2}\right),$$

$$\mu^2\gtrsim 20m^2/7.$$

The nuclear coupling will be predominantly nonderivative.

D. The "Deuteron" States

As in I, we can discuss the nucleon-nucleon states in parallel with the meson states. The interaction may be written conveniently in the form

$$L_{\rm int}=\tfrac{1}{4}g_0[\bar\psi\gamma_\mu\psi^c\bar\psi^c\gamma_\mu\psi-\bar\psi\sigma_{\mu\nu}\psi^c\bar\psi^c\sigma_{\mu\nu}\psi$$
$$+\bar\psi\gamma_\mu\gamma_5\tau_i\psi^c\bar\psi^c\gamma_\mu\gamma_5\tau_i\psi].$$

This is seen to lead to two bound states: a pseudovector, isoscalar ($J=1^+$, $T=0$) coming from the first two interaction terms, and a scalar, isovector ($J+0^+$, $T=1$), coming from the last term. For the $J=1^+$, $T=0$ state (deuteron) the main contribution comes from the attractive tensor interaction, and we get

$$\Gamma_\mu=\sigma_{\mu\nu}q_\nu\Gamma^{A\prime},$$

$$1=\frac{g_0'}{4\pi^2}\int_{4m^2}^{\Lambda^2}\frac{d\kappa^2}{\kappa^2-\mu^2}\left(1-\frac{4m^2}{\kappa^2}\right)^{\frac{1}{2}}\left[\mu^2+\frac{2}{15}(\kappa^2-4m^2)\right],$$

$$\mu^2\gtrsim 17m^2/5.$$

For the $J=0^+$, $T=1$ case we have

$$\Gamma=\gamma_5\gamma\cdot q\Gamma^{S\prime},$$

$$1=\tfrac{4}{3}m^2I^P(-\mu^2),$$

$$\mu^2\gtrsim 16m^2/5.$$

IV. VIOLATION OF γ_5 INVARIANCE

Let us now discuss the violation of the γ_5 invariance as indicated by the finite mass of the real pion. It would be senseless, of course, to talk about the invariance if the observed pion mass implied a large departure from our original Lagrangian, for example, due to a bare nucleon mass as large as the observed mass. So we need to estimate the amount of violation in the Lagrangian.

In general, the bare mass operator, which does not

violate nucleon number conservation,[9] can have the following form

$$M^0=m_1^0+m_2^0\boldsymbol{\tau}\cdot\mathbf{n}+m_3^0i\gamma_5+m_4^0i\gamma_5\boldsymbol{\tau}\cdot\mathbf{n}', \tag{4.1}$$

where $\mathbf{n}$ and $\mathbf{n}'$ are arbitrary unit vectors in the isospin space. The observed mass generated by Eq. (4.1) will also have a similar form. Because of the invariance of the rest of the Lagrangian under the transformations (2.1), we can choose it to be

$$M=m_1+m_2\tau_3+m_3i\gamma_5, \tag{4.2}$$

which gives two eigenmasses

$$m_p\equiv[(m_1+m_2)^2+m_3^2]^{\frac{1}{2}},$$
$$m_n\equiv[(m_1-m_2)^2+m_3^2]^{\frac{1}{2}}. \tag{4.3}$$

The self-consistent self-energy equation to be solved is now

$$M=M^0+g_0\{\mathrm{Tr}S_F^{(M)}(0)-\gamma_5\tau_i\,\mathrm{Tr}[\gamma_5\tau_iS_F^{(M)}(0)]$$
$$+\tfrac{1}{5}\gamma_5\,\mathrm{Tr}[\gamma_5S_F^{(M)}(0)]-\tfrac{1}{5}\tau_i\,\mathrm{Tr}[\tau_iS_F^{(M)}(0)]\}. \tag{4.4}$$

Equating the respective coefficients of both sides, we get

$$\begin{aligned}m_1&=m_1^0+\bar{I}m_1+\tilde{I}m_2,\\ m_2&=m_2^0-\tfrac{1}{5}\bar{I}m_2-\tfrac{1}{5}\tilde{I}m_1,\\ m_3&=m_3^0-\tfrac{1}{5}\bar{I}m_3,\\ 0&=m_4^0+\tilde{I}m_3,\end{aligned} \tag{4.5}$$

where

$$\begin{aligned}\bar{I}&=\tfrac{1}{2}[I(m_p)+I(m_n)],\\ \tilde{I}&=\tfrac{1}{2}[I(m_p)-I(m_n)],\\ I(m)&=-\frac{8ig_0'}{(2\pi)^4}\int\frac{d^4p}{p^2+m^2}F(p,\Lambda).\end{aligned} \tag{4.5'}$$

We are interested in a small change of the non-trivial solution due to M^0. From Eq. (4.5) it is clear that $m_3=0$ unless

$$m_3^0=-(\tfrac{1}{5}+\bar{I}^{-1})m_4^0\neq0.$$

The term m_3 implies a violation of time and space reflections. Since we are not interested in such a violation, we will assume $m_3=m_3^0=m_4^0=0$ from now on. We further note that

$$\bar{I}=I(m_1)+O[(m_2/m_1)^2],\quad \tilde{I}=O[m_2/m_1].$$

In fact, up to the first order in m_2/m_1, we may put

$$m_1=m_1^0+I(m_1)m_1, \tag{4.6a}$$

$$m_2=m_2^0-\tfrac{1}{5}[I(m_1)+2I'(m_1)m_1^2]m_2, \tag{4.6b}$$

where

$$I'(m)=dI(m)/d(m^2)<0.$$

Equation (4.6a) determines m_1 in terms of m_1^0.

The self-consistency condition required, for $m_1^0=0$, is that $I(m)=1$. We may thus expand $I(m)$:

$$I(m_1)=1+I'(m)(m_1^2-m^2),$$

and obtain

$$\Delta m^2=m_1^2-m^2=-m_1^0/[mI'(m)]. \tag{4.7}$$

Since $I'(m)$ is of the order of $-I(m)/m^2$ (see below), this means

$$\Delta m=m_1-m\approx m_1^0. \tag{4.8}$$

From (4.6b), then

$$\begin{aligned}m_2&\approx m_2^0\{1-\tfrac{1}{5}[1+I'(m)m^2]\}^{-1}\\ &\approx m_2^0.\end{aligned} \tag{4.8'}$$

We note that originally there were two solutions $\pm|m|$, which now split into opposite directions according to Eq. (4.7) or (4.8). The meaning of this is as follows. Under the strict γ_5 invariance, there is a complete degeneracy with respect to the transformation (2.1c). The perturbation m_1^0 removes this degeneracy, so that the energy of the vacuum will depend on the orientation of the "γ_5 spin" of the negative energy fermions present in the "vacuum" with respect to this preferred direction. Obviously, the self-consistent procedure, which is similar to the variational method, gives the two extremum configurations corresponding to parallel ($m_0/m>0$) or antiparallel ($m_0/m<0$) γ_5-spin lineup. The parallel case has the larger "gap parameter" $|m|$ than the antiparallel case, so that the former will correspond to the stable ground state. The latter, on the other hand, should correspond to a metastable world.

It is perhaps interesting to see the general behavior of the self-consistency equation for arbitrary magnitude of m_1^0, assuming $m_2^0=0$ for simplicity. The relevant equation,

$$m[1-I(m)]=m^0,$$

is plotted schematically in Fig. 1.

Note that the trivial branch of the solution, which goes through the origin, has $m_0/m<0$. In other words, even in this case the self-consistent solution is qualitatively different from the simple perturbation result. As m^0 increases, it approaches the metastable nontrivial solution, and finally both go into the complex plane.

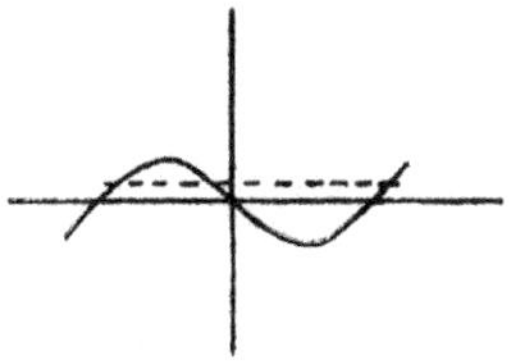

FIG. 1. The three self-consistent mass solutions m (ordinate) as a function of the bare mass m^0 (abscissa).

[9] The most general form of the self-energy Lagrangian (neglecting isospin dependence) is

$$\bar{\psi}[i\gamma\cdot p\,\Sigma_1(p^2)+\Sigma_2(p^2)+i\gamma\cdot p\gamma_5\Sigma_3(p^2)+i\gamma_5\Sigma_4(p^2)]\psi$$
$$+\bar{\psi}^c[i\gamma\cdot p\,\Sigma_5(p^2)+\Sigma_6(p^2)+i\gamma\cdot p\gamma_5\Sigma_7(p^2)+i\gamma_5\Sigma_8(p^2)]\psi$$
$$+\mathrm{H.c.}$$

We do not attempt to study such a problem at this place.

We now come to the meson problem. The pion mass will be determined from

$$\Gamma_j = ig_0'\tau_i \int \mathrm{Tr}[\tau_i S_F(p'+\tfrac{1}{2}q)\Gamma_j S_F(p'-\tfrac{1}{2}q)]d^4p', \quad (4.9)$$

but

$$\underset{\text{isospin}}{\mathrm{Tr}}[\tau_i S_F{}^{(M)}\tau_j S_F{}^{(M)}]$$

$$= \underset{\text{isospin}}{\mathrm{Tr}}\left[\tau_i\left(S_F{}^{(m_p)}\frac{1+\tau_3}{2}+S_F{}^{(m_n)}\frac{1-\tau_3}{2}\right)\right.$$

$$\left.\times\tau_j\left(S_F{}^{(m_p)}\frac{1+\tau_3}{2}+S_F{}^{(m_n)}\frac{1-\tau_3}{2}\right)\right]$$

$$=2\delta_{ij}\frac{S_F{}^{(m_p)}+S_F{}^{(m_n)}}{2}\frac{S_F{}^{(m_p)}+S_F{}^{(m_n)}}{2}$$

$$+2(2\delta_{i3}\delta_{j3}-\delta_{ij})\frac{S_F{}^{(m_p)}-S_F{}^{(m_n)}}{2}\frac{S_F{}^{(m_p)}-S_F{}^{(m_n)}}{2}.$$

The second term yields convergent results, and is $O[(\Delta m/m)^2]$. To the order $\Delta m/m$, therefore, only the first term is important; moreover,

$$(S_F{}^{(m_p)}+S_F{}^{(m_n)})/2\approx S_F{}^{(m_1)}.$$

In other words, there will be no first-order mass splitting of the pion. The mass is then determined from

$$1=J_{p1}(-\mu^2)$$

$$=\frac{g_0'}{4\pi^2}\int_{4m_1^2}^{\Lambda^2}\frac{\kappa^2d\kappa^2}{\kappa^2-\mu^2}\left(1-\frac{4m_1^2}{\kappa^2}\right)^{\frac{1}{2}}. \quad (4.10)$$

For $m_1{}^0=0$, we had originally

$$1=J_p(0)=\frac{g_0'}{4\pi^2}\int_{4m^2}^{\Lambda^2}d\kappa^2\left(1-\frac{4m^2}{\kappa^2}\right)^{\frac{1}{2}},$$

which should now be replaced by

$$1=\frac{m_1{}^0}{m_1}+\frac{g_0'}{4\pi^2}\int_{4m_1^2}^{\Lambda^2}d\kappa^2\left(1-\frac{4m_1^2}{\kappa^2}\right)^{\frac{1}{2}}, \quad (4.11)$$

according to Eq. (4.6a).

From Eqs. (4.10) and (4.11) follows

$$\frac{m_1{}^0}{m_1}=\mu^2\frac{g_0'}{4\pi^2}\int_{4m_1^2}^{\Lambda^2}\frac{d\kappa^2}{\kappa^2-\mu^2}\left(1-\frac{4m_1^2}{\kappa^2}\right)^{\frac{1}{2}}$$

$$\approx\mu^2\frac{g_0'}{4\pi^2}\int_{4m_1^2}^{\Lambda^2}\frac{d\kappa^2}{\kappa^2}\left(1-\frac{4m_1^2}{\kappa^2}\right)^{\frac{1}{2}}$$

$$\lesssim\frac{\mu^2}{4m_1^2}\left(1-\frac{m_1{}^0}{m_1}\right). \quad (4.12)$$

For the observed value of $\mu^2/4m_1{}^2\approx1/200$ we then have, for the stable solution

$$m_1{}^0\lesssim m_1/200\approx5\text{ Mev}. \quad (4.13)$$

The amount of bare mass needed to produce the pion mass is thus surprisingly small.

On the other hand, the metastable solution $(m_1{}^0/m<0)$ produces an imaginary pion mass, indicating the unphysical nature of the solution.

The pion-nucleon coupling constant at the pion pole becomes [see Eq. (2.3)]

$$G_P{}^2/4\pi\approx g_0'[4\pi I_P(-\mu^2)]^{-1},$$

which is changed from the old one only by an order $\mu^2/m_1{}^2\sim m_1{}^0/m_1$.

The other heavy meson states can be treated similarly. We see easily that the changes induced by $m_1{}^0$ are quite small: In general $\Delta\mu^2/m_1{}^2=O(m_1{}^0/m_1)$ and $\Delta G^2/G^2=O(m_1{}^0/m_1)$. Thus the effect of M^0 shows up dramatically only in the pion mass because it was originally zero.

Finally we remark that instead of a bare mass, we could assume slightly different coupling constants g_s and g_p $(<g_s)$ for the scalar and pseudoscalar interaction terms in the Lagrangian (2.3). The nature of the solution is somewhat different from the previous case because the Lagrangian still retains the mass reversal invariance $\psi\rightarrow\gamma_5\psi$, and the solution is twofold degenerate $(\pm m)$. The fractional change of the coupling necessary to produce the pion mass is again small: $|\Delta g/g|\approx\mu^2/4m_1{}^2$.

V. IMPLICATIONS OF THE MODEL

Let us now discuss the relevance of our present model to the physical realities of the nucleons and mesons.

1. We have seen that our Lagrangian (2.3) leads to the nucleon of isospin $\frac{1}{2}$ and the pion of isospin 1. The pion-nucleon coupling constant (pseudoscalar) depends on the cutoff parameter. For the observed large value (≈15) of $G_P{}^2/4\pi$, we see from Eqs. (1.5) and (2.3) that Λ must be of the same order of magnitude as the nucleon mass itself. This is not unreasonable, since the effective nucleon-nucleon interaction in higher approximations would proceed with the exchange of nucleon pairs.

A third parameter, the bare mass, enters our picture in order to make the meson mass finite. It would seem rather unsatisfactory and embarrassing that after all one has to break the postulated symmetry in an *ad hoc* manner. In order to clear up this point, the origin of the effective bare mass then becomes an interesting and important question. Since the required bare mass [Eq. (4.13)] seems to be quite small, a tempting possibility suggests itself that the bare nucleon field is the same as the electron-neutrino field. The electron mass itself could be either intrinsic or of electromagnetic origin.[10] Under this assumption, the bare mass operator would have the form $M^0 = m_e^0(1+\tau_3)/2$, where the word "bare" is used relative to the interaction under consideration. According to the results of the previous section, it is only the isoscalar part of M^0 that produces the large shift of the pion mass, and the amount of violation of the isospin invariance will remain small.

2. Besides the pion, we have also derived vector mesons of both isoscalar ($T=0$) and isovector ($T=1$) types, and a scalar isoscalar meson which is actually unbound. No state corresponding to the isoscalar pion ($\pi_0{}^0$) is found. Of course, these results should depend sensitively on the choice of the interaction in the first place, and to a lesser extent also on the degree of approximation. At any rate, it seems to be a rather interesting and satisfactory feature of the model that these same vector mesons have been anticipated theoretically from various grounds,[11] even though there do not seem to be convincing experimental indications of their existence as yet.[12]

The mass values obtained here are rather high, and these mesons should actually decay into pions very quickly. The coupling constants are generally of the same order as the pion coupling constant, which means a very strong interaction for the vector and scalar mesons. These results, however, may be considerably altered in a better approximation. For one thing, the heavy mesons are coupled strongly to many-pion states which would make the former mere resonances of the latter. Moreover, the nucleon-nucleon and meson-meson interactions can go through long-range forces due to the exchange of these same mesons, which would in turn change the meson states themselves. These processes (the so-called left-hand cuts in the language of the dispersion theory) have not been taken into account in our ladder approximation.

This is a highly cooperative mechanism, and if one wants to handle it in a systematic way, one may be led to the same dispersion theoretical approach that is now widely pursued in pion physics. As a result of such effects, it is then conceivable that the masses of the vector mesons, for example, may come down.[13] Alternatively, it is also conceivable that we have more than one resonance having the same quantum numbers, of which we have obtained the higher ones. These high-energy poles may in turn determine the low-energy resonances.

In addition to the vector mesons, we expect a $T=0$, $J=0^+$ resonance, which has also been postulated by some people.[14] We should try to check these predictions against experimental evidence, such as the characteristic Q-value distributions and angular correlations in meson production processes.

Turning to the nucleon number 2 states, we expect two bound states ($T=0$, $J=1^+$ and $T=1$, $J=0^+$) with comparable masses to those for the vector mesons. This is a qualitatively satisfactory feature in view of the observed deuteron and the singlet virtual states, even though the actual binding is considerably weaker.[15]

3. As was already mentioned in I, our particular model was motivated by the approximate axial vector conservation observed in the nuclear β decay and the role of the pion in it.[5,16] The only difference from I is that (a) we now have the conservation of the isovector axial vector current $i\bar{\psi}\gamma_\mu\gamma_5\tau_i\psi$ instead of the simple axial vector current $i\bar{\psi}\gamma_\mu\gamma_5\psi$, and (b) a small violation of conservation is explicitly introduced. The general treatment of the problem will be completely analogous to the previous case.

Assuming that the β decay occurs through an additional term in the Lagrangian

$$L_\beta = g_\beta\bar{\psi}\gamma_\mu(1+\gamma_5)\tau_+\psi l_{\mu-} + \text{H.c.} \quad [\tau_+ = \tfrac{1}{2}(\tau_1 + i\tau_2)],$$

where l_μ refers to the lepton current, the nuclear β-

[10] The electromagnetic interaction is invariant under the simple γ_5 transformation, but not under the Gürsey transformation since it fundamentally distinguishes between the charged and neutral components. Thus there is a built-in violation which can eventually produce the pion mass.

[11] W. R. Frazer and J. R. Fulco, Phys. Rev. **117**, 1609 (1960); Y. Nambu, *ibid.* **106**, 1366 (1957); G. Chew, Phys. Rev. Letters 4, 142 (1960); J. J. Sakurai, Ann. Phys. **11**, 1 (1960).

[12] J. A. Anderson, Vo X. Bang, P. G. Burke, D. D. Carmony, and N. Schmitz, Phys. Rev. Letters **6**, 365 (1961); A. Abashian, N. Booth, and K. M. Crowe, *ibid.* **5**, 258 (1960).

[13] A crude way to see the general tendency will be to argue as follows: The $T=1$ vector meson is coupled to the nucleon mainly through tensor coupling, so that it will cause a nucleon-antinucleon interaction of the type $-g^2(\boldsymbol{\sigma}_1\cdot\boldsymbol{\sigma}_2\boldsymbol{\tau}_1\cdot\boldsymbol{\tau}_2)e^{-\mu r}/r$. This tends to raise $T=1$, $J=0^+$ and $T=0$, $J=1^-$ meson states, and lower $T=0$, $J=0^+$ and $T=1$, $J=1^-$ states. Any change in the binding force, however, will be offset by the corresponding change in the nucleon mass, which automatically adjusts the pion mass to lie where it should be. The exchange of the $T=0$, $J=1$, and $J=0$ mesons, therefore, would not be so important in determining the relative shift of the meson levels.

[14] J. Schwinger, Ann. Phys. **2**, 407 (1957); M. Gell-Mann and M. Lévy, Nuovo cimento **16**, 705 (1960); S. Gupta, Phys. Rev. **111**, 1436 (1958), Phys. Rev. Letters **2**, 124 (1959); M. H. Johnson and E. Teller, Phys. Rev. **98**, 783 (1955); H. P. Duerr and E. Teller, *ibid.* **103**, 469 (1956). The σ meson mass obtained here is independent of the cutoff Λ, so that there may be some point in arguing that it is more reliable than for the vector mesons. If so, we may expect a nucleon-antinucleon resonance near zero kinetic energy (taking account of the mass shift due to M^0). The width may be quite broad.

[15] In fact, both $T=0$ and $T=1$ vector meson exchanges work in the direction to reduce the binding relative to the nucleon-antinucleon case.

[16] S. Bludman, Nuovo cimento **9**, 433 (1958); F. Gürsey, Ann. Phys. **12**, 91 (1961); Y. Nambu, reference 1; M. Gell-Mann and M. Levy, reference 14; J. Bernstein, N. Gell-Mann, and L. Michel, Nuovo cimento **16**, 560 (1960); J. Bernstein, S. Fubini, M. Gell-Mann, and W. Thirring, *ibid.* **17**, 757 (1960); Chou Kuang-Chao, J. Exptl. Theoret. Phys. (U.S.S.R.) **39**, 703 (1960) [Soviet Phys.—JETP **12**, 492 (1961)].

decay vertex becomes

$$\Gamma_\mu = g_\beta[i\gamma_\mu\tau_+ F_{V1}(q^2) - i\sigma_{\mu\nu}q_\nu\tau_+ F_{V2}(q^2) + \{i\gamma_\mu\gamma_5\tau_+ + [2m_1\gamma_5 q_\mu\tau_+/(q^2+\mu_\pi^2)]f(q^2)\}F_A(q^2)],$$

where q is the momentum change. In the ladder approximation, $F_{V1}(q^2)$ arises from the vector-type nucleon pairs, and $F_{V1}(0)=1$ (in accordance with the Ward identity, applicable to the isospin current, which shows that $F_{V1}(0)=1$ in general.[17])

In the axial vector part, $F_A(q^2)=1$ in our approximation. $f(q^2)$ arises because of the violation of the γ_5 invariance, but it deviates from 1 only to the order $m_1^0/m_1 \sim \mu^2/m_1^2$, as was already seen in the previous section. For practical purposes, therefore, the axial vector current has the desired form which would lead to the Goldberger-Treiman relation[18]

$$2m_1 g_A \approx \sqrt{2}G_\pi g_\pi,$$

where $g_A = g_\beta F_A(0)$ and G_π, g_π are, respectively, pion-nucleon and pion-lepton couplings.

In higher orders, however, $F_A(q^2)$ will be present, and in general $F_A(0)\neq 1$ even under the strict γ_5 invariance. People have conjectured in the past that $F_A(0)=g_A/g_V=1$ as $\mu_\pi \to 0$, but this does not seem to be easily guaranteed. The generalized Ward identity for the axial vector current[19] suffices to prove the Goldberger-Treiman relation, but is not enough to make $F_A(0)=1$. In order that the latter should come out rigorously, we would need a more subtle mechanism. Nevertheless, we can try a working hypothesis that $g_A/g_V=1$ under the strict invariance, and then estimate the deviation due to the violation. This scheme is carried out in the Appendix.

VI. FURTHER PROBLEM

We will consider here some of the general problems which have not been explored, but which seem to be important in a more comprehensive understanding of the elementary particles.

1. The hyperons. In order to incorporate the strange particles into our picture we would have to increase the dimensions of the fundamental field unless we do further unconventional things (see below). The simplest possibility from the point of view of quantum numbers would be to add a bare Λ-particle field as was originally proposed by Sakata.[20] We would then postulate, in addition, the generalized γ_5 symmetry, which would mean the invariance of the left-handed and right-handed components separately under the unitary transformation among the three fields or some subgroups of it. The mass splitting of the three baryons will be obtained from bare masses of similar magnitude, which destroys the otherwise rigorous symmetry.

This approach will produce easily the pions and K mesons and probably more, and their masses can again be related to the baryon bare masses. But we do not yet have a comparable dynamical method to predict Σ and Ξ particles. Consequently, we shall not be able to say whether or not the present model is dynamically satisfactory in this respect.

2. The leptons. In connection with the above model we are naturally led to the lepton problem. Gamba, Marshak, and Okubo[21] have pointed out an interesting parallelism between the $pn\Lambda$ and $\nu e\mu$ triplets. As was remarked in the beginning, our theory gives a special incentive for speculation about this relation because we have obtained two solutions: one ordinary and one extraordinary, differing in masses. Could they both be realized in nature simultaneously? According to our results in I, the answer is no because they belong to different Hilbert spaces. Moreover, the trivial solution gives rise to unphysical mesons at least under the assumption of fixed cutoff, with a large mass $(-\mu^2 \gtrsim \Lambda^2)$ but not necessarily a weak coupling $(G^2 \lesssim \Lambda^4/\mu^4)$. Nevertheless, it would seem too bad if Nature did not take advantage of the two solutions. A straightforward way to make the two solutions co-exist in the same world is obviously to postulate that the world is represented by the direct product of two Hilbert spaces[22]:

$$\mathcal{H} = \mathcal{H}^{(0)} \otimes \mathcal{H}^{(m)}, \tag{6.1}$$

built upon the vacuum state

$$\Omega = \Omega^{(0)} \otimes \Omega^{(m)}. \tag{6.1'}$$

It is true that this is effectively the same as doubling the fields, but here the choice of the two solutions (particles) is dictated by the dynamics of the original nonlinear theory. In order to describe this situation, we may adopt an *effective* Lagrangian

$$L = L^{(1)} + L^{(2)}, \tag{6.2}$$

where each of the $L^{(i)}$ has the same form, only differing in the charge assignments of the respective triplet fields. The Lagrangian obviously yields four subspaces

$$\mathcal{H}_1^{(0)} \otimes \mathcal{H}_2^{(m)}, \quad \mathcal{H}_1^{(m)} \otimes \mathcal{H}_2^{(0)}, \quad \mathcal{H}_1^{(0)} \otimes \mathcal{H}_2^{(0)}$$

and

$$\mathcal{H}_1^{(m)} \otimes \mathcal{H}_2^{(m)}.$$

According to our plan, we must say that we happen to live in the first subspace. [In the second space, the masses of $\nu e\mu$ and $pn\Lambda$ are interchanged, whereas in the third (fourth) case we have two kinds of leptons (baryons).]

[17] R. P. Feynman and M. Gell-Mann, Phys. Rev. **109**, 193 (1958).

[18] M. L. Goldberger and S. B. Treiman, Phys. Rev. **111**, 356 (1958).

[19] J. Bernstein *et al.*, reference 16.

[20] S. Sakata, Progr. Theoret. Phys. (Kyoto) **16**, 686 (1956).

[21] A. Gamba, R. E. Marshak, and S. Okubo, Proc. Natl. Acad. Sci. U. S. **45**, 881 (1959); Z. Maki, M. Nakagawa, Y. Ohnuki, and S. Sakata, Progr. Theoret. Phys. **23**, 1174 (1960).

[22] S. Okubo and R. E. Marshak, Nuovo cimento **19**, 1226 (1961), have independently proposed a similar idea. We thank the authors for valuable communications.

So far there is no interaction between leptons and baryons (except the electromagnetic, which is trivial). To introduce the weak interactions, we may, for example, add to Eq. (5.2) a third nonlinear term involving all the (left-handed) fields. This would complete our program of dealing with the strong and weak interactions.

But, of course, it is not yet a truly unified theory; the weak interaction is introduced only as an *ad hoc* additional process. Moreover, we do not know the mathematical consistency of such a procedure, because the additional interaction, if taken seriously, may qualitatively affect the baryon and lepton solutions we already have.

There is an alternative, but less drastic scheme; namely, to assume six different fields from the beginning, of which three (becoming eventually the baryon fields) have additional strong interactions in the Lagrangian. This may not be devoid of elegance if the interaction is mediated by a vector Bose field coupled to the baryon charge. The intermediate bosons, including the photons and possibly also the weak bosons, could then be interpreted as the agents that distinguish between different components of the bare fermions, which otherwise would enjoy a high degree of symmetry.

We would like to throw in another remark here that there may be also a possibility of utilizing the ordinary and extraordinary solutions in distinguishing between electron and muon, or baryons of different strangenesses.

3. The γ_5 invariance for general systems. In our theory the γ_5 invariance is a very essential element. It is a particular symmetry which exists in the Lagrangian, but is masked in reality because of the (approximate) degeneracy of the vacuum with respect to that symmetry. We have used the pion and the β decay in support of the assumption. In order to firmly establish its validity, however, we must try to find more evidences. For one thing, the induced pseudoscalar terms in nucleon β decay and μ capture should be examined more closely.

Furthermore, if such a symmetry is to have a general meaning, we must be able to consider partially conserved currents for processes such as

$$\begin{aligned} &H^3 \to He^3,\\ &He^6 \to Li^6,\\ &C^{14} \to N^{14},\\ &\Sigma^- \to \Sigma^0,\\ &\Sigma^- \to \Lambda. \end{aligned} \tag{6.3}$$

An elementary definition of the γ_5 transformation for the general system is obvious: When the wave function of a system is expressed in terms of the fundamental (bare) spinors obeying the rules Eq. (2.1), the transformation is unambiguously defined for each component, and thereby the total axial vector current is determined.

For superallowed transitions with spin $\frac{1}{2}$, the problem is particularly simple, since it is the same as for the neutron case. Thus for $H^3 \to He^3$ we have the same Goldberger-Treiman relation

$$(M_H+M_{He})g_A(\text{H,He})/\sqrt{2}G_\pi(\text{H,He}) \approx g_\pi, \tag{6.4}$$

where g_A, G_π now characterize the β decay and the (unknown) pion coupling for the transition under consideration.

Similar relations hold for the Σ decays.[23] For the Σ-Σ case, we have

$$2m_\Sigma g_A(\Sigma\Sigma)/G_\pi(\Sigma\Sigma) \approx g_\pi. \tag{6.5}$$

For the Σ-Λ case, the axial vector vertex becomes[24]

$$\begin{aligned} \Gamma_A \approx [i\gamma_\mu\gamma_5 + (m_\Sigma+m_\Lambda)\gamma_5 q_\mu/(q^2+\mu_\pi^2)]F_{1A}(q^2)\\ +i\gamma_5\sigma_{\mu\nu}q_\nu F_{2A}(q^2)\\ (m_\Sigma+m_\Lambda)F_{1A}(0)/G_\pi(\Sigma\Lambda) \approx g_\pi, \end{aligned} \tag{6.6}$$

if the relative Σ-Λ parity is even. The vector current conservation is also violated because of the Σ-Λ mass difference, and it looks as though this would predict a corresponding scalar meson term. However, the analogy is rather superficial. Firstly, the violation disappears if $m_\Sigma = m_\Lambda$, in which case there would be no need for a scalar meson. The Σ-Λ mass difference itself might be due to the breakdown of the γ_5 symmetry. Secondly, it is an "unfavored" transition ($\Delta T=1$), so that the vector part, corresponding to the off-diagonal element of the isospin current, should vanish in the ideal limit of strict isospin invariance and $q \to 0$. In other words, we expect

$$\Gamma_V \approx [q^2 i\gamma_\mu - (m_\Sigma - m_\Lambda)q_\mu]F_{1V}(q^2) + \sigma_{\mu\nu}q_\nu F_{2V}(q^2). \tag{6.7}$$

In case the Σ-Λ parity is odd,[25] the vector and axial vector parts will interchange their roles. The vector part, which now looks like the axial vector current, would have the form

$$\begin{aligned} \Gamma_V \approx [q^2 i\gamma_\mu\gamma_5 + (m_\Sigma+m_\Lambda)\gamma_5 q_\mu]F_{1V}(q^2)\\ +i\gamma_5\sigma_{\mu\nu}q_\nu F_{2V}(q^2). \end{aligned} \tag{6.8}$$

The axial vector part can similarly be put in the form

$$\begin{aligned} [i\gamma_\mu - q_\mu(m_\Sigma - m_\Lambda)/(q^2+\mu_\pi^2)f(q^2)]F_{1A}(q^2)\\ +\sigma_{\mu\nu}q_\nu F_{2A}(q^2). \end{aligned} \tag{6.9}$$

But $f(q^2)$ need not be ≈ 1 if the Σ-Λ mass difference is also due to the violation of the γ_5 symmetry.

There are other processes for which the chirality conservation can be tested in a direct way. Although

---

[23] L. B. Okun', *Ann. Rev. Nuclear Sci.* **9**, 61 (1959); M. Gell-Mann, *Proceedings of the 1960 Annual International Conference on High-Energy Physics at Rochester* (Interscience Publishers, Inc., New York, 1960), p. 522.

[24] We have in this case three independent terms.

[25] See S. Barshay, Phys. Rev. Letters **1**, 97 (1958); Y. Nambu and J. J. Sakurai, *ibid.* **6**, 377 (1961).

extraordinary solutions are in general not eigenstates of chirality (even under strict γ_5 invariance), the conservation law should still apply to the expectation values of chirality. In fact, we can express the chirality conservation law $\langle X_i \rangle = \langle X_f \rangle$ for any reaction $i \to f$; for example

$$p+\pi \to p+\pi, \quad p+\pi+\pi', \text{ etc.,}$$
$$p+p \to p+p, \quad p+p+\pi, \text{ etc.,}$$

as a relation between the change of nucleon chirality and the magnitude of the pion production amplitude.

The ideas outlined in this section will be taken up in more detail elsewhere.

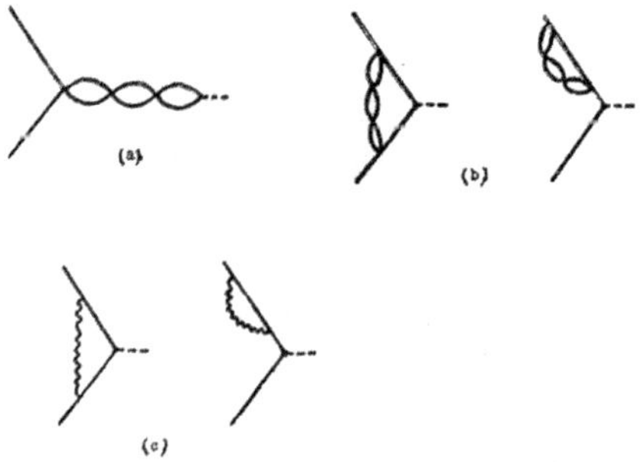

FIG. 2. Typical graphs considered in the evaluation of the axial vector vertex.

APPENDIX

We calculate here the renormalization of the axial vector (Gamow-Teller) coupling constant g_A for nuclear β decay under the following assumptions:

(1) Under strict γ_5 invariance (Gürsey type), there is no renormalization, namely $g_A = g_{A0}$ $(=g_{V0}=g_V)$, where g_{A0} is the bare coupling constant.

(2) The violation of the invariance gives rise to the finite pion mass as well as the deviation of the ratio $R = g_A/g_{A0} = g_A/g_V$ from unity, so that there is a functional relation between the two quantities.

Let us first consider the isovector axial vector vertex Γ_A in the usual perturbation theory. In our model, it consists of various graphs, some of which are shown in Fig. 2(a) and (b). The "ladder" graphs 2(a) have been considered in I as well as in the present paper, since they are intimately related to the γ_5 gauge transformation. In I (Appendix) we found that $R>1$ when both pseudoscalar and pseudovector type interactions are present.[26] The graphs 2(b) have not been considered yet. These will come into our consideration as soon as we take corresponding higher-order approximations for the self-energy, which was briefly discussed in I. The chain of bubbles in these graphs will act like a meson when there is such a dynamical pole [Fig. 2(c)].

The (divergent) renormalization effect due to intermediate mesons is always negative,[27] irrespective of the type of the meson, so that the effect of these meson-like bubble graphs is also expected to be similar. When the chain does not produce a pole, however, the effect can be opposite.

Combining all these effects, we have no way to predict the resultant magnitude and sign of the renormalization correction. So we simply assume these contributions to cancel out under strict γ_5 invariance.

Next let us suppose that the invariance is slightly violated. This will cause changes in the propagators in all these graphs. Most of these changes are, however, quite small, being of the order of $m^0/m \approx \mu^2/4m^2$, as will be clear from the results of Sec. IV. The largest effect is naturally expected to come from the "pion" contribution in Fig. 2(b), as this is a change from zero mass (infinite range) to a finite one.

Let us accordingly take the effective pion graph from Fig. 2(c) with an arbitrary pion mass μ. Call its contribution to the vertex renormalization (for zero momentum transfer) $\Lambda(\mu)$. Then according to the above assumption

$$R = \Gamma_A(\mu_\pi)/g_A = \Gamma_A(\mu_\pi)/\Gamma_A(0) \approx 1 + \Lambda(\mu_\pi) - \Lambda(0). \tag{A1}$$

The difference $\Lambda(\mu) - \Lambda(0)$ is convergent, which turns out to be

$$\Lambda(\mu)-\Lambda(0) = \frac{G^2}{16\pi^2}\frac{\mu^2}{m^2}\left\{\left(3-\frac{5\mu^2}{2m^2}\right)\ln\frac{m^2}{\mu^2}-5 + \frac{16\mu}{\sqrt{3}m}\left[\tan^{-1}\left(\frac{4m^2}{3\mu^2}-\frac{2}{3}\right)+\tan^{-1}\left(\frac{2}{3}\right)\right]\right\}, \tag{A2}$$

where G is the phenomenological pion coupling constant.

As was expected, this goes like $(\mu^2/m^2)\ln(m^2/\mu^2)$ for small μ, which is more important than the contributions from the neglected processes behaving like μ^2/m^2.

With $(G^2/4\pi)(\mu^2/4m^2) = f^2/4\pi = 0.08$, Eq. (A2) gives

$$R-1 = \begin{cases} 0.18 \\ 0.24. \end{cases} \tag{A3}$$

The first figure is the entire contribution from Eq. (A2), while the second is the contribution from the leading logarithmic term alone. Experimentally, R is estimated to be ≈ 1.25.[28]

[26] See also Z. Maki, Progr. Theoret. Phys. (Kyoto) 22, 62 (1959).

[27] We owe Dr. J. de Swart the mathematical check on this point.

[28] M. T. Burgy, V. E. Krohn, T. B. Novey, G. R. Ringo, and V. L. Telegdi, Phys. Rev. 120, 1829 (1961).

Chapter 6: Nambu and the Prelude to QCD

In 1965, Nambu wrote three papers which set the stage for Quantum Chromodynamics, the theory of strong interactions:

January 1965: "Dynamical Symmetries and Fundamental Fields"[76]

April 1965, with Moo-Young Han: "Three-Triplet Model with Double $SU(3)$ Symmetry"[77]

May 1965: "A Systematics of Hadrons in Subnuclear Physics"[78]

Background

1964 was a harvest year for fundamental physics, the discovery of the microwave background radiation, as well as Murray Gell-Mann (1929–2019) and George Zweig (1937–) hypothesis[79] of fractionally-charged constituents (quarks/aces) of mesons and baryons.

In July, Feza Gürsey (1921–1992) and Luigi Radicati Di Brozolo[80] (1919–2019) generalized Gell-Mann's and Ne'eman's eightfold way by adding spin,[81] $SU_3 \rightarrow SU_6 = SU_3 \times SU_2$. Although it was a non-relativistic union, pseudoscalars and vector mesons were fitted together in one SU_6 multiplet: all spin 3/2 and spin 1/2 hadrons in the symmetric 56-dimensional SU_6 multiplet. This contradicted the Gell-Mann–Zweig picture of the lowest lying hadrons as s-wave bound states of three quarks, which according to Pauli should be antisymmetric.

By Fall 1964, three concerns faced theoretical physicists — one experimental, fractionally-charged particles had never been observed, the other two theoretical:

[76] *Proceedings of the Second Coral Gables Conference on Symmetry Principles at High Energy*, (W. H. Freeman and Company, San Francisco 1965).

[77] M.-Y. Han and Y. Nambu, *Phys. Rev.* **139**, 1006–1010 (1965).

[78] A. De-Shalit, H. Feshback, and L. Van Hove (eds.), *Preludes in Theoretical Physics*, (North Holland 1966) pp. 133–142.

[79] M. Gell-Mann, *Phys. Lett.* **8**, 214 (1964); G. Zweig, *CERN Preprint*, Feb 24 1964.

[80] F. Gürsey and L. A. Radicati, *Phys. Rev. Lett.* **13**, 173 (1964); A. Pais, *Phys. Rev. Lett.* **13**, 175 (1964); B. Sakita, *Phys. Rev.* **136**, B1765 (1964).

[81] In the spirit of E. Wigner and E. Stückelberg supermutiplets.

the relativistic unification of space-time with flavor symmetries, and the violation of the Pauli exclusion principle.

The first had motivated integrally-charged variants of Gell-Mann–Zweig quarks; three days after Zweig's CERN preprint, Henri Bacry (1928–2010), Jean Nuyts (1936–2017) and Leon Van Hove (1924–1990),[82] introduced two flavor $SU(3)$-triplets with integral charges, T and Θ that they called "trions".

The strange Gürsey–Radicati baryon assignment was further validated by the startling agreement with experiment of the calculation of the ratio of neutron to proton magnetic moments.[83]

The first solution to the Gürsey–Radicati statistics puzzle was proposed in the Fall by O. W. Greenberg (1932–) *"Spin and Unitary-Spin Independence in a Paraquark Model of Baryons and Mesons"*.[84]

For each $SU(6)$ label λ, Greenberg introduced three independent sets of fermion ladder operators, $a_\lambda^{(\alpha)\dagger}$, $a_\lambda^{(\alpha)}$ $\alpha = 1, 2, 3$ with the usual anticommutation relations,

$$\{a_\lambda^{(\alpha)\dagger},\, a_\mu^{(\alpha)}\} = \delta_{\lambda,\mu},\ \{a_\lambda^{(\alpha)\dagger},\, a_\mu^{(\alpha)\dagger}\} = 0,$$

and commuted for different values of α. He used H. S. Green's (1920–1999) Ansatz[85] of order three, to introduce a symmetric linear combination,

$$a_\lambda^\dagger = \sum_{\alpha=1}^{3} a_\lambda^{(\alpha)\dagger},$$

with only $SU(6)$ labels. The symmetric cubic combination,

$$f_{\lambda\mu\nu}^\dagger = \{a_\lambda^\dagger\,,\, a_\mu^\dagger\,\}\,,\, a_\nu^\dagger\,\} = \epsilon^{\alpha\beta\gamma} a_\lambda^{(\alpha)\dagger}\, a_\mu^{(\beta)\dagger}\, a_\nu^{(\gamma)\dagger},$$

yielded 56. It is also a singlet under a new $SU(3)$, implied by the Levi-Civita tensor.

Greenberg's algebraic construction restored the Pauli principle with three quark varieties, later known as color, but he did not offer any dynamical scheme.

In January 20–22 1965, at the "Second Coral Gables Conference on Symmetry Principles at High Energy", Nambu presented a model,[86] which answered both conundrums by considering two Gell-Mann–Zweig-like triplets, and suggesting a scalar or Abelian interaction to bind them. The quarks had integral charges and the $SU(6)$ statistics impasse was resolved, an answer to two of the concerns are mentioned above.

There was scarce mention of parastatistics from the invitation-only participants: only one word from J. Robert Oppenheimer (1904–1967) in the discussion sessions, and a detailed description by Yuval Ne'eman's (1925–2006) presentation.

[82] H. Bacry, J. Nuyts and L. Van Hove, *Phys. Lett.* **9**, 279 (1964).
[83] M. A. B. Bég, B. W. Lee and A. Pais, *Phys. Rev. Lett.* **13**, 514 (1964).
[84] O. W. Greenberg, *Phys. Rev. Lett.* **13**, 598 (1964).
[85] H. S. Green, *Phys. Rev.* **90**, 270 (1953).
[86] *Proceedings of the Second Coral Gables Conference on Symmetry Principles at High Energy*", (W. H. Freeman and Company, San Francisco 1965).

The next paper of Nambu that we discuss is his collaboration with Moo-Young Han (1934–2016). Born in South Korea, Han traveled to the United States at the end of the Korean war. He attended Carroll College in Waukesha, Wisconsin, where he started a successful academic career: graduated in 1957 with a B.S. in electrical engineering, and was accepted into the graduate program in Physics at the University of Rochester. Under the tutelage of E. C. G. Sudarshan (1931–2018) he obtained his Ph.D. in 1963. It was followed by a postdoctoral position at Syracuse University from 1964 to 1966.

In March 1965, Han listed at the SLAC Preprint Server a paper entitled *"On the Introduction of Nambu Triplets with Integral Charges."* Unfortunately we could not find a copy, but the title suggests it was inspired by Nambu's paper at Coral Gables. While Han did not attend, we believe that his adviser Sudarshan, who presented a talk at that conference, informed him of Nambu's two-triplet paper.

A month later, Han submitted in collaboration with Nambu the seminal paper "Three-Triplet Model with Double $SU(3)$ Symmetry." We could not find how their collaboration began, but we believe it was initiated by the more senior author.

In 1967, M. Y. Han joined the faculty at Duke University in North Carolina. An esteemed teacher, he was awarded in 1972 Duke University's "Alumni Distinguished Undergraduate Teaching Award." An author of many books, he created an electronic newsletter, the "Information Exchange for Korea-American Scholars" (IEKAS) and served as its editor for many years. He retired as full professor in 2011, and passed away five years later.

A month later, Nambu sent his contributions to "Preludes in Theoretical Physics," the Festschrift volume[87] in honor of Viki Weiskopf (1908–2002) who was retiring as "Director Général du CERN." It mostly repeats the two and three-triplet models.

The Papers

Coral Gables

In January 1965 at Coral Gables, Nambu spoke on *"Dynamical Symmetries and Fundamental Fields."*

The first section begins with the Gell-Mann–Zweig picture of fundamental triplets underlying known particles and resonances, even though these triplets had not been observed. Further into the paper, Nambu speculated that fractionally-charged quarks were unlikely to exist — at least one of these quarks must be stable and would have been *"observed quite easily among cosmic rays, for example."*

He surveyed pre-$SU(6)$ models based on Gürsey *et al.*,[88] and A. Salam's (1926–1996) summary at the 1964 Dubna conference:

[87]op. cit.

[88]T. D. Lee, F. Gürsey, and N. Nauenberg, *Phys. Rev.* **135**, B467 (1964).

1. A single fundamental triplet (quark-ace model)
2. A fundamental quartet ($SU(4)$ model)
3. Two fundamental triplets ($U_3 \times U_3$ or W_3 models)

He remarked that in July 1964, F. Gürsey and L. Radicati had combined particles of different spin to enlarge the symmetry from $SU(3)$ to $SU(6)$, with *"striking consequences which are in agreement with known facts"* for baryons as three-body systems of fundamental triplets.

After these preambles, Nambu presented a remarkable extension of the Gell-Mann–Zweig quark model to three fundamental triplets,[89] t_1, t_1' and t_2, with integral electric charges,

$$Q_{t_1} = (z_1 + 1,\, z_1,\, z_1), \quad Q_{t_1'} = (z_1' + 1,\, z_1',\, z_1'), \quad Q_{t_2} = (z_2 + 1,\, z_2,\, z_2),$$

and required that the sum of all the charges vanish, $z_1 + z_1' + z_2 + 1 = 0$.

He assigned each triplet a *"charm number"*,[90] C with $C_1 = 3z_1+1$, $C_1' = 3z_1'+1$, $C_2 = 3z_2 + 1$ for the triplets. This way, the physical three-triplet combinations automatically have zero "Nambu Charm", $C_1 + C_1' + C_2 = 0$, the same as the sum of all nine charges. $SU(6)$ symmetry treats all three triplets on the same footing, preserving the nucleons magnetic moment ratio. This simple model of integrally-charged quarks restored the Pauli principle by inventing a new quantum number.

Motivated perhaps by the trion model of Bacry *et al.*, Nambu presented a simpler example with two triplets,

$$t_1 = t_1' \neq t_2, \quad z_1 = z_1' = 0,\, z_2 = -1, \quad C_1 = C_1' = 1,\, C_2 = -2$$

with integral charges $Q_{t_1} = (1, 0, 0)$, $Q_{t_2} = (0, -1, -1)$. The three-quark bound states with $C = 0$ were of the form $t_1 t_1 t_2$, which correctly described the 56 baryons.

He mentioned a number of possible new symmetries: *"Since we have two triplets, the symmetry may be extended to $U(3)_1 \times U(3)_2$ and then $U(6)_1 \times U(6)_2$,"* with three ways to choose baryon number which ensures $B = 1$ for the three-triplet combinations:

(1) $B = 1/3$ for all triplets,
(2) $B = 1$ for t_1, -1 for t_2,
(3) $B = 0$ for t_1, 1 for t_2,
and B and C can be identified with charges of $U(1)_1 \times U(1)_2$.

He next turned to dynamics: *"Just as an electrically neutral system tends to be the most stable, a neutral charm* (Nambu's) *system will have a tendency to have the lowest potential energy in the first approximation (no terms quadratic in the number of particles)."* Nambu reasoned that $C = 0$ suggested a strong field coupled to Nambu's Charm, although he did not mention the type (scalar or vector).

[89] Nambu did not seem to be aware of O. W. Greenberg's Parastatistics.

[90] Not to be confused with the charmed quark of B. J. Bjorken and S. L. Glashow, *Phys. Lett.* **11**, 255 (1964).

It is implied but left unsaid until the last section how charm resolves the statistics problem by restricting the three quark composites to $t_1t_1t_2$. "*The baryons belong to* $(\mathbf{21}, \mathbf{6})$ *of* $SU(6)_1 \times SU(6)_2$*, which reduces to* $\mathbf{56}+\mathbf{70}$ *of SU(6),*" $\subset SU(6)_1 \times SU(6)_2$. This section concludes with a semi-empirical mass formula for the bound states in terms of baryon number and Nambu's charm.

This paper contains one of Nambu's seminal contributions. Although derived from integrally charged quarks, he extended the number of flavor triplets, solved the statistics conundrum, and introduced a new quantum number to explain the dynamics of the bound states. As we will see later, Nambu's charm is one the color charges, soon to be introduced in the paper with Han.

The next section *"SU(6) and its Interpretation"* discusses how $SU(6)$ could be realized into a relativistic theory. While a detailed presentation is not needed for our purposes, we pick out several Nambu-like remarks:

"... in order to obtain a chiral $SU(6)_L \times SU(6)_R$*, the effective interaction Lagrangian must contain energy momentum vectors so that the* $SU(6)$ *symmetry can be discussed only in the rest frame".*

"If $SU(6)$ *is good, then either (1) somehow the kinetic energy of the triplets is suppressed compared with the potential energy, or (2) there is no spin-orbit coupling."* Among possible resolutions,

"$\hbar \sim 0$ *the kinetic part of the Lagrangian is much smaller than the potential energy, and the triplets are classical particles. We may also say that it is strongly coupled to some field* $(g^2 \to \infty)$*".*

Judging from the discussions after Nambu's presentation, the focus of the Coral Gables participants were on the dynamical details, which he presented mostly in the second and third sections. Perhaps confused by Nambu's reasoning by analogies, no attendee directly addressed his two-triplet model.

Han–Nambu's Paper

The introduction of the *"Three-Triplet Model with Double SU(3) Symmetry"*[91] highlighted three perceived challenges of the quark picture: fractionally-charged quarks, the failure of the Pauli exclusion principle and an unknown binding mechanism for the meson and baryon constituents.

Oddly, it contains no reference to Gürsey and Radicati nor to Sakita although SU_6 is specifically mentioned in the text, and nor to Greenberg's parastatistics.

Han and Nambu summarized the Coral Gables paper with two integrally-charged quark flavor triplets that resolved the statistics problem, and identified the low-lying baryons and mesons as "zero-charm" states.

[91] M.-Y. Han and Y. Nambu, *Phys. Rev.* **139**, 1006–1010 (1965).

Their initial focus is on models that respect the Nishijima (1926–2009) Gell-Mann relation with integrally-charged quarks.

Three Triplets

In the original Gell-Mann–Zweig quark model, the electric charge is written in terms of the hypercharge Y and the third component of isospin I_3

$$Q = I_3 + \frac{Y}{2},$$

is expressed in terms of three diagonal traceless flavor $SU(3)$ generators,

$$Q = -B_1^1, \quad Y = -B_3^3 = -B_1^1 - B_2^2, \quad I_3 = \frac{1}{2}(B_2^2 - B_1^1).$$

Acting on the (u, d, s) quark flavor triplet,

$$B_1^1 = \frac{1}{3}\mathrm{Diag}(-2, 1, 1), \quad B_2^2 = \frac{1}{3}\mathrm{Diag}(1, -2, 1), \quad B_3^3 = \frac{1}{3}\mathrm{Diag}(1, 1, -2),$$

with one constraint $B_1^1 + B_2^2 + B_3^3 = 0$. These three quark flavors are fractionally charged.

In the $Y - I_3$ plane, the three quarks lie on the vertices of an inverted equilateral triangle centered at the origin. The same electric charges lie along the side at $-45°$.

S. Okubo (1930–2015) and collaborators[92] built models with integrally-charged quarks which satisfy the Nishijima–Gell-Mann relation with a minimal extension of the symmetry, $U(3) = SU(3) \times U(1)$. They simply replaced the traceless B_j^i generators with,

$$A_j^i = B_j^i + \frac{\tau}{3}\delta_j^i, \quad \tau = A_1^1 + A_2^2 + A_3^3,$$

where τ is the $U(1)$ operator. Setting $\tau = -1$ yields,

$$A_1^1 = \frac{1}{3}\mathrm{Diag}(1, 0, 0), \quad A_2^2 = \frac{1}{3}\mathrm{Diag}(0, 1, 0), \quad A_3^3 = \frac{1}{3}\mathrm{Diag}(0, 0, 1).$$

The new electric charges were obtained by substitution,

$$Q \to \tilde{Q} = -A_1^1 = Q, \quad Y \to \tilde{Y} = -A_1^1 - A_2^2,$$

leaving I_3 unchanged. The new quark charges satisfy a new Nishijima–Gell-Mann formula,

$$\tilde{Q} = I_3 + \frac{\tilde{Y}}{2},$$

but they are now integrally charged, two neutral quarks and one with positive charge.

Graphically, this quark triplet is also represented by an inverted equilateral triangle, but its center is raised to 2/3, the average hypercharge. This center can be thought of as the new quark flavor triplet.

[92]S. Okubo, C. Ryan and R. E. Marshak, *Il Nuovo Cimento* **XXXIV**, 759 (1964).

Nambu's two-triplet model appeared in the same plane as two inverted equilateral triangles, the first with center vertically displaced from the origin by $+2/3$, the second with center displaced by $-4/3$.

Han and Nambu proceeded to generalize the Okubo *et al.* construction. They found two more quark triplets with integral charges that satisfy a Nishijima–Gell-Mann relation, with the same $U(3)$ symmetry.

Their key observation was the three inequivalent ways to embed $SU(2)$ into $SU(3)$: the first is the weak isospin, the other two are called U-spin and V-spin,[93] each with its different Nishijima–Gell-Mann relation,

$$Q_I = I_3 + \frac{Y_I}{2}, \quad Q_U = A_2^2 + A_3^3 = U_3 + \frac{Y_U}{2}, \quad Q_V = -A_1^1 = V_3 + \frac{Y_V}{2},$$

with

$$I_3 = \frac{1}{2}(A_1^1 - A_2^2), \quad U_3 = \frac{1}{2}(A_3^3 - A_2^2), \quad V_3 = \frac{1}{2}(A_1^1 - A_3^3).$$

These three quark triplets, t_I, t_U, t_V, have respective electric charges Q_I, Q_U, Q_V.

They represent the three triplets on the same $\tilde{Y} - \tilde{I}_3$ plane. To each corresponds an equilateral triangle, anchored with one vertex at the origin symmetrically distributed at 120° from one another. Their centers are located at their respective values of $(\tilde{I}_3, \tilde{Y})$,

$$t_I: \quad (0, 2/3); \quad t_U: \quad (-1/2, -1/3); \quad t_V: \quad (+1/2, -1/3).$$

They also noticed something interesting: the three centers form an upright equilateral triangle — it looks as if each flavor triplet t_I, t_U, t_V is one state of an antitriplet of a new $SU(3)$!

Then Han and Nambu postulated a new $SU(3)''$, a "simple" step for them but a huge step towards our understanding of the strong interactions!

Double SU(3) Symmetry

The nine quarks are now represented by triplet–triplets with two indices $t_{i,\alpha}$; the flavor $SU(3)'$ acts on $\alpha = 1, 2, 3$ and the second $SU(3)''$ on the index $i = 1, 2, 3$, known years later as the color index.

Under $SU(3)' \times SU(3)''$, the nine quarks transform as $(\mathbf{3}, \bar{\mathbf{3}})$. The rest is group theory,

$$\text{Mesons}: \quad (\mathbf{3}, \bar{\mathbf{3}}) \times (\bar{\mathbf{3}}, \mathbf{3}) = (\mathbf{8}, \mathbf{1}) + (\mathbf{1}, \mathbf{1}) + \cdots,$$

$$\text{Baryons}: \quad (\mathbf{3}, \bar{\mathbf{3}}) \times (\bar{\mathbf{3}}, \bar{\mathbf{3}}) \times (\bar{\mathbf{3}}, \bar{\mathbf{3}}) = (\mathbf{1}, \mathbf{1}) + 2(\mathbf{8}, \mathbf{1}) + (\mathbf{10}, \mathbf{1}) + \cdots,$$

where dots denote $SU(3)''$ non-singlets.

[93] C. A. Levinson, H. J. Lipkin, and S. Meshkov, *Phys. Rev. Lett.* **10**, 361 (1963), and references therein.

They noticed that the $SU(3)''$ singlets correspond to the meson nonet, together with the baryon octets and decuplet of the eight-fold way. This suggested a simple dynamical principle: the binding potential is minimum for $SU(3)''$-singlet states. It is easy to write such a potential in terms of the quadratic and cubic $SU(3)''$ Casimir operators.

It gets better: *The advantage of the three-tripet model is that the* $SU(6)$ *symmetry can be easily realized with s-state triplets.* The symmetry is generalized to $SU(6) \times SU(3)''$, so that

"Since an $SU(3)''$ *singlet is anti-symmetric, the over-all Pauli principle requires the baryon states to be the symmetric* $SU(6)$ *56-plet."*

It is clear that Han and Nambu have the Yang–Mills theory in mind:

Instead of the Charm field, we introduce now eight gauge vector fields which behave as $(1, 8)$*, namely as an octet of* $SU(3)''$*, but as singlets of* $SU(3)'$*. Since their coupling to the individual triplets is proportional to* λ_i'' *(the generators of* $SU(3)''$*) the interaction energy arising from the exchange of these vector fields will yield ...",* that is a linear mass formula of the Casimir operators.

In today's language these are the gluons. Their $SU(3)''$ Yang–Mills is the theory of strong interaction, although the dynamical details were lacking. How could they have foreseen Asymptotic Freedom?
To conclude,

The Han–Nambu path to color was motivated from integrally charged quarks.

The Greenberg path to color was motivated by Green's parastatictics.

Both were part of the scaffolding often required in building a correct theory.

Neither approach seems to have been followed by the community which paid more attention to the scaffolding than to the beautiful emerging structure!

Preludes in Theoretical Physics

Three weeks after submitting the paper with M. Y. Han, Nambu contributed to "Preludes in Theoretical Physics", the Festschrift volume in the honor of Viki Weiskopf (1908–2002) who was retiring as "Directeur Général du CERN".

His paper *"A Systematics of Hadrons in Subnuclear Physics"*,[94] summarized his January two-triplet model and the April three-triplet model with M. Y. Han.

[94] A. De-Shalit, H. Feshback, and L. Van Hove (eds.), *Preludes in Theoretical Physics*, (North Holland 1966, pp. 133–142).

Most of it was on deriving the binding potential and semi-empirical mass formulas. It does not add to the discussion of the previous papers. The problems are challenging, and without resolution, and we do not discuss them any further.

We do find in it Nambu's explicit proposal of a Yang–Mills theory underlying the Strong Interactions:

An important difference from the two-triplet case is that instead of the charm gauge group $U(1)$, we have the group $SU(3)''$. The charm gauge field C must be replaced by an octet of gauge fields G_μ, ..., coupled to the infinitesimal $SU(3)''$ generators (currents) λ''_μ of the triplets with a strength g.

We do not know if the Weisskopf Festschrift volume was accompanied by a conference, but Murray Gell-Mann[95] regretted not having attended his advisor's Festschrift. He was not aware of the structure Han–Nambu were building, and thought in retrospect that he would have pointed out that it did not require integrally charged quarks.

95 "Professor Nambu", as Gell-Mann called him, was the first visitor he had invited soon after joining Caltech.

DYNAMICAL SYMMETRIES AND FUNDAMENTAL FIELDS

Y. Nambu

University of Chicago, Chicago, Illinois

I. Fundamental Triplets

At the moment, the success of the SU(3) symmetry is only partial because we have not understood the absence of the fundamental triplet representation among the known particles and resonances. Although it may be possible or even useful to rationalize this fact by slightly modifying the SU(3) group, in my opinion it is only a superficial approach. The most important message to be read from the partial success of the SU(3) seems to be that there are quite probably more fundamental fields and associated particles than those known to us. In other words, the baryons and mesons are by no means elementary, but are actually composite systems built up from more basic units, just as atoms and nuclei are themselves composite systems. Such a viewpoint is, of course, not new but actually dates back at least to Sakata. Especially within the past year or so, numerous attempts have been made to understand the Eightfold Way in terms of the symmetries of a few fundamental fields. They may be classified into the following possibilities[1]

1. A single fundamental triplet (quark-ace model)
2. A fundamental quartet (SU(4) model)
3. Two fundamental triplets ($U_3 \times U_3$ or W_3 model)

As noted above, any model except the first leaves room for a symmetry higher than the SU(3), which must be rather badly broken. In addition, we can accommodate the γ_5-type symmetry in each of these models.

Now a recent attempt to combine particles of different spin in the SU(6) symmetry[2] has led to several striking consequences which are in agreement with known facts. If these agreements are not an accident, it suggests strongly that the baryon is a three-body system of fundamental triplets. We observe, however, that it is not necessary to use fractionally-charged "quark" fields.[3] Suppose we take three triplets t_1, t_1', t_2 with the charge assignments (z_1+1, z_1, z_1), $(z_1'+1, z_1', z_1')$, (z_2+1, z_2, z_2) to construct a baryon. Then it is necessary that

$$z_1 + z_1' + z_2 + 1 = 0$$

or in terms of the "charm" number[4] $C = 3z + 1$,

$$C_1 + C_1' + C_2 = 0 \tag{1}$$

The predictions of SU(6) symmetry, such as the characteristic F:D ratio for the magnetic moment, remain unchanged if we treat the three triplets on an equal footing. Therefore, we will take the attractive solution[5]

$$C_1 = C_1' = 1 \quad C_2 = -2 \text{ or } t_1 = t_1' = (1,0,0),\ t_2 = (0,-1,-1) \tag{2}$$

which gives only integral charges 0, ± 1 for the members. We now have two distinct triplets of fundamental fermions, of which we use two of one kind and one of another. It is conceivable that they have different masses.

The reason we prefer this model to the quark model is that fractional charges are unlikely to exist. At least one of the quarks must be stable and could have been observed rather easily among cosmic rays, for example. (We reject the idea that quarks are mere mathematical fiction. If they exist, they ought to have a finite mass of observable magnitude.) Moreover, we would have to say that leptons are also composite. Since we have two triplets, a natural symmetry group of the model might be $U(3)_1 \times U(3)_2 \sim SU(3)_1 \times SU(3)_2 \times U(1)_1 \times U(1)_2$. Its SU(6) generalization is then $SU(6)_1 \times SU(6)_2 \times U(1)_1 \times U(1)_2$.

There remains the question of baryon number (B) assignment. We mention three simple choices:

1) $B = 1/3$ for all triplets
2) $B = 1$ for t_1, -1 for t_2 (3)
3) $B = 0$ for t_1, 1 for t_2

In 1), we may redefine the baryon number and call $B' = 1$ for the triplets, and $B' = 3$ for the baryons.

We remark that the two conservation laws (B and C) are, of course, due to the conservation of what we may call triplet charges N_1, N_2 for t_1 and t_2 respectively ($U(1)_1 \times U(1)_2$). Namely,

$$\begin{aligned} N_1 &= n_1 - \bar{n}_1,\ N_2 = n_2 - \bar{n}_2 \\ B &= \alpha N_1 + \beta N_2,\ C = N_1 - 2N_2 \end{aligned} \tag{4}$$

where n_i and $\bar{n}_i$ are the number of particles belonging to t_i and $\bar{t}_i$, while α and β depend on the choice (3).

We shall next tackle the problem of stabilizing the charm zero states (baryons) against other possible states in a purely dynamical and qualitative way. Here we can draw lessons from atomic and nuclear physics. Somehow the three-particle configuration with C = 0 must have saturation characteristics. This can be understood if there is a strong field (or force) coupled to the charm number. Just as an electrically neutral system tends to be most stable, a neutral charm system will have a tendency to have the lowest potential energy in the first approximation (no terms quadratic in the number of particles).

However, we have also to consider meson states (B=C=0) and many-baryon states ($B \geqslant 2$, $C = 0$), and make sure that the mass increases with B. This may be achieved by two additional mechanisms:

1) A mass difference between the triplets, and/or
2) Another field coupled to the baryon number.

In this connection we observe that the three possible meson configurations $\bar{t}_1 t_1$, $\bar{t}_2 t_2$ and $\bar{t}_1 t_2$ ($\bar{t}_2 t_1$) may have large mass and potential energy differences so that only one of them corresponds to the known low-lying mesons. (The charm attraction is largest for $\bar{t}_2 t_2$.)

With the interplay of these counteracting elements, we may anticipate arriving at a "semi-emprical mass formula"

$$M = m_o \left[|B| + x\, C^2 + y \right], \quad x \geqslant 1, \; y \geqslant 0^* \tag{5}$$

*Analytically, a more satisfactory form might be

$$M^2 = m_o^{\,2} \left[B^2 + xC^2 + y \right]$$

With B and C expressed by (4), Eq. (5) has the correct property that C = 0 states lie in a valley in the B-C plane, with the mass increasing linearly with B.

II. SU(6) and its Interpretation

To go on further, we have to consider seriously the implications of SU(6) symmetry. Here again there is ample room for speculation ranging from very sophisticated to very naive. We shall ramble over this topic for a while before settling down to a particular model.

The trouble with SU(6) is, as is well known, that it clashes with relativity in the sense that the free Dirac equation cannot be reconciled with it because of the inherent spin-orbit coupling. On the other hand, it is easy to construct a local non-linear interaction which is invariant under a relativistic extension of the SU(6) group.

Since SU(6) contains SU(2) x SU(3) where the first one is the spin rotation group, a minimum relativistic extension comprising the Lorentz group SL(2) is SL(6), as was noted by Michel and Sakita. This group leaves the form $\bar{U}_R U_L$, $\bar{U}_L U_R$ invariant, where the sextets U_L and U_R transform contragradiently to each other. To conserve parity, however, we have to deal with 12 component fields ψ and $\bar{\psi}$. (We will not distinguish the two triplets). This means introducing the "ρ-spin", where $\rho_3 = \pm 1$ for U_L and U_R respectively. The group which leaves $\bar{\psi}\psi = \bar{U}_R U_L + \bar{U}_L U_R$ invariant is the M(12) discussed by Bardakci et al.[6]

Another possible extension of SU(6) is $SU(6)_L \times SU(6)_R$ or $U(6)_L \times U(6)_R$ which leave $U_L^+ U_L$ and $U_R^+ U_R$ invariant; but they are not invariant under the Lorentz group. It is a natural extension of $SU(3)_L \times SU(3)_R$ which contains the γ_5 group. Incorporating the Lorentz group, we are led to $SL(6)_L \times SL(6)_R$ or $GL(6)_L \times GL(6)_R$. Eventually we can go up to GL(12) which contains the following interesting subgroups:

$$\begin{aligned} &GL(12) \supset U(12) \supset SU(12) \supset SU(6)_L \times SU(6)_R \supset SU(6) \\ &GL(12) \supset SL(12) \supset M(12) \supset SL(6) \supset SU(6) \end{aligned} \tag{6}$$

Note that the first chain involves only unitary groups but does not include the Lorentz group. The second chain (except SU(6)) includes the Lorentz group, but not a unitary γ_5 group.

It is always possible to write down a quantized local non-linear interaction term which is parity conserving and invariant under any subgroup of GL(12). For example

$$L \sim \det\,(\psi_\alpha \psi_\beta^\dagger),\ \alpha,\beta = 1, \ldots 12 \tag{7}$$

is invariant under SL(12) x U(1), where U(1) stands for the baryon or charm gauge group. For $SL(6)_L \times SL(6)_R$ we can take

$$\begin{aligned} L \sim g_1 &\left[\det\,(U_{L\alpha} U_{L\beta}^+) + \det\,(U_{R\alpha} U_{R\beta}^+)\right] \\ + g_2 &\left[\det\,(U_{L\alpha} U_{R\beta}^+) + \det\,(U_{R\alpha} U_{L\beta}^+)\right] \end{aligned} \tag{8}$$

We observe here that there is a minimum degree n of non-linearity ($L \sim U^n\, U^{+n}$) for each group. For all unitary groups, n = 1. For a Lorentz invariant group, n is as follows:

$$
\begin{array}{ll}
SL(12) & n = 12 \\
M(12) & 1 \\
SL(6) & 1 \\
SL(6)_L \times SL(6)_R & 6 \\
SU(3)_L \times SU(3)_R & 2
\end{array}
\qquad (9)
$$

unless we introduce extra contragredient fields. In order that we have an effective two-body and three-body interaction, n must be $\leqslant 2$. This leaves only M(12) (SL(6)) or $SU(3)_L \times SU(3)_R$. A significant point is that we cannot have a γ_5 group, SU(6) group and Lorentz group at the same time since they lead to $SL(6)_L \times SL(6)_R$. So we conclude that in order to have γ_5 and SU(6) together, namely, $SU(6)_L \times SU(6)_R$, the <u>effective</u> interaction Lagrangian must contain energy momentum vectors so that the SU(6) symmetry can be discussed only in the rest frame.

As is well known, the free Dirac Lagrangian is not invariant under any of the above groups. The kinetic energy violates SU(6) (spin-orbit coupling), and the mass violates γ_5 symmetry. If SU(6) is good, then either 1) somehow the internal kinetic energy of the triplets is suppressed compared with the potential energy, or 2) there is no spin-orbit coupling. Possible field theoretical interpretations of the above statement are:

1a. $\hbar \sim 0$ (since $\hbar$ appears only in the kinetic part of the Lagrangian). The triplets behave like classical particles. We may also say that it is strongly coupled to some field ($g^2/\hbar c \to \infty$).

1b. $Z_2 \sim 0$, which might mean that the triplets are not elementary (spontaneous breakdown or bootstrap?)

2. The fundamental fermions obey a Klein-Gordon equation:

$$L_o = -\partial_\mu \bar{\psi} \partial_\mu \psi - m^2 \bar{\psi} \psi$$

In this last case L_o has the symmetry group M(12). However, the degree of freedom is doubled (parity doublet), and besides we are led to the well known difficulty of indefinite metric. Assuming that one of these mechanisms is operating, the internal symmetry will depend on the form of the interaction terms. In this sense, it is a dynamical symmetry. These interaction terms we are talking about include the charm and baryonic fields as well as weaker forces, including those mediated by ordinary mesons.

We have no clearcut preference for a group larger than SU(6), except that at least $U(6)_L \times U(6)_R$ is desirable in view of the known but not well understood existence of γ_5 symmetry. For the sake of fun, let us take SU(12), the largest of the SU(n) groups, and classify the states in terms of three spins $\lambda^i(SU_3)$, σ and ρ. The intrinsic spin is σ, the chirality is ρ_1, and the intrinsic parity is given by $(-1)^{\rho_3}$. An interesting outcome is that, for the baryons, the totally symmetric three-particle states (dimensionality 364) of SU(12) contain representations 56 ($|\rho| = 3/2$) and 70 ($|\rho| = 1/2$) of SU(6). The ρ-spin value indicates their multiplicity (4 and 2 respectively, counting both positive and negative energy states). Actually, the meaning of ρ spin is obscure since some states correspond to a mixed product of creation and annihilation operators in terms of free fields. For the mesons (representation 143 of SU(12)), we have the ρ assignment

	$\lvert\rho\rvert$	ρ_1	$SU(3)_L \times SU(3)_R$
S+P	1	1	(3,3*) + (3*,3)
S-P	1	-1	
V_k	1	0	(1,8) + (8,1)
V_o	0	0	
T_L	1	1	(3,3*) + (3*,3)
T_R	1	-1	
A_k	0	0	(1,8) + (8,1)
A_o	1	0	

We can assign $|\rho| = 1$ to the known vector and pseudoscalar mesons at rest. A discussion of the possibility of ρ spin conservation is being attempted.

III. An Extreme Non-relativistic Model

We present here a most simple-minded model of baryons based on the previous considerations. In order to have SU(6), the internal velocity v of the triplets is assumed to be small. Let M be the common mass of the triplets, R be the size of the composite baryon, and V be the depth of the potential in which the triplets move. The condition $v \ll 1$ means

$$MR \gg 1 \quad (\hbar = c = 1). \tag{10}$$

Now R cannot be larger than the known size of the nucleon. Namely, we estimate

$$10^{-13}\ \text{cm} \gtrsim R \gtrsim 10^{-14}\ \text{cm} \tag{11}$$

Then the baryon mass m is given by

$$m = 3M - V \ll M \tag{12}$$

So the triplets are much heavier than the baryons, and there is a near cancellation between the potential and kinetic energy.

The system now looks like the triton or a helium atom, except in the differences of scale. There are two like particles (charm +1) and one unlike particle (charm -2). If ℓ_1 is the relative orbital angular momentum of the like particles, and ℓ_2 is the angular momentum of the pair against the third particle, then the total orbital angular momentum L is

$$L = |\ell_1 + \ell_2|, \; |\ell_1 - \ell_2| -1, \ldots |\ell_1 - \ell_2|$$

and the parity $P = (-1)^{\ell_1+\ell_2}$. If the baryon is in the symmetric 56 of SU(6), ℓ_1 must be odd. With $\ell_1 = 1$, we have

$$\ell_2 = 0; \quad L = 1$$

$$\ell_2 = 1; \quad L = 0,\ 1,\ 2; \text{ etc.}$$

Perhaps $\ell_1 = 1$ (odd) is favored compared to $\ell_1 = 0$ (even) because of the charm repulsion between the two like particles, (as in the 1S 2S states of the helium atom!). Using the same argument, then, ℓ_1 should be 0 (attraction between unlike particles). If, on the other hand, a short range repulsive baryon number force is important, ℓ_2 might also be 1. At any rate, the total j is now

$$\begin{aligned}
&L = 0;\ s = j = 3/2 \text{ (decuplet)},\ 1/2 \text{ (octet)}, \text{ parity } + \\
&L = 1: \quad j = 3/2,\ 1/2 \text{ for } s = 1/2 \text{ (octet)} \\
&\qquad\quad\ \ j = 5/2,\ 3/2,\ 1/2 \text{ for } s = 3/2 \text{ (decuplet)} \\
&\qquad\qquad\qquad \text{parity } + \text{ or } -
\end{aligned} \tag{13}$$

and so on.

The energy levels in our dynamical model are primarily determined by the spatial configuration, and the SU(6) symmetry operating among all three particles influences the groupings in a given configuration. There will be a large splitting between levels of different ℓ_1, ℓ_2 and L assignments. Then there will be spin-orbit coupling, and finally spin-spin coupling, (perhaps with the relative order of magnitude $\sim M\, v/c$ and $\sim M(v/c)^2$ respectively). The SU(6) symmetry correlates levels with different spin and SU(3) spin. Namely the baryon belongs to (21,6) of $SU(6)_1 \times SU(6)_2$ which reduces to 56 + 70 of SU(6).

Let us, as an example, take the non-trivial possibility that the baryons belong to L = 1. The spin orbit coupling $\sim 2g\, S\cdot L$ will split these levels according to

		s = 1/2	s = 3/2
j =	1/2	-2g	-5g
	3/2	g	-2g
	5/2	-	3g

Thus the two levels j = 1/2, s = 1/2 and j = 3/2, s = 3/2 are degenerate as in the simple SU(6)! However, they are not the lowest levels. (Of course, we would not have the trouble if we chose L = 0 ($\ell_1 = \ell_2 = 1$) as the ground state configuration.) A highly speculative way out of the above difficulty is to assume that 1) the S-L splitting is larger than 2m, and 2) a negative mass state is reinterpreted as an opposite parity state. For example, with m_0 (common mass) = m, g = m, we have the following picture

$$\begin{aligned} m(8,\ 1/2^+) &= m(10,\ 3/2^+) = m \\ m(8,\ 3/2^-) &= 2m \\ m(10,\ 1/2^+) &= 4m \\ m(10,\ 5/2^-) &= 4m \end{aligned} \tag{14}$$

where we may identify (8, 3/2⁻) with the γ octet ($N^*(1512)$, $Y_1^*(1660)$, $Y_0^*(1680?)$, $\Xi^*(1810)$). With such an interpretation, we find in a very crude estimation

$$M\, v/c \sim 1000 \text{ Mev},\ M(v/c)^2 \sim 100 \text{ Mev so that } M \sim 10 \text{ Bev.},\quad v/c \sim 1/10.$$

We conclude with a few comments concerning our model.
1) The triplets themselves and all $C \neq 0$ states are supposed to be very massive (in the above example, $M \sim 10$ Bev), and in general will quickly decay via weak interactions. Only in the case B = 1/3 (for all triplets) are the triplets really stable against decay into ordinary matter.
2) There may be a significant mass difference between the two triplets.

3) The γ_5 symmetry, if it exists at all, cannot be explained unless the large triplet mass itself is of dynamical origin. This does not prevent us from computing the F/D ratio of the axial vector current according to SU(6).
4) The very strong potential with long range ($\sim 1/m$) might conflict with ordinary field theoretical interpretation because it means $g^2/\hbar c \sim M/m \gg 1$ for a Yukawa potential. The charm and baryon number mesons, if they exist as particles, will have spin 1, no electric charge, and a relatively low mass ($\sim m$), and will decay very fast.
5) The remarkable ratios obtained for the baryon magnetic moments do not yet explain their absolute magnitude unless the triplets themselves, singly or collectively, have large anomalous moments. This brings back the problem of the role played by the ordinary meson cloud. Such a cloud must exist. The only plausible interpretation will be that the properties of the core particles are faithfully amplified by the cloud. Since the mesonic excitations are small in our scale, their role in dynamics is expected to be important. There is no conflict with the bootstrap theory in this sense.
6) We should expect to keep on observing baryon resonances up to energies of the order V or M, at which point $C \neq 0$ particles begin to be produced.

References

1. See e.g. Lee, Gürsey and Nauenberg, Phys. Rev. 135, B467 (1964). A. Salam, Dubna Conference Report.
2. Gürsey and Radicati, Phys. Rev. Letters 13, 173 (1964). A. Pais, Phys. Rev. Letters 13, 175 (1964). B. Sakita, preprint.
3. M. Gell-Mann, Phys. Letters 8, 214 (1964). G. Zweig, CERN preprint.
4. See Salam, Ref. 1; B.J. Bjørken and S.L. Glashow, Phys. Letters 11, 255 (1964).
5. H. Bacry, J. Nuyts, L. van Hove, Phys. Letters 9, 279 (1964).
6. K. Bardakci, J.M. Cornwall, P.G.O. Freund and B.W. Lee, preprint.

DISCUSSION

Dashen - Isn't it true that when you have baryon number zero for a fermion, that is stable also, absolutely stable?
Nambu - No, it could decay into leptons.
Dashen - That would be violating the lepton number conservation. Right?
Nambu - I haven't assigned the lepton number as yet.
Dashen - Ok, I'll buy that. I wonder did you run into any difficulty with the -1 for the baryon number just in conservation of baryon number?
Nambu - Well in the electric charge you have the universal electromagnetic field. In the case of baryon number we don't have that kind, so I don't take it quite objectionable to happen. It's just a redefinition of the baryon number.
Dashen - And you can assign lepton number to these triplets?
Nambu - Yes.
Coester - How do you reconcile the first and the last inequalities on the (10) and (12)? It seems to me when the binding is large compared to the rest masses of the particles involved you will have large velocities.
Nambu - No. If you take the square well, it can be as deep as you like.
Coester - I notice that I still get the large velocity inside.
Nambu - No, not necessarily. I'm thinking of a purely static point of view.
Marshak - One point that you have made about interpreting a negative mass particle as a particle of positive mass and opposite parity is very sticky. I just would like to say that in our discussion of the $\bar{W}(3)$ model which is the U(3) X U(3) parity doubling theory where you can define parity in a conventional way, we just could not reconcile your interpretation with our method of constructing baryons. This is why in our paper with Mukunda, we had to work with the baryon octet because in the W(3) limit, if you take a baryon nonet and then split it by the mass term, you must get a negative mass for the singlet, and positive mass for the octet, since you start with zero mass. We could not give a meaning to the negative mass. But I know Gell-Mann was willing to accept your recipe and identify the 9th baryon as the 1405 $J^P=\frac{1}{2}^-$ resonance. This disturbs me; I think you've provided a recipe, but could you justify it?
Nambu - No, I cannot fully justify this. There is a difficulty that exists: And if you want to make an

interpretation, of course, you have to redefine the vacuum somehow. I really do not know.

Tanaka - What assumptions went into obtaining your mass formula which depends on C and B?

Nambu - This formula is drawn on electrostatic analogy. The electrostatic repulsion goes like the square of the charge. The mass of the nucleus increases linearly, more or less, with the baryon number. Then these charm number non-zero states are highly unstable when you have a large mass compared to the ordinary one, so the x must be large compared to 1.

Nuyts - It is not absolutely necessary to stick to the W(3) group. You may also take the SP(6) group or the SU(6) group. We also tried to incorporate the spin into that, then you get either SP(12) or SO(12). Then Professor Van Hove told me four days ago that he had found some nice property with SP(12), so that also SP(12) or SO(12) could give nice results.

Zumino - What is the difficulty with interpreting the negative mass as an opposite parity? If one just takes now, as the field of the particle, γ_5 times the original field, and if the theory has some sort of a γ_5 invariance in the interaction, then where does the difficulty arise?

Nambu - The problem seems to boil down to this: If what you are computing is the mass (or the mass operator), then negative sign may be reinterpreted as negative parity. But if you are computing the energy, a negative sign means that we have not defined the vacuum properly. In a static theory it is difficult to tell what one is doing in this respect.

(Editors Note:
An interpretation of negative mass states as positive mass particles with the parity reversed was made from the Regge theory point of view in the following reference. Possible Connections between the Fermion Trajectories, Ismail A. Sakmar, Phys. Rev. 135, B249 (1964)).

PHYSICAL REVIEW VOLUME 139, NUMBER 4B 23 AUGUST 1965

Three-Triplet Model with Double $SU(3)$ Symmetry*

M. Y. HAN
Department of Physics, Syracuse University, Syracuse, New York

AND

Y. NAMBU
The Enrico Fermi Institute for Nuclear Studies, and the Department of Physics, The University of Chicago, Chicago, Illinois

(Received 12 April 1965)

With a view to avoiding some of the kinematical and dynamical difficulties involved in the single-triplet quark model, a model for the low-lying baryons and mesons based on three triplets with integral charges is proposed, somewhat similar to the two-triplet model introduced earlier by one of us (Y. N.). It is shown that in a $U(3)$ scheme of triplets with integral charges, one is naturally led to three triplets located symmetrically about the origin of I_3-Y diagram under the constraint that the Nishijima–Gell-Mann relation remains intact. A double $SU(3)$ symmetry scheme is proposed in which the large mass splittings between different representations are ascribed to one of the $SU(3)$, while the other $SU(3)$ is the usual one for the mass splittings within a representation of the first $SU(3)$.

I. INTRODUCTION

ALTHOUGH the $SU(6)$ symmetry strongly indicates that the baryon is essentially a three-body system built from some basic triplet field or fields, the quark model[1] is not entirely satisfactory from a realistic point of view, because (a) the electric charges are not integral, (b) three quarks in s states do not form the symmetric $SU(6)$ representation assigned to the baryons, and (c) a simple dynamical mechanism is lacking for realizing only zero-triality states as the low-lying levels.

These difficulties may be avoided if we introduce more than one basic triplet. Recently one of us (Y. N.) has attempted a two-triplet model[2] where the members of the triplets t_1 and t_2 had the charge assignment $(1,0,0)$ and $(0,-1,-1)$, as had been proposed earlier by Bacry *et al.*[3] The baryon would be represented by the combination $t_1t_1t_2$, whereas the mesons would correspond to some combination $\sim a t_1\bar{t}_1' + b t_2\bar{t}_2'$. The triplets are assumed to have masses large compared to the baryon mass, which would mean that baryons and mesons have very large binding energies. A dynamical mechanism for this is provided by a neutral field coupled strongly to the "charm number"[4] C, which is 1 for t_1 and -2 for t_2, and therefore $C=0$ for baryons and mesons. In analogy with electrostatic energy, we can argue that the potential energy due to the charm field would be lowest when the system is "neutral," namely, $C=0$. Thus all other unwanted configurations with $C\neq 0$, which include among others triplet, sextet, etc. representations, would have high masses, and hence would not be easily observed.

There have been proposed two different ways in which to introduce basic triplet or triplets with integral charges. One approach essentially involves a modification of the Nishijima–Gell-Mann relation by way of introducing an additional quantum number, the triality quantum number,[5] and this has led to considerations of higher symmetry schemes based on rank-three Lie groups.[6] On the other hand, Okubo *et al.*[7] have recently shown that the minimal group required for this purpose is actually the group $U(3)$.[8] It is shown that a triplet scheme may be defined in $U(3)$ such that the triplet always possesses integral values of charge and hypercharge and satisfies the Nishijima–Gell-Mann relation without a modification. The $U(3)$ triplet considered by Okubo *et al.* is of Sakata type; i.e., it consists of an isodoublet and an isosinglet. Actually, the $U(3)$ scheme is much more appealing than those of the rank-three Lie groups on two accounts: firstly, the Nishijima–Gell-Mann relation is satisfied universally by triplets as by octets and decuplets, and secondly as far as the hitherto realized representations are concerned, $U(3)$ is equivalent to $SU(3)$.[9]

In what follows, we show that the $U(3)$ scheme, when fully utilized as described below, naturally and uniquely

* Work supported in part by the U. S. Atomic Energy Commission under the Contract No. AT(30-1)-3399 and No. AT(11-1)-264.

[1] M. Gell-Mann, Phys. Letters **8**, 214 (1964); G. Zweig, CERN (to be published).

[2] Y. Nambu, *Proceedings of the Second Coral Gables Conference on Symmetry Principles at High Energy* (W. H. Freeman and Company, San Francisco, 1965).

[3] H. Bacry, J. Nuyts, and L. van Hove, Phys. Letters **9**, 279 (1964).

[4] This name was originally used in connection with the $SU(4)$ symmetry. B. J. Bjørken and S. L. Glashow, Phys. Letters **11**, 255 (1964); A. Salam, Dubna Conference Report, 1964 (unpublished).

[5] G. E. Baird and L. C. Biedenharn, *Proceedings of the First Coral Gables Conference on Symmetry Principles at High Energy* (W. H. Freeman and Company, San Francisco, 1964); C. R. Hagen and A. J. Macfarlane, Phys. Rev. **135**, B432 (1964) and J. Math. Phys. **5**, 1335 (1964).

[6] For example, see I. S. Gerstein and M. L. Whippmann, Phys. Rev. **137**, B1522 (1965). Earlier references are given in this paper.

[7] S. Okubo, C. Ryan, and R. E. Marshak, Nuovo Cimento **34**, 759 (1964).

[8] The use of $U(3)$ in this connection has also been remarked by I. S. Gerstein and K. T. Mahanthappa, Phys. Rev. Letters **12**, 570, 656(E) (1964).

[9] S. Okubo, Phys. Letters **4**, 14 (1963).

leads to a set of three basic triplets with integral charges, namely an I-triplet (isodoublet and isosinglet), a U-triplet (U-spin doublet and U-spin singlet) and a V-triplet (V-spin doublet and V-spin singlet).[10] These triplets arise from three different ways of defining charge Q, hypercharge Y, and a displaced isospin I_3 in the $U(3)$ group as opposed to the $SU(3)$, in such a way that the charge and hypercharge have integral values, while keeping the Nishijima–Gell-Mann relation intact, and they differ from each other in their quantum-number assignments as well as in their transformation properties under the Weyl reflections.[11] This is described in Sec. II. In Sec. III, a double $SU(3)$ symmetry scheme is proposed based on the three-triplet model in which the large mass splittings between different representations are ascribed to one of the $SU(3)$, and the other $SU(3)$ is, as usual, responsible for the mass splittings within a representation. The low-lying baryon and meson states may be taken as singlets with respect to one of the $SU(3)$. The extended symmetry group with respect to the $SU(6)$ symmetry is briefly discussed.

II. THREE TRIPLETS

We shall denote the infinitesimal generators of $U(3)$ by $A_\nu{}^\mu$ which satisfies the following commutation relations:

$$[A_\beta{}^\alpha, A_\nu{}^\mu] = \delta_\beta{}^\mu A_\nu{}^\alpha - \delta_\nu{}^\alpha A_\beta{}^\mu, \tag{1}$$

where all indices take on the values 1, 2, and 3. The corresponding infinitesimal generators $B_\nu{}^\mu$ of $SU(3)$ are then given by

$$B_\nu{}^\mu = A_\nu{}^\mu - \tfrac{1}{3}\delta_\nu{}^\mu A_\lambda{}^\lambda \tag{2}$$

which satisfy the following equations:

$$[B_\beta{}^\alpha, B_\nu{}^\mu] = \delta_\beta{}^\mu B_\nu{}^\alpha - \delta_\nu{}^\alpha B_\beta{}^\mu \tag{3}$$

and

$$B_\lambda{}^\lambda = 0. \tag{4}$$

Furthermore, the unitary restriction gives

$$(A_\nu{}^\mu)^\dagger = A_\mu{}^\nu, \quad (B_\nu{}^\mu)^\dagger = B_\mu{}^\nu. \tag{5}$$

Let us now briefly summarize the relevant results of Okubo *et al.* In the $SU(3)$ scheme, the charge Q, the hypercharge Y and the third component of isospin I_3 are identified as follows[12]:

$$Q = -B_1{}^1, \tag{6a}$$

$$Y = B_3{}^3 = -B_1{}^1 - B_2{}^2 \quad [\text{by the relation (4)}], \tag{6b}$$

$$I_3 = \tfrac{1}{2}(B_2{}^2 - B_1{}^1). \tag{6c}$$

In the $U(3)$ scheme, the corresponding quantities $\tilde{Q}$, $\tilde{Y}$, and $\tilde{I}_3$ are defined as follows:

$$\tilde{Q} = -A_1{}^1 = Q - \tfrac{1}{3}\tau, \tag{7a}$$

$$\tilde{Y} = -A_1{}^1 - A_2{}^2 = Y - \tfrac{2}{3}\tau, \tag{7b}$$

$$\tilde{I}_3 = \tfrac{1}{2}(A_2{}^2 - A_1{}^1) = I_3, \tag{7c}$$

where

$$\tau = A_1{}^1 + A_2{}^2 + A_3{}^3. \tag{8}$$

With these definitions, the Nishijima–Gell-Mann relation is seen to be equally satisfied by the $U(3)$ and $SU(3)$ theories, i.e.,

$$Q = I_3 + \tfrac{1}{2}Y \tag{9}$$

and

$$\tilde{Q} = \tilde{I}_3 + \tfrac{1}{2}\tilde{Y}, \tag{10}$$

respectively. Since the generators $A_1{}^1$, $A_2{}^2$, and $A_3{}^3$ possess integral eigenvalues in any representation,[13] the identifications of $\tilde{Q}$ and $\tilde{Y}$ to be the charge and the hypercharge, respectively, in $U(3)$ theory shall always lead to integral values for the charge and the hypercharge. In particular, in the three-dimensional representation, the $U(3)$ triplet has the eigenvalues

$$\tilde{Q} = \begin{bmatrix} 1 & 0 & 0 \\ 0 & 0 & 0 \\ 0 & 0 & 0 \end{bmatrix}, \quad \tilde{I}_3 = \begin{bmatrix} \frac{1}{2} & 0 & 0 \\ 0 & -\frac{1}{2} & 0 \\ 0 & 0 & 0 \end{bmatrix}, \quad \tilde{Y} = \begin{bmatrix} 1 & 0 & 0 \\ 0 & 1 & 0 \\ 0 & 0 & 0 \end{bmatrix}. \tag{11}$$

This triplet corresponds to the Sakata triplet which we call an I triplet for short.

We can now generalize the above constructions of the $U(3)$ triplet in the following way. Comparing (6b) and (7b), we see that a particular choice has been made for $\tilde{Y}$. Had we defined $\tilde{Y}$ to be $A_3{}^3$, it would still have integral eigenvalues but the relation (10) would have been violated. This is because $B_\lambda{}^\lambda = 0$ in $SU(3)$ but $A_\lambda{}^\lambda \neq 0$ in general in $U(3)$ and thus some care is needed in defining corresponding quantities in $U(3)$. Making use of (4), the definition in (6) can be written more generally as

$$Q = -B_1{}^1 = B_2{}^2 + B_3{}^3, \tag{12a}$$

$$Y = B_3{}^3 = -B_1{}^1 - B_2{}^2, \tag{12b}$$

$$I_3 = \tfrac{1}{2}(B_2{}^2 - B_1{}^1) = \tfrac{1}{2}(2B_2{}^2 + B_3{}^3) = -\tfrac{1}{2}(2B_1{}^1 + B_3{}^3). \tag{12c}$$

As in (7), replacing $B_\nu{}^\mu$'s in (12) by corresponding $A_\nu{}^\mu$'s, we list all possible candidates for the corresponding quantities in $U(3)$ which are now however not equivalent to each other [they are equivalent, of course, when reduced to $SU(3)$], i.e.,

$$\tilde{Q}: \quad -A_1{}^1, \quad A_2{}^2 + A_3{}^3, \tag{13a}$$

$$\tilde{Y}: \quad A_3{}^3, \quad -A_1{}^1 - A_2{}^2, \tag{13b}$$

$$\tilde{I}_3: \quad \tfrac{1}{2}(A_2{}^2 - A_1{}^1), \quad \tfrac{1}{2}(2A_2{}^2 + A_3{}^3), \quad -\tfrac{1}{2}(2A_1{}^1 + A_3{}^3). \tag{13c}$$

[10] C. A. Levinson, H. J. Lipkin, and S. Meshkov, Nuovo Cimento **23**, 236 (1961); Phys. Letters **1**, 44 (1962) and Phys. Rev. Letters **10**, 361 (1963).
[11] A. J. Macfarlane, E. C. G. Sudarshan, and C. Dullemond, Nuovo Cimento **30**, 845 (1963).
[12] We use the sign convention of S. P. Rosen, J. Math. Phys. **5**, 289 (1964).
[13] For a derivation of this result, see Eq. (7) of Ref. 7.

To start with, the alternative choices in (13) provide twelve inequivalent ways in which to choose a set of three quantities $\tilde{Q}$, $\tilde{Y}$ and $\tilde{I}_3$ for the $U(3)$ scheme. In every choice $\tilde{Q}$ and $\tilde{Y}$ will have integral eigenvalues, but as can be easily checked the Nishijima–Gell Mann relation will not be valid for all of them. In fact, there are only three cases for which it is valid and we are thus naturally led to three inequivalent triplets in the $U(3)$ scheme; they are defined by the following three choices:

$$t_I:\quad \tilde{Q}=-A_1{}^1,\qquad \tilde{Y}=-A_1{}^1-A_2{}^2,\qquad \tilde{I}_3=\tfrac{1}{2}(A_2{}^2-A_1{}^1),\tag{14a}$$

$$t_U:\quad \tilde{Q}=A_2{}^2+A_3{}^3,\qquad \tilde{Y}=A_3{}^3,\qquad \tilde{I}_3=\tfrac{1}{2}(2A_2{}^2+A_3{}^3),\tag{14b}$$

$$t_V:\quad \tilde{Q}=-A_1{}^1,\qquad \tilde{Y}=A_3{}^3,\qquad \tilde{I}_3=-\tfrac{1}{2}(2A_1{}^1+A_3{}^3).\tag{14c}$$

Now the first one, t_I, for which

$$\tilde{Y}=-A_1{}^1-A_2{}^2,\tag{15}$$

$$\tilde{I}_3=\tfrac{1}{2}(A_2{}^2-A_1{}^1)=\tfrac{1}{2}(B_2{}^2-B_1{}^1)=I_3\tag{16}$$

corresponds to the I triplet mentioned above.

The structure of the remaining triplets t_U and t_V can be brought to much more transparent and symmetric forms in terms of the U-spin and V-spin subalgebras.[10] As in the case of relations (9) and (10) for $SU(3)$ and $U(3)$, we define the U and V spin of $U(3)$ in exactly the same forms as in $SU(3)$ except that all quantities are tilded quantities. From the $SU(3)$ definitions,[12] we then have

$$\tilde{Y}_U=-\tilde{Q}=-A_2{}^2-A_3{}^3,\tag{17}$$

$$\tilde{U}_3=\tilde{Y}-\tfrac{1}{2}\tilde{Q}=\tfrac{1}{2}(A_3{}^3-A_2{}^2)=\tfrac{1}{2}(B_3{}^3-B_2{}^2)=U_3\tag{18}$$

for (14b), and

$$\tilde{Y}_V=\tilde{Q}-\tilde{Y}=-A_3{}^3-A_1{}^1,\tag{19}$$

$$\tilde{V}_3=-\tfrac{1}{2}(\tilde{Y}+\tilde{Q})=\tfrac{1}{2}(A_1{}^1-A_3{}^3)=\tfrac{1}{2}(B_1{}^1-B_3{}^3)=V_3\tag{20}$$

for (14c). They correspond, therefore, to a U triplet and a V triplet, respectively, and hence the notations t_I, t_U, and t_V. With respect to the $SU(3)$ triplet (quark), these $U(3)$ triplets have their respective "hypercharges" (i.e., Y, Y_U, and Y_V) shifted by the amount of $\frac{2}{3}$ and as such they have quite different transformation properties under the Weyl reflections W_1, W_2, and W_3[11] which are reflections about the axis $I_3=0$, $U_3=0$, and $V_3=0$, respectively. Whereas the $SU(3)$ triplet is invariant under all three Weyl reflections, the $U(3)$ triplets are not. They transform according to

$$W_1:\quad t_I\rightarrow t_I,\quad t_U\leftrightarrow t_V;\tag{21a}$$

$$W_2:\quad t_U\rightarrow t_U,\quad t_I\leftrightarrow t_V;\tag{21b}$$

$$W_3:\quad t_V\rightarrow t_V,\quad t_I\leftrightarrow t_U.\tag{21c}$$

Figure 1 and Table I(a) list the quantum numbers $\tilde{I}_3$ and $\tilde{Y}$ for the single triplet (quark) model; a possible assignment implied by the two-triplet model[3] is shown in Fig. 2 and Table I(b); the corresponding quantum numbers for the three-triplet model are given in Fig. 3 and Table I(c).

TABLE I. Quantum-number assignments for (a) the quark model, (b) the two-triplet model, and (c) the three-triplet model.

(a)

	quark		
$\tilde{I}_3$	$\frac{1}{2}$	$-\frac{1}{2}$	0
$\tilde{Y}$	$\frac{1}{3}$	$\frac{1}{3}$	$-\frac{2}{3}$
$\tilde{Q}$	$\frac{2}{3}$	$-\frac{1}{3}$	$-\frac{1}{3}$

(b)

	t_1			t_2		
$\tilde{I}_3$	$\frac{1}{2}$	$-\frac{1}{2}$	0	$\frac{1}{2}$	$-\frac{1}{2}$	0
$\tilde{Y}$	1	1	0	-1	-1	-2
$\tilde{Q}$	1	0	0	0	-1	-1

(c)

	$t_1(t_I)$			$t_2(t_U)$			$t_3(t_V)$		
$\tilde{I}_3$	$\frac{1}{2}$	$-\frac{1}{2}$	0	0	-1	$-\frac{1}{2}$	1	0	$\frac{1}{2}$
$\tilde{Y}$	1	1	0	0	0	-1	0	0	-1
$\tilde{Q}$	1	0	0	0	-1	-1	1	0	0

III. DOUBLE $SU(3)$ SYMMETRY

Let us call the three triplets $t_1(=t_I)$, $t_2(=t_U)$, and $t_3(=t_V)$. Each triplet may be characterized in general by the average values, $\bar{I}_3$ and $\bar{Y}$, of $\tilde{I}_3$ and $\tilde{Y}$ for its three members. This specifies the location of the center of the triplet in the $\tilde{I}_3-\tilde{Y}$ diagram. Since $\bar{A}_1{}^1=\bar{A}_2{}^2=\bar{A}_3{}^3=\tilde{\tau}/3=\tau/3$, Eq. (14) gives for the three definitions of

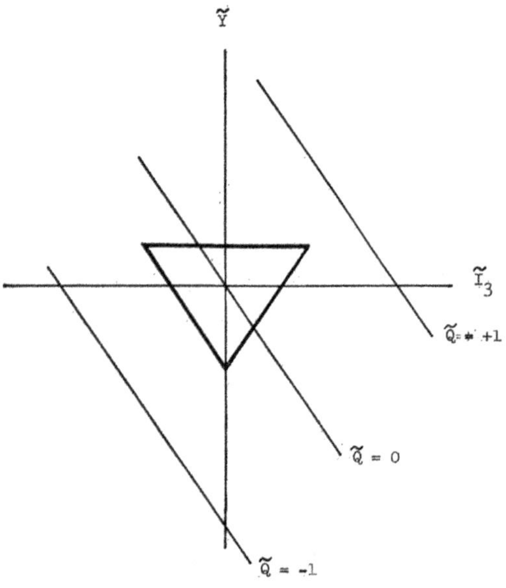

FIG. 1. The single-triplet (quark) model.

$\bar{I}_3$ and $\bar{Y}$,

$$\begin{aligned}\bar{I}_3&=0,\ \tfrac{1}{2}\tau,\ -\tfrac{1}{2}\tau,\\ \bar{Y}&=-\tfrac{2}{3}\tau,\ \tfrac{1}{3}\tau,\ \tfrac{1}{3}\tau,\end{aligned} \tag{22}$$

respectively, where $\tau=-1$ for all the triplets. We may define new quantities I_3, Y and $Q=I_3+\frac{1}{2}Y$ by the relations:

$$\begin{aligned}\tilde{I}_3&=\bar{I}_3+I_3,\\ \tilde{Y}&=\bar{Y}+Y,\\ \tilde{Q}&=\bar{I}_3+\tfrac{1}{2}\bar{Y}+I_3+\tfrac{1}{2}Y=\bar{Q}+Q.\end{aligned} \tag{23}$$

It is clear that I_3 and Y play the role of $SU(3)$ generators within each triplet. The charm number C defined in the two-triplet model[2] is then

$$\tfrac{1}{3}C=\bar{Q}=\bar{I}_3+\tfrac{1}{2}\bar{Y}. \tag{24}$$

Now it is interesting to note that according to Eq. (22) and Fig. 3, the centers of the three triplets form an antitriplet, equivalent to an antiquark, symmetrically located around the origin. Let us suppose that the nine members of the three triplets $t_{1\alpha}$, $t_{2\alpha}$, $t_{3\alpha}$, $\alpha=1, 2, 3$ be combined into a single multiplet $T=\{t_{i\alpha}\}$, $i=1, 2, 3$. We can then imagine two distinct sets of $SU(3)$ operations on T. One is the $SU(3)$ acting on the index α for each triplet, while the other $SU(3)$ acts on the index i, which mixes corresponding members of different triplets. T is then a representation $(3,3^*)$ of this group $G\equiv SU(3)'\times SU(3)''$.[14] The quantum numbers of $SU(3)'$ and $SU(3)''$ are identified as $I_3'=I_3$, $Y'=Y$, $I_3''=\bar{I}_3$ and $Y''=\bar{Y}$ in Eq. (22), so that

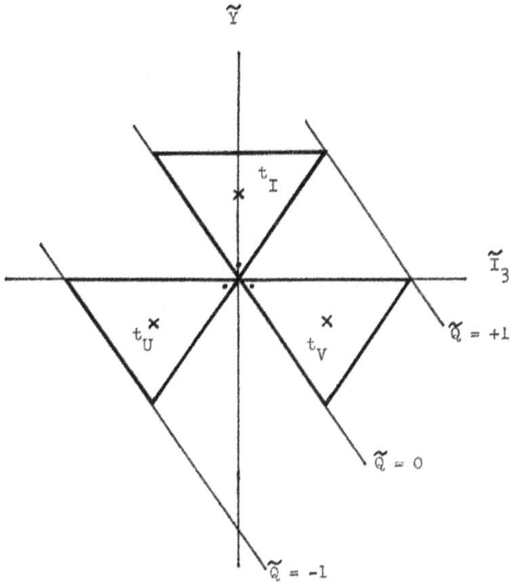

FIG. 3. The three-triplet model.

$$\begin{aligned}\tilde{I}_3&=I_3'+I_3'',\quad \tilde{Y}=Y'+Y'',\\ \tilde{Q}&=I_3'+I_3''+\tfrac{1}{2}Y'+\tfrac{1}{2}Y'',\\ \tfrac{1}{3}C&=I_3''+\tfrac{1}{2}Y''.\end{aligned} \tag{25}$$

A general representation of G may be characterized by four numbers p', q', p'', q'' so that $D(p',q',p'',q'')\sim D(p',q')\times D(p'',q'')$, where $D(p,q)$ is a representation of $SU(3)$. However, in our scheme where the nonet T is the fundamental field, we do not get all the possible representations of G. This can be illustrated by means of the triality numbers[5] $t'=p'-q' \bmod(3)$, $t''=p''-q'' \bmod(3)$. The nonet T has $t'=1$, $t''=-1$. All representations constructed out of T and T^* then satisfy $t'=-t''$.

Let us next consider the meson and baryon states $\sim TT^*$ and $\sim TTT$. The $SU(3)'\times SU(3)''$ contents of these 81- and 729-plets are

$$\begin{aligned}(3,3^*)\times(3^*,3)&=(8,1)+(1,1)+(1,8)+(8,8),\\ (3,3^*)\times(3,3^*)\times(3,3^*)&=(1,1)+2(8,1)+2(1,8)\\ &\quad+(1,10^*)+(10,1)+2(8,10^*)+2(10,8)\\ &\quad+4(8,8)+(10,10^*).\end{aligned} \tag{26}$$

It is an attractive possibility to postulate at this point that the energy levels are classified according to $SU(3)''$. The masses will then depend on the Casimir operators of $SU(3)''$. For example, a simple linear form will be

$$m=m_0+m_2C_2''+m_3C_3'', \tag{27}$$

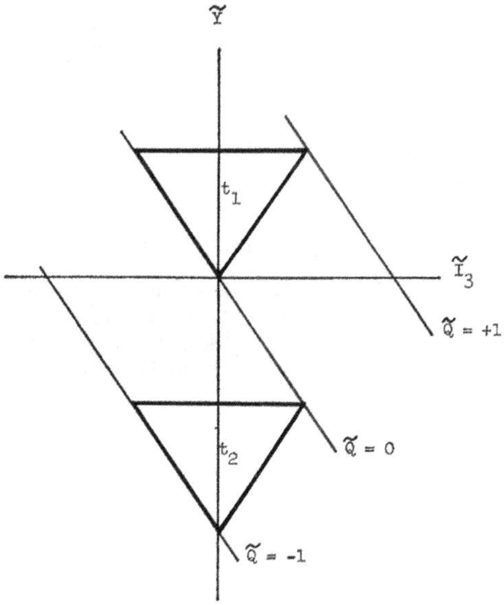

FIG. 2. The two-triplet model.

[14] Such a nonet provides a natural basis for the symmetry of $SU(9)$. However, we will not consider it here.

where C_2'', C_3'' are the eigenvalues of quadratic and cubic Casimir operators of $SU(3)''$. In particular, we may assume that the main mass splitting comes from C_2''. Since this increases with the dimensionality of representation, the lowest mass levels will be $SU(3)''$ singlets. This selects the low-lying meson and baryon states to be (8,1), (1,1) and (8,1), (1,1), (10,1), respectively. In general, all low-lying states will have triality zero, $t'=t''=0$.

As for the baryon number assignment to the triplets, the simplest possibility would be to assign an equal baryon number, i.e., $B=\frac{1}{3}$, to them. In this case the triplets themselves would be essentially stable, and their nine members would behave like an octet plus a singlet of "heavy baryons" as may be seen from Fig. 3. Another simple possibility may be $B=\frac{1}{3}+Y''$, namely $B=(1,0,0)$ for (t_1,t_2,t_3). We expect a mass splitting depending on B or Y'', which may be the origin of the Okubo-Gell-Mann mass formula.

The advantage of the three-triplet model is that the $SU(6)$ symmetry can be easily realized with s-state triplets. The extended symmetry group becomes now $SU(6)'\times SU(3)''$. Since an $SU(3)''$ singlet is antisymmetric, the over-all Pauli principle requires the baryon states to be the symmetric $SU(6)$ 56-plet. Other $SU(6)$ representations such as the 70, will be obtained by bringing in either the orbital angular momentum or the "ρ spin" of the Dirac spinor triplets.

As in the two-triplet model mentioned in the Introduction, the mass formula of the type (27) may be derived dynamically. Instead of the charm number field, we introduce now eight gauge vector fields which behave as (1,8), namely as an octet in $SU(3)''$, but as singlets in $SU(3)'$. Since their coupling to the individual triplets is proportional to λ_i'' [the generators of $SU(3)''$], the interaction energy arising from the exchange of these vector fields will yield the first and second terms of Eq. (27). If these mesons obey again a similar type of mass formula, they will be expected to be massive compared to the ordinary mesons. However, it is not clear whether the resulting short-range character of the interaction can be readily reconciled with the postulated largeness of the interaction energy.

We may characterize the hierarchy of interactions and their symmetries implied by the above model as follows. First, the ***superstrong*** interactions responsible for forming baryons and mesons have the symmetry $SU(3)''$, and causes large mass splittings between different representations. The scale of mass involved would be comparable or large compared to the baryon mass, namely $\gtrsim 1$ BeV. The lowest states, i.e., $SU(3)''$ singlet states, would split according to $SU(3)'$, which would be the $SU(3)$ group observed among the known baryons and mesons, with their ***strong*** interactions. The scale of mass splitting would then be $\lesssim 1$ BeV.

When we go to the massive $SU(3)''$ nonsinglet states, there may very well be coupling between the two $SU(3)$ groups similar to the $L\cdot S$ coupling. The levels should be classified in terms of the three sets of Casimir operators formed out of λ_i', λ_i'', and $\lambda_i=\lambda_i'+\lambda_i''$, respectively. The splitting due to the coupling would naturally be intermediate between the above two splittings, namely ~ 1 BeV. Because of this coupling, the separate conservation of the two $SU(3)$ spins, I_3' and Y' on the one hand, and I_3'' and Y'' on the other, would be destroyed, and only the sums $I_3=I_3'+I_3''$ and $Y=Y'+Y''$ would be conserved under ***strong*** interactions. This in turn would mean that all the massive states are in general highly unstable, and decay strongly to the low-lying states. (In the two-triplet model, we considered only weak decays of $C\neq 0$ states. But strong decays are also a possibility as is contemplated here.)

We have discussed here a possible model of baryons and mesons based on three triplets. How can we distinguish this and other different models mentioned already? Certainly different models predict considerably different structure of massive states. These states are characterized by the triality for the quark model, by the charm number for the two-triplet model and by the $SU(3)''$ representation for the present three-triplet model. If we restrict ourselves to the low-lying states only, however, it seems difficult to distinguish them without making more detailed dynamical assumptions.

ACKNOWLEDGMENTS

One of us (M. Y. H.) wishes to thank Professor E. C. G. Sudarshan and Professor A. J. Macfarlane for their encouragement and useful discussions and Professor L. O'Raifeartaigh and J. Kuriyan for helpful comments.

A SYSTEMATICS OF HADRONS IN SUBNUCLEAR PHYSICS

YOICHIRO NAMBU

The Enrico Fermi Institute for Nuclear Studies and the Department of Physics, The University of Chicago, Chicago, Illinois

(*Received May 3, 1965*)

1.

With the recognition that the SU(3) symmetry is the dominant feature of the strong interactions, the main concern of the elementary particle theory has naturally become directed at the understanding of the internal symmetry of particles at a deeper level. An immediate question that arises in this regard is whether there are fundamental objects (such as triplets or quartets) of which all the known baryons and mesons are composed. These fundamental objects would be to the baryons and mesons what the nucleons are to the nuclei, and the electrons and nuclei are to the atoms. If that was really the case, it would certainly precipitate a new revolution in our conceptual image of the world. At the moment we can only hope that the question will be answered within the next ten to twenty years when the 100 GeV to 1000 GeV range accelerators will have been realized.

Even now, the amusing and rather embarassing success of the SU(6) theory [1] lends support to the existence of those fundamental objects. It is embarassing because this is basically a non-relativistic and static theory, and we do not know exactly how this can cover the realm of high energy relativistic phenomena.

Putting aside those theoretical difficulties mainly associated with relativity, let us make the working hypothesis that there are fundamental objects which are heavy ($\gg 1$ GeV), though not necessarily stable, and that inside each baryon or meson they are combined with a large binding energy, yet moving with non-relativistic velocities. Though this might look like a contradiction, at least it does not violate the uncertainty principle in non-relativistic quantum mechanics since the range of the binding forces ($10^{-14}-10^{-13}$ cm) is large compared

Reprinted from Preludes in Theoretical Physics, eds. A. De-Shalit, H. Feshbach, L. van Hove, © 1966 North-Holland, pp. 133–142.

to the Compton wave lengths of those constituents, and the strength of the forces can be arbitrarily adjusted. In other words, we have a model very similar to the atomic nuclei except for large binding energies. Theoretical justification of such a hypothesis must await future investigation.

In a previous article [2], we have put forward such a model with the following characteristic features.

1) There exist two fundamental fermion triplets t_1 and t_2 with charge assignments (1, 0, 0) and (0, -1, -1) for their three members. The baryons have the structure $\sim t_1 t_1 t_2$, and the mesons $\sim a t_1 \bar{t}_1 + b t_2 \bar{t}_2$.

2) To t_1 and t_2 are assigned "charm" charge $C = +1$ and -2 respectively. Thus the baryons and mesons (zero triality states) have $C = 0$. The primary binding forces acting on them are proportional to C. Let us imagine these forces to be mediated by a field (C-field). The resulting Coulomb-like energy though probably of finite range, then stabilizes the $C = 0$ ("uncharmed") systems against the $C \neq 0$ ("charmed") states, such as the triplets themselves.

3) The SU(6) symmetry can be brought in, with the Pauli principle taken into account, since the constituent particles are non-relativistic. In another paper, we also considered a three-triplet model, in which t_1, t_2 and t_3 have charge assignments (1, 0, 0), (1, 0, 0) and (0, -1, -1) respectively. This has the advantage that the baryon states (the 56-dimensional representation of SU(6)) may be realized with s-state triplets as $\sim t_1 t_2 t_3$.

The reasoning that has gone into the above stability problem is similar to the one used in nuclear physics in deriving the semi-empirical formula of Weizsäcker. The purpose of the present paper is to put this idea into a more precise form, even though the outcome should still be called at best semi-quantitative.

2.

Let us first consider states composed of an arbitrary number of t_1 and t_2, but without antiparticles $\bar{t}_1$ and $\bar{t}_2$. Their masses are M_1 and M_2, respectively, and the "charm" numbers 1 and -2, as was mentioned already. The pairwise interaction energy through the C-field will depend on the spatial configurations of the particles, but we will represent it, in the first approximation, by a constant V_c, as long as the size of the system is comparable with the range of the force. If the number of t_1's and t_2's are n_1 and n_2, respectively, the total energy of the system is

$$\begin{aligned} E(n_1, n_2) &= M_1 n_1 + M_2 n_2 + \\ &\quad + V_c \tfrac{1}{2} n_1(n_1 - 1) + 4V_c \tfrac{1}{2} n_2(n_2 - 1) - 2V_c n_1 n_2 \\ &= M_1 n_1 + M_2 n_2 + \tfrac{1}{2} V_c(n_1 - 2n_2)^2 - \tfrac{1}{2} V_c(n_1 + 4n_2) \qquad (1) \\ &= (M_1 - \tfrac{1}{2} V_c) n_1 + (M_2 - 2V_c) n_2 + \tfrac{1}{2} V_c C^2, \\ C &= n_1 - 2n_2. \end{aligned}$$

As expected, the leading quadratic term depends only on the total charm C. If V_c is sufficiently large, this will favor $C = 0$ as the lowest states, which means $n_1 = 2n_2$. Restricting ourselves to $C = 0$ states now, the remaining terms are linear in n_1 and n_2, implying a saturation property. With $n_1 = 2n_2$, we have

$$E(2n_2, n_2) = (2M_1 + M_2 - 3V_c) n_2. \qquad (2)$$

From the physical requirement that this increases with n_2 and that the baryon ($n_2 = 1$) be lighter than the triplets, we further need

$$M_1, M_2 > 2M_1 + M_2 - 3V_c > 0. \qquad (3)$$

Thus the energy surface in the $n_1 - n_2$ plane has a valley running along the line $C = n_1 - 2n_2 = 0$, and its level rises linearly with increasing coordinates. However, it will be further necessary to make sure that the $C = 0$ states are actually lower than their neighbors even for small n's. Namely

$$\begin{aligned} E(2n_2 \pm 1, n_2) &> E(2n_2, n_2), \\ E(2n_2, n_2 \pm 1) &> E(2n_2, n_2). \end{aligned} \qquad (4)$$

This gives two more conditions

$$V_c - M_1 > 0, \qquad 4V_c - M_2 > 0. \qquad (5)$$

Combining Eqs. (3) and (5), we obtain

$$3V_c - 2M_1 > M_2 - M_1 > 3(V_c - M_1) > 0. \qquad (6)$$

The second triplet, therefore, must be heavier than the first, but not too much heavier. This is because we have to maintain a balance between the energy due to rest masses and that due to interaction.

Eq. (1) may be expressed in terms of C and the baryon number B if we make an appropriate assignment: $B = x$ for t_1 and $B = y$ for t_2. Since the baryon $\sim t_1 t_1 t_2$ has $B = 1$, we require $2x+y = 1$. Possible choices given in ref. [2] are

$$(x, y) = (\tfrac{1}{3}, \tfrac{1}{3}) \quad \text{or} \quad (0, 1) \quad \text{or} \quad (1, -1). \tag{7}$$

The numbers n_1 and n_2 may be then expressed in terms of C and B as

$$\begin{aligned} n_1 &= 2B+yC \\ n_2 &= B-xC \end{aligned} \tag{8}$$

and thus

$$\begin{aligned} E(B, C) = \tfrac{1}{2}V_c C^2 + (2M_1+M_2-3V_c)B \\ +[(M_1-\tfrac{1}{2}V_c)-(2M_1+M_2-3V_c)x]C. \end{aligned} \tag{9}$$

At this point we should add a reservation that the linear terms in the above mass formula are not as meaningful as the leading quadratic terms since the effects depending on spatial configurations, such as those due to the finite range character of the C-field and the exchange energy, can be of the same order as the former.

3.

In order to consider the meson states, we will next bring in antiparticles as well in the picture. We make the basic assumption that a system consists of definite numbers of $n_1, \bar{n}_1, n_2, \bar{n}_2$ of $t_1, \bar{t}_1, t_2$ and $\bar{t}_2$. This means that *we regard pair creation and annihilation as forbidden processes*, which is consistent with our basic non-relativistic approach.

The formula corresponding to Eq. (1) becomes

$$\begin{aligned} E(n_1, \bar{n}_1, n_2, \bar{n}_2) = (M_1-\tfrac{1}{2}V_c)(n_1+\bar{n}_1)+(M_2-2V_c)(n_2+\bar{n}_2)+ \\ +\tfrac{1}{2}V_c C^2, \end{aligned} \tag{10}$$

$$C = n_1-\bar{n}_1-2(n_2-\bar{n}_2).$$

The requirement that $E > 0$ demands

$$M_1 - \tfrac{1}{2}V_c > 0, \qquad M_2 - 2V_c > 0 \tag{11}$$

in contrast to Eq. (5), which was derived for the special case $\bar{n}_1 = \bar{n}_2 = 0$. We find, together with Eqs. (3) and (5),

$$M_1 > V_c - M_1 > M_2 - 2V_c > 0 \tag{12}$$

which replaces Eq. (6).

We will now relate the constants M_1, M_2 and V_c to the baryon $(t_1 t_1 t_2)$ and meson $(t_1 \bar{t}_1$ and $t_2 \bar{t}_2)$ masses m, μ_1 and μ_2:

$$\begin{aligned} m &= 2M_1 + M_2 - 3V_c, \\ \mu_1 &= 2M_1 - V_c, \\ \mu_2 &= 2M_2 - 4V_c, \end{aligned} \tag{13}$$

from which we obtain an identity

$$2\mu_1 + \mu_2 = 2m. \tag{14}$$

Because of this, we cannot determine the three unknowns M_1, M_2, V_c uniquely. Instead, we can express Eq. (10) in terms of μ_1 and μ_2:

$$E(n_1, \bar{n}_1, n_2, \bar{n}_2) = \tfrac{1}{2}\mu_1(n_1 + \bar{n}_1) + \tfrac{1}{2}\mu_2(n_2 + \bar{n}_2) + \tfrac{1}{2}V_c C^2. \tag{15}$$

Turning to the relation (14), we put $m \sim 1.2$ GeV, $\mu_1 \sim 600$ MeV $= \tfrac{1}{2}m$ corresponding to the average baryon and meson masses, and *predict* a value

$$\mu_2 \sim m \sim 2\mu_1 \tag{16}$$

for the second meson. This is not an unreasonable value in view of the fact that a large number of unidentified meson resonances seem to exist in this energy range. Eq. (15) reduces then to the simple form

$$E(n_1, \bar{n}_1, n_2, \bar{n}_2) = \tfrac{1}{2}\mu_1[n_1 + \bar{n}_1 + 2(n_2 + \bar{n}_2)] + \tfrac{1}{2}V_c^2. \tag{17}$$

It is rather surprising that such a naive picture as ours can yield non-trivial and qualitatively reasonable results.

By way of a remark, we note from Eq. (13) that

$$\begin{aligned} M_1 &= \tfrac{1}{2}V_c + \tfrac{1}{2}\mu_1 \sim \tfrac{1}{2}V_c, \\ M_2 &= 2V_c + \tfrac{1}{2}\mu_2 \sim 2V_c \sim 4M_1 \end{aligned} \tag{18}$$

since $V_c \gg \mu_1, \mu_2$ by assumption. Interestingly enough, the above relation admits the interpretation that the mass of each triplet is made up of a self-energy due to the C-field plus a small "bare mass" $\frac{1}{2}\mu$.

4.

We will now turn to the three-triplet model [3] proposed as an alternative to the two-triplet model. The three triplets t_1, t_2 and t_3 altogether contain nine fermions $T_{i\alpha}$, $i, \alpha = 1, 2, 3$, where the index i distinguishes different triplets, and α the different members of a triplet. Two different SU(3) operations, called SU(3)′ and SU(3)″, are introduced, acting respectively on α and i, and in these spaces $T_{i\alpha}$ behave as a representation (3, 3*). The electric charge is assigned to each particle according to

$$Q = I_3' + \tfrac{1}{2}Y' + I_3'' + \tfrac{1}{2}Y'' \tag{19}$$

which takes integral values. In fact both $T_{1\alpha}$ and $T_{2\alpha}$ have the assignment (1, 0, 0), and $T_{3\alpha}$ have (0, −1, −1), exactly like t_1 and t_2 of the previous two-triplet model.

An important difference from the two-triplet case is that instead of the charm gauge group $U(1)$, we have the group SU(3)″. The charm gauge field C must then be replaced by an octet of gauge fields G_μ, $\mu = 1, \ldots, 8$, coupled to the infinitesimal SU(3)″ generators (currents) λ_μ'' of the triplets, with a strength g. For a system containing altogether N particles, the exchange of such fields between a pair then results in an interaction energy

$$V_G = +g^2 \sum_{n>m} \lambda_\mu''^{(n)} \cdot \lambda_\mu''^{(m)} = \tfrac{1}{2}g^2 \Big[\sum_{n=1}^{N} \lambda_\mu''^{(n)}\Big]\Big[\sum_{m=1}^{N} \lambda_\mu''^{(m)}\Big] - \tfrac{1}{2}g^2 \sum_{n=1}^{N} \lambda_\mu''^{(n)} \lambda_\mu''^{(n)}$$
$$= \tfrac{1}{2}g^2 [C_2 - NC_{20}], \tag{20}$$

where $\lambda_\mu^{(n)}$ refers to the n-th particle, C_2 is the quadratic Casimir operator of SU(3), and $C_{20} = 4/3$ is its value for a triplet representation $D(1, 0)$ or $D(0, 1)$. In general C_2 is given by

$$C_2(l_1, l_2) = \tfrac{1}{3}(l_1^2 + l_1 l_2 + l_2^2) + (l_1 + l_2) \tag{21}$$

for a representation $D(l_1, l_2)$.

Note that the only dependence on the total number N of constituents appears in the second term of Eq. (17).

We add to V_G the rest masses (M = common mass), and obtain the total energy

$$E = (M - \tfrac{1}{2}C_{20}g^2)N + \tfrac{1}{2}g^2C_2. \tag{22}$$

Bound states are characterized by $V_G < 0$, and the low lying states by the smallest value of C_2, namely $C_2 = 0$ for the singlet $D(0, 0)$. For the latter, E is simply proportional to the total number N of constituents, starting with the meson $(N = 2) \sim t_1\bar{t}_1 + t_2\bar{t}_2 + t_3\bar{t}_3$ and the baryon $(N = 3) \sim t_1t_2t_3$ (antisymmetric combination). Their masses are thus related by

$$\mu = 2(M - \tfrac{1}{2}C_{20}g^2) = \tfrac{2}{3}m, \tag{23}$$

and Eq. (22) becomes

$$E = \tfrac{1}{2}\mu N + \tfrac{1}{2}g^2C_2. \tag{24}$$

These are to be compared with Eqs. (14) and (15). Because of the high symmetry among the three triplets, we have found only one set of mesons with $N = 2$. In any case, the energy is simply proportional to the total number of constituents as long as $C_2 = 0$, as if it were made up of non-interacting basic units of mass $\tfrac{1}{2}\mu$.

5.

Having disposed of the gross mass spectrum of many-triplet compound systems, we now turn our attention to the "fine structure" of low lying states, which in our view comprise all the mesons and baryon resonances known so far. In all probability, however, our crude qualitative arguments aie not really satisfactory for discussing these details. We will therefore restrict ourselves to general remarks only.

Because of our basic assumptions about the superstrong interactions and the static behaviour of particles, the dynamics we have been dealing with so far does not depend on the spin and the SU(3) spin variables, therefore the system possesses the symmetry of superstrong interactions, the SU(6) symmetry of combined spin and SU(3) spin, and the symmetry of orbital angular momentum. The overall Pauli principle imposes constraints among these symmetries, and thereby single out certain SU(6) and orbital states for the lowest configuration with respect to the superstrong interaction. The general classification of these states can be done as in the case of nuclear and atomic physics, but this will be beyond the scope of the present paper.

In the three-triplet model, however, the problem is relatively simple if we take only s-state triplets. The low lying three particle configuration is a SU(3)″ singlet, so the baryon must go into a complete symmetric SU(6) representation 56. No other states are possible without changing the spatial configuration, but this will cause some change in the superstrong interaction. For the mesons, we obviously obtain 36 = 35+1 SU(6) states which are degenerate. These results are in accordance with those of the original SU(6) theory, as well as its "relativistic" version.

We must next discuss the two additional effects which do exist and tend to upset the symmetries. One arises from the internal motion of particles, and the other from the presence of virtual mesons. Contrary to the prevalent view, we regard the mesons as perturbing forces rather than the decisive factors in the physics of hadrons. Since the strong interactions are then merely first forbidden processes, so to speak, the meson and baryon resonances are really bound states decaying via violation of superstrong interaction symmetry. Nevertheless, these secondary effects can affect, and may even decide, the "fine structure" of low lying states. Perhaps we may compare the situation to the electronic levels of an atom where the main spectrum is determined by the static Coulomb force, and both the fine structure and the photon emission processes are higher order effects. In this sense, we do not necessarily find a contradiction between the present approach and the conventional strong interaction theory as far as the low lying states are concerned.

The reason we consider the strong interaction as generally symmetry breaking is that the virtual exchange of 36 virtual mesons do not possess an SU(6) symmetric form. An ideal SU(6) symmetric interaction would involve the 35 generators χ_μ as in Eq. (20):

$$V \sim \pm g^2 \sum_{n>m} \chi_\mu^{(n)} \chi_\mu^{(m)} = \pm \frac{g^2}{8} \sum_{n>m} [\tfrac{2}{3}\sigma_i^{(n)}\sigma_i^{(m)} + \lambda_\alpha^{(n)}\lambda_\alpha^{(m)} + \sigma_i^{(n)}\sigma_i^{(m)}\lambda_\alpha^{(n)}\lambda_\alpha^{(m)}]. \tag{25}$$

Viewed as a static force, this requires an exchange of 35 scalar and axial vector mesons.

if the relative signs of the various terms are to be correctly maintained for both particle-particle and particle-antiparticle interactions. [For processes involving meson-baryon scattering, however, Capps [4] and Belinfante and Cutkosky [5] have shown their compatibility with SU(6).]

Next consider the effect of the internal motion. This disturbs the basic symmetry in two senses. It mixes the Dirac spinor components, introducing corrections to the static superstrong forces. Further it simply adds the kinetic energy of orbital motion to the system. As far as the symmetry is concerned, these perturbations act like adding a neutral singlet meson with a suitable spin-parity. Its order of magnitude will depend on the internal velocity v of the particles, which should be of the order $1/MR$ where R is the size of the system. If we take this correction to be of the order $Mv^2 \sim 100$ MeV, and $R \sim 1/M$, we obtain the estimation

$$M \sim 10m, \qquad \frac{v}{c} \sim \frac{1}{10}$$

as we did before [2].

6.

Finally we would like to comment on some obvious difficulties and intriguing problems concerning our model of the subnuclear structure of hadrons.

a) What is the origin of superstrong interactions?

Are these another kind of vector fields or something entirely new? If they are ordinary fields, their range must be at least of the order of the baryon size, and moreover sufficiently smooth and well-behaved in order to keep the kinetic energy small. It is conceivable that no single or a relatively few well defined meson states are responsible for this. A direct confirmation of such interactions would be difficult.

b) The magnetic moments of baryons, for example, agree closely with the SU(6) symmetry, yet obviously the bulk of contributions come from the meson cloud. This means that regardless of whether the meson cloud obeys SU(6) symmetry or not, the baryon should not be considered as composed of three bare triplets without structure. How, then, can we justify our picture that each system, including the mesons, is composed of a definite number of triplets? The answer to this probably should be that the quantities like charm are at any instant well localized at a definite number of centers in space, and these centers are accompanied by large concentrations of energy, moving with slow velocities; whereas the quantities like ordinary charge are more uniformly spread out and carried by faster moving matter. In order to test such a picture experimentally, we would have to use some phenomena which depend on the energy distribution, the correlation functions of charges and energies at different points, the internal velocity of particles, etc.

c) The notion that decays and resonances are actually forbidden processes was first recognized as a surprising paradox in the process of adapting SU(6) to relativity. In our view, this is not only natural, but also simplifies the whole picture. We should be able to discuss the classes of first forbidden, second forbidden, etc. transitions, and they will be accessible to experimental test [6]. For this we should look especially for small, inconspicuous bumps in cross sections, many particle decay modes, and relatively rare events.

d) It has been widely speculated that an axial vector current conservation as relativistic chival symmetry has physical significance. If this is actually the case, it is probably beyond the capacity of our extreme static approach, since we have first to explain away the large masses of triplets, even though we can formally apply group theoretical arguments and the Goldberger-Treiman type relations to individual problems.

REFERENCES

1) F. Gürsey and L. A. Radicati, Phys. Rev. Letters, **13** (1964) 173;
A. Pais, Phys. Rev. Letters, **13** (1964) 175;
B. Sakita, Phys. Rev. **136** (1964) B1765.
2) Y. Nambu, Proc. of the Second Coral Gables Conference on Symmetry Principles at High Energy, University of Miami, January, 1965.
3) M. Y. Han and Y. Nambu, Syracuse University preprint 1206-SU-31.
4) R. Capps, Phys. Rev. Letters, **14** (1965) 31.
5) J. G. Belinfante and R. E. Cutkosky, Phys. Rev. Letters, **14** (1965) 33.
6) A more detailed study of this problem will be done elsewhere.

Chapter 7: Nambu and the String

From June 1969 to August 1970, three works of art established Nambu as one of the main architects of String Theory:

"Quark model and the Factorization of the Veneziano Amplitude.[96]*"*

"Dual Model of Hadrons.[97]*"*

"Duality and Hadrodynamics.[98]*"*

Background

In the 1960s Strong Interactions remained a puzzle — perturbation theory did not apply and alternative approaches had to be considered. A promising venue, advocated by Geoffrey Chew (1924–2019) and Steve Frautschi[99] among others, was to constrain the S-matrix that describes the strong interactions between mesons and hadrons; it had to satisfy Lorentz invariance, causality, locality, crossing symmetry, and unitarity. It was hoped that these would be enough to determine the strongly interacting amplitudes, or even "bootstrap" some particles from a smaller set. An additional ingredient was Tullio Regge's[100] (1931–2014) description of amplitudes as functions of complex angular momentum, resulting in the high energy behavior of amplitudes in terms of the exchange of families (Regge trajectories) of resonances in a cross-channel. These analyticity requirements resulted in "superconvergence relations" of the form,

$$\int d\nu \, Im \, A(s,t) = 0,$$

valid under broad conditions.

[96]R. Chand (ed.), *Int. Conf. Symmetries and Quark Models*, Wayne State Univ., June 1969, 1970 Gordon and Breach.
[97]American Physical Society Meeting, January 1970, *EFI* 70-07.
[98]High Energy Symposium, Copenhagen, August 1970, *EFI* 69-64.
[99]op. cit.
[100]T. Regge, *Il Nuovo Cimento* **10**, vol. 14, (1959).

In 1967, R. Dolen, D. Horn (1937–), and C. Schmid (1937–) considered[101] $\pi - N$ scattering as a laboratory for testing superconvergence, where the s-channel integral was summed over two ingredients. There, the amplitude is dominated at lower energies by the N (Δ) fermion resonances while its asymptotic behavior is expressed from the Regge behavior, dominated by the ρ meson trajectory in the t-channel. Their careful analysis found these two contributions cancelling, confirming superconvergence.

Furthermore, Regge's asymptotic behavior could be analytically continued to the lower energy region where the Δ resonances dominate, and they discovered the unexpected: the continued Regge contribution traces out the average over the resonances! Since Regge's contribution is dominated by the ρ meson in t-channel, this implies a relation to the fermion hadron resonances, a relation they dubbed "duality."

Marco Ademollo (1936–2022), Hector Rubinstein (1933–2009), Gabriele Veneziano (1942–), and Miguel Virasoro (1940–2021)[102] suggested a dual bootstrap for Regge trajectories:

"the amplitude in the resonance region in the direct channel can be obtained by use of crossing as the analytic continuation of the Regge amplitude describing scattering at high energy in the crossed channel."

They proceeded to construct the simplest amplitude to study Dolen–Horn–Schmid duality. Their amplitude of choice was entirely bosonic,

$$\pi + \pi \longrightarrow \pi + \omega,$$

where ω is the vector boson predicted by Nambu a decade earlier. Its S-matrix, described by a single amplitude $A(\nu, t)$ with $\nu = (s - u)/4$, was found to have a simple Regge asymptotic behavior,

$$A(\nu, t) \longrightarrow \beta(t)\xi(\alpha)\left[\frac{\nu}{\nu_1}\right]^{\alpha(t)-1}$$

where $\alpha(s)$ is a linear Regge trajectory, $\beta(t) = \bar{\beta}(t)/\Gamma(\alpha(t))$ the residue, and $\xi(\alpha) = [1 - e^{-i\pi\alpha(t)}]/\sin \pi\alpha(t)$.

On his way to his next postdoc position at CERN, the sea air inspired Veneziano[103] to seek a crossing symmetric version of this formula. He wrote it in a form

$$A(s, t) \longrightarrow \frac{\bar{\beta}}{\pi}\Gamma\big[1 - \alpha(t)\big]\big[-\alpha(s)\big]^{\alpha(t)-1} + (s \leftrightarrow u),$$

that highlights the lack of crossing symmetry: the $\Gamma(1 - \alpha(s))$ produces poles in the s-channel, which are not duplicated in the t-channel, and the asymptotic form

[101] R. Dolen, D. Horn, and C. Schmid, *Phys. Rev. Lett.* **19**, 402 (1967); *Phys. Rev.* **166**, 1768 (1968).

[102] M. Ademollo, H. R. Rubinstein, G. Veneziano, and M. A. Virasoro, *Phys. Rev.* **176**, 1904 (1968).

[103] G. Veneziano, *Nuovo Cimento* **LVII-A**, 190 (1968), submitted July 29.

$(-\alpha(s))^{\alpha(t)-1}$ singles out the s-channel. His brilliant step was to replace the last term by $\Gamma(1-\alpha(t))$ divided by $\Gamma(2-\alpha(s)-\alpha(t))$ to reproduce the same asymptotic behavior and the same poles in the t- and s-channels. The result was the well-known (to mathematicians) beta function,

$$B(s,t) = \frac{\Gamma(1-\alpha(s))\Gamma(1-\alpha(t))}{\Gamma(2-\alpha(s)-\alpha(t))},$$

resulting in a unique crossing-symmetric amplitude,

$$A(s,t,u) = \frac{\bar{\beta}}{\pi}\Big[B(1-\alpha(s),1-\alpha(t)) + B(1-\alpha(t),1-\alpha(u)) + B(1-\alpha(s),1-\alpha(u))\Big],$$

valid for all s, t and u. Veneziano's seminal paper was the first step to the Dual Resonance Model, followed by Strings, Superstrings, and Supersymmetry.

Veneziano's four-point amplitude was soon generalized to five legs by Korkut Bardakçi (1936–) and Henri Ruegg[104] (1930–2020) and also by Virasoro;[105] by December 1968, it had been derived for any number of legs.[106]

Extracting physics from the general amplitude proceeded at a frenetic pace. In April 1969, Keiji Kikkawa (1935–2013), Bunji Sakita (1930–2002) and Miguel Virasoro, interpreted the dual amplitudes as tree-level Born approximations[107] of a more complete theory. The same month, two groups, Sergio Fubini (1928–2005) and Veneziano, as well as Korkut Bardakçi and Stanley Mandelstam (1928–2016) put forward compelling arguments[108] for the full factorization of the amplitudes, and residues with imaginary couplings, although with possible cancellations analogous to the Ward-like identity of QED.

• In June 1969, Nambu was invited to the "International Conference on Symmetries and Quark Models" at Wayne State University, where he presented *"Quark Model and the Factorization of the Veneziano Amplitude"*.

The topic was not yet at the center of attention of the community; of the twenty-one talks delivered at Wayne State, Nambu's was the only one on the Veneziano amplitudes. As usual, he was well ahead of the field.

Nambu noted the compatibility between duality and the quark picture of mesons and, especially by the interpretation of duality of quark diagrams advocated by Haim Harari (1940–) and Jonathan Rosner (1941–).[109]

---

[104] K. Bardakçi, H. Ruegg, *Phys. Lett.* **28B**, 342 (1968).

[105] M. A. Virasoro, *Phys. Rev. Lett.* **22**, 37 (1969).

[106] H. M. Chan and T. S. Tsou, *Phys. Lett.* **28B**, 485 (1969); C. Goebel and B. Sakita, *Phys. Rev. Lett.* **22**, 257 (1969); Z. Koba and H. B. Nielsen, *Nucl. Phys.* **B10**, 633 (1969).

[107] K. Kikkawa, B. Sakita, M. A. Virasoro, *Phys. Rev.* **184**, 1701 (1969).

[108] S. Fubini and G. Veneziano, *Nuovo Cimento* **LXIV A**, 811 (1969); K. Bardakçi and S. Mandelstam, *Phys. Rev.* **D184**, 1640 (1969).

[109] H. Harari, *Phys. Rev. Lett.* **22**, 562 (1969); J. Rosner, *Phys. Rev. Lett.* **22**, 689 (1969).

It contained two important results:

(i) A general proof of factorization of the Veneziano amplitudes by introducing an infinite number of space-time vector harmonic oscillators. Similar conclusions were obtained by others.[110]

(ii) With his instinct for the fundamental, Nambu identified the oscillators as vibration modes of a quantum string (or a cavity resonator), the structure behind Veneziano's "Dual Resonance Model" amplitudes.

By December 1969, Virasoro had derived an infinite number of subsidiary conditions, one for each oscillator. They were compatible with the equations of motion only for an unphysical value of the intercept, $\alpha(0) = 1$, which implied a spectrum with a massless vector boson and a tachyon. Virasoro did not succeed in getting around this restriction, and eventually published,[111] without writing the famous algebra that bears his name.

• In January 1970, at the American Physical Society Meeting, Nambu expanded on his previous talk at Wayne State, and of course, introduced new concepts:

(i) Mesons are open strings with quarks and antiquarks at their end points. They are $SU(3)''$-singlets under the three-triplet model M. Y. Han and Nambu had introduced[112] in 1965. The three quarks are whimsically given names: D, N, and A, so that mesons $\sim (D\bar{D} + N\bar{N} + A\bar{A})$ and baryons $\sim D\,N\,A$, are all singlets.

(ii) Meson interactions proceed by the creation of a quark–antiquark pair on the string. Veneziano tree amplitudes are generated from a pair at the end of the string.

(iii) The string, tracing a two-dimensional world sheet, naturally displays the infinite-dimensional symmetry of the Laplace equation. Nambu suggested it enforced duality and generated Virasoro's infinite number of conditions; Nambu's footnote explicitly displayed the "Virasoro algebra"!

• In August 1970, Nambu was invited to lecture at the High Energy Symposium at Copenhagen.

A week or so before the meeting, Nambu and his son John were driving back to Chicago. Near the Nevada–Utah border they were passed by an obnoxious driver. At John's urging, the Professor gunned his older car trying to pass him, and ... broke his car's transmission. They spent a week in Wendover waiting for parts; by

[110] S. Fubini, D. Gordon and G. Veneziano, *Phys. Lett.* **29B**, 679 (1969); L. Susskind, *Phys. Rev. Lett.* **23**, 545 (1969); Z. Koba and H. B. Nielsen, *Nucl. Phys.* **B10**, 633 (1969); *Nucl. Phys.* **B12**, 517 (1969).
[111] M. A. Virasoro, *Phys. Rev.* **D1**, 2933 (1970).
[112] op. cit.

then it was too late to attend the Symposium in Copenhagen. Undelivered, this lecture exists only as a preprint, reprinted in the "Yellow Book"[113] by Tohru Eguchi (1948–2019) and Kazuhiko Nishijima (1926–2009).

Entitled *"Duality and Hadrodynamics"*, it summarized the new insights he had presented at the Wayne State and APS meetings:

(a) Factorization,
(b) Strings,
(c) Han-Nambu quarks at the end of open strings,
(d) Quark-antiquark pairs as seeds for string interactions.

He proposed a geometrical action for the relativistic string, as the area was spanned by its evolving world-sheet — independently later derived by Tetsuo Goto[114] (1931–1982), known today as the Nambu–Goto action.

Nambu mused on a possible link with the Ising model with *"its glorious Onsager solution"*: the dual string as an infinite array of aligned quark–antiquark pairs. Duality would require the same array in the "euclidean time" direction, resulting in a square polarizable medium with "$SU(3)' \times SU(3)''$" flavor and color charges on each site.

He reiterated that perhaps dual strings are Dirac strings with monopoles and antimonopoles attached at their extremities.

Nambu's Three Presentations

Wayne State Talk

Nambu began by pointing out areas of compatibility between the quark picture of mesons and the duality of Dolen, Horn, and Schmid (DHS).

One is dynamical — mesons in the $q\bar{q}$ configuration are described by wavefunctions that depend on the relative coordinates of the quarks, with not only orbital but also radial excitations with Regge trajectories *"forming $U(6)$ 36-plets"*. With a complex mesonic structure beyond the orbital excitations, it is consistent with DHS duality with *its elegant mathematical realization embodied in the Veneziano amplitude* and its infinite number of resonances.

The second hint of a link was the Harari–Rosner interpretation of duality in terms of planar quark diagrams.

Nambu first wanted to find *"a self-consistent, factorizable set of Veneziano-type amplitudes for all hadronic processes"*.

He started from the generalized form of the N-point Veneziano amplitude, with the external momenta $(p_1, p_2, \ldots p_N)$ lined up in a multiperipheral configuration.

[113] op. cit.
[114] T. Goto, *Prog. Theor. Phys.* **46**, 1560 (1971).

It is an integral over $N-1$ parameters with products of external momenta in the exponent terms of the form $(1-x)$, $(1-y)$, $(1-xy)$, etc. where $x, y, \ldots$ are parameters to be integrated over.

Factorization would be proved if the generalized amplitude can be written as multiple sums over intermediate states like,

$$\sum \langle N | \cdots | l \rangle \langle l | \Gamma | m \rangle \langle m | \Gamma | k \rangle \langle k | \Gamma | 1 \rangle.$$

A step to achieve the result is to examine the original Veneziano four-point function,

$$\int_0^1 dx (1-x)^{-\alpha(t)} x^{-\alpha(s)}$$

where $s = (p_1 + p_2)^2$, $t = (p_2 + p_3)^2$ are the Mandelstam variables. The t-channel poles are hiding in

$$(1-x)^{-2p_2 \cdot p_3 - c_2} = \exp\left[(2p_2 \cdot p_3 + c_2) \sum_{r=1}^{\infty} \frac{x^r}{r}\right],$$

where c_2 is the intercept of the Regge trajectory. Factorization requires the coefficient of r^n to be of the form,

$$\sum_k \langle 4 | \Gamma(p_3) | k \rangle \langle k | \Gamma(p_2) | 1 \rangle.$$

In order to show that, Nambu introduced an infinite number of vector creation and annihilation operators, $a_\beta^{(r)}$, $a_\gamma^{(r)\dagger}$, $r = 1, 2 \ldots$, where $\beta, \gamma = 1, 2, 3, 4, 5$ are the four space-time indices; the fifth is a device to account for different Regge intercepts as needed. They satisfy the usual commutator equations,

$$\left[a_\beta^{(r)}, a_\gamma^{(s)\dagger}\right] = \delta^{rs} g_{\beta\gamma}.$$

Since their commutators are c-numbers, the well known formula $e^A e^B = e^B e^A e^{[A,B]}$, applies and,

$$\langle 0 | e^{ik' \cdot a/\eta} e^{ik \cdot a^\dagger \xi} | 0 \rangle = e^{k \cdot k' \xi/\eta}.$$

Applying to the Veneziano formula, it yielded the vertex at level $N = \sum_r r a^{(r)\dagger} \cdot a^{(r)}$,

$$\langle N | \Gamma(k) | 0 \rangle = \left\langle N \right| \exp\left[i \sum_r \sqrt{\frac{2}{r}} k \cdot a^{(r)\dagger}\right] \left| 0 \right\rangle.$$

This is followed by a technical but straightforward exposition which proves for the first time the full factorization of the N-point Veneziano amplitude.

It was Nambu at his technical best, but what set him apart was his instinct for an underlying structure. He assembled the ladder operators into a periodic coordinate field in a circle, $\varphi_\mu(\xi) = \varphi_\mu(\xi + 2\pi)$, and momentum $\pi_\mu(\xi)$,

$$\varphi_\mu(\xi) = \sum_{r=1}^{\infty} \frac{1}{\sqrt{2r}} [a_\mu^{(r)} + a_\mu^{(r)\dagger}] \cos(r\xi), \qquad \pi_\mu(\xi) = i \sum_{r=1}^{\infty} \sqrt{\frac{r}{2}} [a_\mu^{(r)} - a_\mu^{(r)\dagger}] \cos(r\xi)$$

and expressed N as

$$N = -\frac{1}{\pi}\int_0^{2\pi} :(\partial_\xi \varphi(\xi)\cdot\partial_\xi \varphi(\xi) + \pi(\xi)\cdot\pi(\xi)): d\xi.$$

To Nambu, it seemed that,

"the internal energy of a meson is analogous to that of a quantized string of finite length (or a cavity resonator for that matter) whose displacements are described by the field $\varphi_\mu(\xi)$."

Moreover, the vertex operator is the normal-ordered operator,

$$\Gamma =: e^{2ik\cdot\varphi(0)} :,$$

which looks like the form factor of the naive quark model in which one of the quarks interacts with an external field carrying momentum k.

From the Harari–Rosner planar graphs, he concluded that the interaction looked like three resonances merging in a non-local interaction.

Throughout the talk, he acknowledged unsolved problems, ghosts from the vector operators, how to avoid multiple counting of diagrams, and the actual interaction of the strings. This will come in his next talk.

American Physical Society Talk

Delivered in January 1970, in front of a mixed audience at the Chicago American Physical Society winter meeting, this talk was more conceptual than technical.

Nambu integrated his string picture of the Veneziano model with the three-triplet model he had devised in collaboration[115] with Moo-Young Han in 1965: quarks (antiquarks) at the ends of open strings. He also suggested another possibility, magnetic monopoles (antimonopoles) at the end of open Dirac strings.

An introduction explained the excitement generated by Dolen, Horn, and Schmid duality, and its mathematical realization, the Veneziano model. It pointed to new ideas about the nature of hadrons and their interactions, with linear Regge trajectories and many resonances yet to be discovered.

"Much of what I am going to say is not original with me, or otherwise have been conceived by others independently."

Factorization of the Veneziano amplitudes had been shown[116] in special cases. Nambu then gave a complete proof at Wayne State, and typicall did not refer to himself.

[115]op. cit.

[116]K. Bardakci and S. Mandelstam, *Phys. Rev.* **184**, 1640 (1968); S. Fubini and G. Veneziano, *Nuovo Cimento* **A64**, 811 (1969).

He proposed a simple pictorial interpretation of hadron dynamics which incorporated the factorization property and the quark diagram selection rules of Harari and Rosner. To implement this idea, Nambu used the Han–Nambu integrally-charged three-triplet quark model. The triplets span the three-dimensional representation of a new $SU(3)''$ (later recognized by others as the color group). Nambu gave the triplets evocative names, D, N, A and antiparticles $\bar{D}$, $\bar{N}$, $\bar{A}$. The observed hadrons are composites,

$$\text{Mesons} \sim (D\bar{D} + N\bar{N} + A\bar{A})/\sqrt{3}; \qquad \text{Baryons} \sim D\,N\,A$$

all singlets under their $SU(3)''$. One of its charges was the "Nambu Charm" he had introduced at the Coral Gables conference. Nambu also liked it because it was proportional to the quark's integral electric charges.

For Nambu, Veneziano amplitudes are generated by the vibrations of the "rubber string" he had proposed at Wayne State. At one end of an open string lives a quark, at the other an antiquark. Flavor $SU(3)$ and spins are carried by quarks while the energy comes from the string itself,

The string sweeps out a two-dimensional world-sheet governed by a wave equation

$$\frac{\partial^2 y_\mu}{\partial \eta^2} - \omega^2 \frac{\partial^2 y_\mu}{\partial \xi^2} = 0, \quad \mu = 0, 1, 2, 3,$$

where η and ξ are time-like and space-like parameters, respectively. ξ runs between 0 and 1 with boundary $\partial y_\mu / \partial \xi = 0$ at the ends.

The energy is the infinite sum of the quadratic contributions with ever increasing frequencies,

$$H_0 = p^2\mu + \omega \sum_{n=1}^{\infty} \sum_{\mu=0}^{3} n a_\mu^{n\dagger} a_\nu^n g^{\mu\nu} = p^2\mu + \omega N,$$

with a finite number of states at each mass, and exponentially-rising degeneracy.

Nambu offered a qualitative description of the interactions:

a string may be cut in two by creating a pair of "quarks" at the points where it is cut."

Imagining a string stretched between quark and antiquarks yields the duality diagrams of Harari and Rosner: Applied to mesons, the interaction proceeds as,

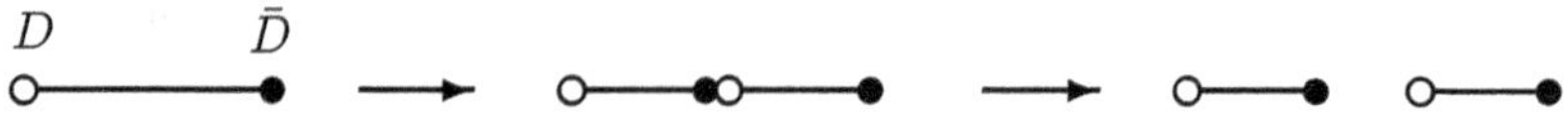

Cutting can be applied to baryons with one simple graphical depiction, but, as he noted in a footnote, it does not produce an $SU(3)''$ singlet; this would require other configurations.

The Hamiltonian that generated the cutting would depend on the overlap between initial and final configurations. Nambu considered a process where a meson string splits into two, and a supposed one is created in its ground state. With the cut at the end of the initial string, it looks as if an external force hits the string at its end. Applying perturbation theory to compute the peripheral tree diagram, Nambu obtained the Veneziano amplitude, but gave no details of his assertion. While admitting that

"... it is not obvious that duality comes out of the rubber string model, at least at first glance. On second thought, however, one might say it is very natural, because one can imagine a two-dimensional world sheet stretched four ways by four external lines..."

he argued for a symmetry between s- and t-channels in this diagram, and found it in the two-dimensional Laplace equation which *"manifestly exhibits the duality properties of Veneziano amplitudes"*, referring to Ziro Koba and Holger Bech Nielsen's[117] *"elegant formulation"*.

He connected this symmetry as a possible solution of ghost states in the spectrum of Veneziano amplitudes. Space-time vector operators generate negative norm states which must be eliminated. This problem had been encountered and solved in QED using $U(1)$ gauge symmetry. Applied to the dual model, Fubini and Veneziano[118] found a similar gauge condition for one oscillator. Virasoro then found an infinite number of gauge conditions, one for each vector oscillator capable of decoupling unwanted states, at the cost of a tachyon and a massless vector state.

Nambu suggested in a footnote that Virasoro's conditions were enabled by the symmetries of a two-dimensional system — the string. It would take several years until Peter Goddard (1945–) and Charles Thorn (1946–) provided a proof,[119] valid only in ten space-time dimensions.

At the very end of his talk, Nambu speculated on a new connection between the dual three-triplet model and Dirac's magnetic monopole. Although very different, one could attach magnetic monopoles and antimonopoles at the end of open strings, where *"...the charm number in our DNA model can be interpreted as the magnetic charge."* He mentioned Julian Schwinger's (1918–1994) dyons[120] which have both electric and magnetic charges as possible ingredients. Their electric dipole moments

[117] Z. Koba and H. B. Nielsen, *Nucl. Phys.* **B12**, 517 (1969).
[118] op. cit.
[119] P. Goddard and C. B. Thorn, *Phys. Lett.* **B40**, 235 (1972).
[120] J. Schwinger, *Science* **165**, 757 (1969); **166**, 690 (1969).

are given by $\sum_n < \sigma_n >< \lambda''_n >$, and vanish since these strings are $SU(3)''$ singlets: his Charm number is the combination $\lambda''_3 + \lambda''_8\sqrt{3}$, one of the two $SU(3)''$ (color) generators.

This conceptual talk had introduced several interpretations for strings and made new connections, a window of Nambu's imagination and instinct for simplicity.

Copenhagen Lecture (undelivered)

I. Introduction

Nambu's stated purpose is to *"...guess the internal structure and dynamics of hadrons which underlie the Veneziano model."* He emphasized its crucial property, the appearance of the same resonances in the factorization of all its amplitudes, and listed some of its predictions and concerns:

- Linearly rising Regge trajectories, and regularly spaced daughters.

- The existence of ghosts in the residues is a serious problem, but the papers[121] of Fubini and Veneziano, and of Virasoro hinted at their possible decouplings.

He whimsically mentioned "en passant" that one may have to live with ghosts, and refers to E. C. G. Sudarshan, Tsung-Dao Lee (1926–) and Gary Feinberg (1933–1992), but did not believe it.

II. Factorized Veneziano Model

Factorization motivated Nambu's string picture. He began with the well-known proper time formalism of a free particle of mass m, with coordinate $x_\mu(\tau)$. Its properties can be derived from two actions,

$$I_{\text{length}} = -m \int d\tau \sqrt{-dx_\mu dx^\mu}, \qquad I = \frac{1}{2} m \int \left(\frac{dx_\mu}{d\tau}\frac{dx^\mu}{d\tau} - 1 \right) d\tau.$$

I_{length} the geometrical length of a world line trajectory. I is its linearized version with two translation invariances, $\tau \to \tau + c$, $x^\mu \to x^\mu + a^\mu$ that lead to constants of the motion, the Hamiltonian and the momenta:

$$H = \frac{1}{2m}(p^\mu p_\mu + m^2), \qquad p^\mu = m\frac{dx^\mu}{d\tau}.$$

Imposing the constraint $\frac{dx_\mu}{d\tau}\frac{dx^\mu}{d\tau} = -1$ normalizes τ to become the proper time, and yields $H = 0$. Transition to quantum mechanics postulates the canonical commutation relations and the Schrödinger equation,

$$[x^\mu, p^\nu] = ig^{\mu\nu}, \qquad i\frac{\partial}{\partial\tau}\Psi = H\Psi,$$

[121] op. cit.

yielding the wave function

$$\Psi(\tau) = \exp\left[-i\frac{(p^2+m^2)\tau}{2m}\right]\Psi(0),$$

and the transition amplitude

$$\big(\Psi(x',\tau),\Psi(x,0)\big) = -\frac{im^2}{4\pi^2}\exp\left[im\tau((x/\tau)^2-1)/2\right].$$

Integration over τ from 0 to ∞, yields the Feynman propagator,

$$\frac{-i}{p^2+m^2-i\epsilon}.$$

Having set the stage for the proper-time description of the elastic string, Nambu imagined it as an infinite number of points, each tracing a world line, a world sheet $x_\mu(\xi,\tau)$ with two coordinates, $\xi, 0 \leq \xi \leq \pi$, and $\tau, -\infty \leq \tau \leq \infty$.

Analogy with the linearized point particle action I suggests the string action,

$$I_{\text{string}} = \frac{1}{4\pi}\int\int\left(\frac{\partial x_\mu}{\partial\tau}\frac{\partial x^\mu}{\partial\tau} - \frac{\partial x_\mu}{\partial\xi}\frac{\partial x^\mu}{\partial\xi}\right)d\xi d\tau,$$

designed to yield the equation of motion,

$$\left(\frac{\partial^2}{\partial\tau^2}-\frac{\partial^2}{\partial\xi^2}\right)x^\mu(\xi,\tau) = 0,$$

with boundary conditions $\partial x^\mu/\partial\xi = 0$ at $\xi = 0, \pi$.

Nambu thought of duality as a symmetry between ξ and τ, except for differences in domain and metric. In computing scattering amplitudes a "euclidean" version $\tau \to i\eta$ brings out duality explicitly.

The same translations of the point particle yield two constants of the motion,

$$H = \int\left[\pi p^\mu p_\mu + \frac{1}{4\pi}\left(\frac{\partial x^\mu}{\partial\xi}\right)\left(\frac{\partial x_\mu}{\partial\xi}\right)\right]d\xi,$$

$$P^\mu = \frac{1}{2\pi}\int d\xi\, p^\mu, \qquad p^\mu = \frac{1}{2\pi}\left(\frac{\partial x^\mu}{\partial\tau}\right).$$

Transition to quantum mechanics is achieved through the normal mode decomposition,

$$x^\mu = x_0^\mu + \sum_{n=1}^{\infty} x_n^\mu \cos n\xi,$$

$$H = p_0^\mu p_{\mu 0} + \frac{1}{2}\sum_{n=1}^{\infty}(p_n^\mu p_{\mu n} + n^2 x_{\mu n}x_n^\mu),$$

with $x_{\mu 0}$ and $p_{\mu 0}$ being the center of mass parameters. Quantization of the normal modes yields an infinite number of states with masses

$$M^2 \sim \sum_{\mu,\, n} n a_{\mu n}^\dagger a_n^\mu + \alpha.$$

His description ended with several remarks:

(1) Whether the system is a rubber string or a rubber band depends on the boundary conditions, and the rubber band has twice as many oscillators;

(2) This I_{string} does not have a geometrical meaning, although it is invariant under the scaling $\tau,\ \xi \to \lambda\tau,\ \lambda\xi$, one aspect of the conformal invariance of the two-dimensional Laplace equation.

The point particle has the length of its world line as its "geometrical" action. It motivated Nambu to seek an analogous geometrical description for the string:

"For curiosity, then, let us try to construct a geometric action as one does in general relativity. Obviously a natural candidate for it is the surface area of the two-dimensional world sheet; ..."

He used the techniques of General Relativity to embed the string in Minkowskian four-space $y_\mu(\xi^0,\xi^1)$, $(\xi^0 \sim \tau,\ \xi^1 \sim \xi)$, with surface element,

$$d\sigma^{\mu\nu} = G^{\mu\nu} d^2\xi, \qquad G^{\mu\nu} = \frac{\partial(y^\mu, y^\nu)}{\partial(\xi^0,\xi^1)}.$$

In terms of the two-dimensional metric and line element in the space spanned by ξ^α,

$$g_{\alpha\beta} = \frac{\partial y_\mu}{\partial \xi_\alpha}\frac{\partial y^\mu}{\partial \xi_\beta}, \qquad ds^2 = g_{\alpha\beta} d\xi^\alpha d\xi^\beta, \qquad (\alpha,\beta = 0,1).$$

He could rewrite the string action in a linearized form,

$$I_{\text{string}} = -\frac{1}{4\pi}\int\int g_{\alpha\beta}\mathring{g}^{\alpha\beta} d^2\xi, \qquad \mathring{g}^{\alpha\beta} = \begin{pmatrix} -1 & 0 \\ 0 & 1 \end{pmatrix}.$$

Nambu then proposed the geometric string action as the area spanned by its world sheet,

$$I_{\text{area}} = \int \left| d\sigma^{\mu\nu} d\sigma_{\mu\nu} \right|^{1/2},$$

with nonlinear equations. Still, it is the simplest, without curvature nor with higher derivative equations. He goes no farther in its analysis. It was left to others to work it out.[122]

From the linearized action Nambu obtained the world sheet energy-momentum in the (τ,ξ) space,

$$T_{\alpha\beta} = \frac{1}{2\pi}\left(g_{\alpha\beta} - \frac{1}{2}\mathring{g}_{\alpha\beta} g_{\gamma\delta}\mathring{g}^{\gamma\delta} \right).$$

The space integrals over a test function $f(\xi)$,

$$\overline{T}_{\alpha\beta}[f] = \int_0^\pi T_{\alpha\beta}(\xi) f(\xi) d\xi$$

[122] J. Goldstone, P. Goddard, C. Rebbi and C. B. Thorn, *Nucl. Phys. B* **56**, 109 (1973).

satisfy a commutator algebra as an integral form of Schwinger's conditions.[123] For test functions $f_n = 1 - e^{-2in\xi}$, $n = 0, 1, 2\ldots$, he defined,

$$L_n^{\pm} = (\overline{T}_{00} \pm \overline{T}_{01})[e^{2in}], \qquad L^{\pm}[f_n] = L_0^{\pm} - L_{-n}^{\pm},$$

and found the commutator algebra,

$$[\, L_n^{\pm}, L_m^{\pm}\,] = 2(n-m)L_{n+m}^{\pm},$$

universally known as the "Virasoro Algebra" (without the c-number). Nambu offered a modest assessment,

"$L^{\pm}[f_n]$ have been found useful in generating various gauge conditions"

Nambu was not unaware of the remaining difficulties: ... *the most serious defect ...is the indefinite metric ...".* There are two possible interpretations:

(1) The wave function of quantum space-time vector harmonic oscillators "explodes" in the time-like direction, with positive mass squared;

(2) If the mass squared are negative, one is dealing with a unitary description with the non-compact group $U(3,1)$ with an infinite degeneracy at each residue.

"The Veneziano model seems to prefer the former", but then there is trouble with the form factors (in inelastic scattering). He emphasized two kinds of trouble, negative probabilities or negative mass-squared, and noted,
"This is a general agony of making the choice, not restricted to Mr Veneziano alone."

III. Quarks and the dual model

In this section Nambu summarized the picture he had proposed six months earlier in Chicago:

(1) quarks and antiquarks at the end of open strings,

(2) interactions occur when a quark–antiquark pair is created somewhere on the string, splitting it into two,

$$A \longrightarrow B + C.$$

The quark model he used are the three triplets under $SU(3)''$, the symmetry he had introduced earlier with M. Y. Han. It contained a natural figure of merit for the known hadrons as states of zero charm ($C = 0$), which explained the $SU(6)$ classification hadrons. Today's alert reader will recognize $SU(3)''$ as the color group, and C, the "Nambu Charm" as one of the color charges.

In detail, the nine quarks transform as $(\mathbf{3}', \bar{\mathbf{3}}'')$ under $SU(3)' \times SU(3)''$. The first is flavor, the second color. Nambu's quarks have the right color, but the wrong

[123] J. Schwinger, *Phys. Rev.* **162**, 324 (1962).

electric charges; part of the scaffolding and does not invalidate the color structure of the hadrons.

"We will also make things a bit more exciting by naming the three different triplets D, N and A", with "Nambu Charm" proportional to $(\lambda_3'' + \lambda_8''/\sqrt{3})$, respectively $(1, -2, 1)$.

Baryons and mesons follow the usual pattern for hadrons and mesons,

$$B \sim DNA, \qquad M \sim D\bar{D} + N\bar{N} + A\bar{A},$$

B is the completely antisymmetric $SU(3)''$ singlet, " *...and this takes care of the Pauli principle"*. A welcome consequence is the zero triality of the Gell-Mann–Ne'eman $SU(3)'$.

Nambu discussed several approaches to build a Hamiltonian for spring splitting hadrons as "molecules" bound by superstrong interactions and the baryons are like *"... a resonating linear structure"*, to preserve his picture of the interaction as the creation of a $T - \bar{T}$ pair where the string breaks, and that each part *"grows into a full grown string, like an earthworm!"*. A string can only be viewed as a large and indefinite number of N mass points. A hadron would be a linear superposition of configurations having different values of N, all proportional to one another.

The interaction Hamiltonian would be the sum of an infinite number of trilinear product of wavefunctions $\Psi_A(x^{(1)}, x^{(2)} \cdots x^{(N)})$ of N points with the selection rule $N_A = N_B + N_C$. In the Veneziano model, the breaking of the strings occurs only at the end point, with $N_i \to \infty$, $N_2/N_3 \to 0$, or 1. This formalism of string-splitting leads to many unanswered questions, and Nambu suggested yet another possibility.

The string as a chain of $T - \bar{T}$ pairs, *"... like a polarized medium with opposite charges at its ends"*. Interactions between neighbors would depend on their $SU(3)'$ and $SU(3)''$ (flavor and color) spins. Duality would be satisfied with a similar pattern in the euclidean "time direction". The result would be a two-dimensional array with flavor and color "spins" at each pair.

$$\begin{array}{l}
T\bar{T}T\bar{T}\cdots\cdots\bar{T}\\
\bar{T}T\bar{T}T\cdots\cdots T\\
T\bar{T}T\bar{T}\cdots\cdots\bar{T}\\
\bar{T}T\bar{T}T\cdots\cdots T\\
\cdots\cdots\cdots\cdots\cdots\\
\cdots\cdots\cdots\cdots\cdots
\end{array}$$

He proposed a nearest neighbor interaction of the form

$$I = g\sum_{n,n'} \gamma_\mu^{(n)}\gamma^{\mu(n')} + \sum_n \gamma_\mu^{(n)}\varphi^{\mu(n)},$$

where $\gamma_\mu^{(n)}$ are the charge matrices at each point, and $\varphi_\mu^{(n)}$ is an external field. The Veneziano amplitudes would be generated by the partition function as the exponential of this action summed over the charges, in the limit of an infinite number of pairs.

"The simplest mathematical model of this type is the well-known Ising model with its glorious Onsager solution. Perhaps we can adopt the Ising model here as a prototype."

IV. Magnetic Monopoles

Nambu returned to his earlier proposal of monopoles at the end of strings. In that case, $SU(3)' \times SU(3)''$ are interpreted as electric and magnetic charges. He constructed the Maxwell field generated by a pair of monopoles,

$$F_{\mu\nu}(x) = g\epsilon_{\mu\nu\rho\sigma} \int\int \delta^{(4)}(x - y(\xi, \tau))d\sigma^{\rho\sigma},$$

where $d\sigma^{\rho\sigma}$ is the surface element of the world sheet spread between the world lines of the two monopoles. The variation of Lagrangian $\int F_{\mu\nu}F^{\mu\nu}$ with respect to $y(\xi, \tau)$ yields the *"correct equations"*.

We do not discuss the remaining sections of the "Copenhagen Lecture". Section V discusses the statistical interpretation of string models and Section VI inelastic electron proton scattering on ways to obtain the observed non-Gaussian form factors.

In closing, we believe these presentations of several new ideas were not yet quite worked out was Nambu's gentle way of motivating its (mostly young) audience.

Reprinted from Proceedings of the International Conference on Symmetries and Quark Models, ed. R. Chand, © 1970 Gordon and Breach, pp. 269–278.

*Quark model and the factorization of the Veneziano amplitude**

YOICHIRO NAMBU

The Enrico Fermi Institute
The University of Chicago, Chicago, Illinois

IF WE TAKE the quark model in the most realistic sense, a meson in the $q\bar{q}$ configuration, for example, would have a wave function which, beside the spin and the $SU(3)$ indices, depends on the relative coordinates of the quarks. We would therefore expect in general excitations in L as well as in the radial mode, and hence a family of Regge trajectories forming 36-plets of $U(6)$. So far we do not have a clearcut evidence for the radial excitations, but the apparent complexity and structure we see in higher mesonic levels suggest that there may well be excitations other than the orbital L type. This is also supported by the successes of the duality concept[1] and its elegant mathematical realization embodied in the Veneziano[2] amplitude. Although we do not yet know the exact meaning of duality, nor do we know whether it is to be regarded as a new fundamental principle, the Veneziano model seems to contain a lot of implications which have yet to be tested.

A close connection between duality and the quark model has been pointed out by a number of people recently.[3] It is based on a diagrammatic interpretation of hadronic reactions in terms of quark exchanges which satisfy simple topological rules, the most important of which is to first consider only "planar diagrams" with no crossed quark lines. These rules make qualitative but powerful predictions, and it should be highly interesting if we could

* This work supported in part by the U.S. Atomic Energy Commission.

make a more quantitative and precise fusion of the quark model and the analytic structure of the Veneziano representation. Work along this line seems to be going on already at various places. Our results which we would like to discuss here are in fact very similar to those of the Berkeley[4] and the MIT groups,[5] but it would be useful to put the problem in a clear, if not perfect, perspective.

The Veneziano amplitude for a two-body reaction is typically represented by the function (or more generally, a sum of functions)

$$V_{lmn}(s, t) = \Gamma(1 - \alpha_s)\,\Gamma(m - \alpha_t)/\Gamma(n - \alpha_s - \alpha_t) \tag{1}$$

with $\alpha_s = s + a_s$, $\alpha_t = s + a_t$ and an appropriate set of integers l, m, n. V has the duality in the sense that it can be decomposed into an infinite series of resonances either in the s or in the t channel, which converges in its respective physical range, and exhibits the right Regge behavior but for the unphysical infinitely sharp resonance structure superimposed on it. This can probably be remedied by unitarizing V through iteration, perhaps at the risk of violating strict duality, but we will not worry about this question since we do not hold duality to be a sacred principle. Of course we expect that the Regge cuts and even the Pomeranchukon may be generated by starting from V as the input Born term.[6]

Our primary concern will be to find a self-consistent, factorizable set of Veneziano-type amplitudes for all hadronic processes. By this we mean that we want to find a set of hadronic states (represented e.g. by a master wave function with infinite components) and their basic coupling among themselves in such a way that if we build up "tree diagrams" one by one using the basic vertices we would obtain amplitudes for arbitrary many-particle processes and that the result will automatically exhibit duality. That the Veneziano amplitudes can be generalized to many-particle processes has been shown,[7] and we will use the general formula as the basis for our program. In order to build a realistic theory from the quark model it is of course important to discuss the way the $SU(3)$ and the spinor indices are to be coupled, but this is a straightforward problem once the basic topological rules are given.[4] On the other hand, the factorization of analytic functions of many variables requires introduction of a large number of internal coordinates, many more than would correspond to a naive bound state picture. A remarkable point, however, is that this can be done with a finite degree of degeneracy for each level, although the solution is not necessarily a unique one.

To focus our attention on the essential part, we will ignore the spins and consider an n-point Veneziano function for neutral scalar particles. Such a function can be written down according to the known rules, and it takes a particularly simple form in the multiperipheral configuration:

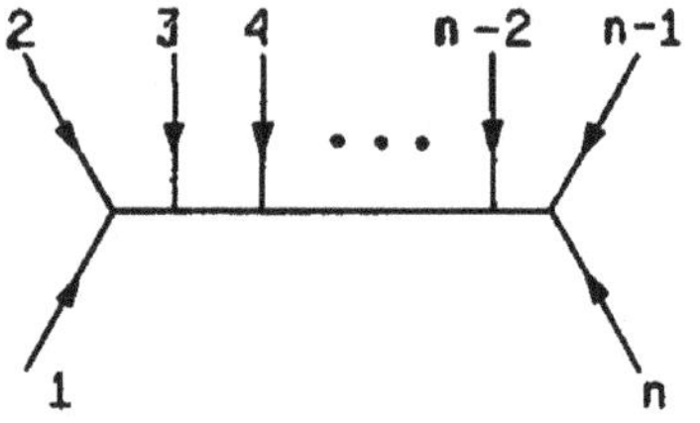

$$
\begin{aligned}
V = \int x^{(12)} y^{(123)} z^{(1234)} \cdots w^{(n-1,n)} \times \\
\times \left(\frac{1-x}{1-xy}\right)^{(23)} \left[\frac{(1-y)(1-xyz)}{(1-xy)(1-yz)}\right]^{(34)} \cdots \times \\
\times \left(\frac{1-xy}{1-xyz}\right)^{(234)} \left[\frac{(1-yz)(1-xyzu)}{(1-xyz)(1-yzu)}\right]^{(345)} \cdots \times \\
\times \cdots \times \frac{dx\, dy\, dz \cdots dw}{(1-xy)(1-yz)\cdots}
\end{aligned}
\tag{2}
$$

where

$$
\begin{aligned}
(12) &\equiv -\alpha_2(s_{12}) - 1 = -s_{12} - a_2 - 1, \qquad s_{12} = (p_1 + p_2)^2, \quad \text{etc.} \\
(123) &\equiv -\alpha_3(s_{123}) - 1 = -s_{123} - a_3 - 1, \quad s_{123} = (p_1 + p_2 + p_3)^2, \quad \text{etc.}
\end{aligned}
\tag{3}
$$

This function is invariant under the cyclic permutation $(12 \cdots n) \to (23 \cdots n1)$ and the inversion $(12 \cdots n) \to (n \cdots 21)$. The a's are the intercepts for the i-body Regge trajectories, which need not be equal to each other. Regrouping the individual factors and expanding the variables: $s_{12} = (p_1 + p_2)^2 = 2p_1 \cdot p_2 + 2m^2$, etc., we get

$$
\begin{aligned}
V = \int x^{(12)} y^{(123)} \cdots dx\, dy \cdots \times \\
\times (1-x)^{-2p_2 \cdot p_3 - c_2} (1-y)^{-2p_3 \cdot p_4 - c_2} \cdots \times \\
\times (1-xy)^{-2p_2 \cdot p_4 - c_3} (1-yz)^{-2p_3 \cdot p_5 - c_3} \cdots \times \\
\times (1-xyz)^{-2p_2 \cdot p_5 - c_4} (1-yzu)^{-2p_3 \cdot p_5 - c_3} \cdots \times \\
\times \cdots \times \\
\times (1-xyz \cdots w)^{-2p_2 \cdot p_{n-1} - c_{n-2}}
\end{aligned}
\tag{4}
$$

where

$$c_2 = a_2 + 2m^2 + 1$$
$$c_3 = a_3 - 2a_2 - m^2 \tag{5}$$
$$c_n = a_n - 2a_{n-1} + a_{n-2}, \quad n \geq 4.$$

The last relations may be written in a uniform fashion for all c_n's by defining

$$a_0 \equiv 1, \quad a_1 \equiv -m^2. \tag{6}$$

Thus the external particles belong to the ground state $\alpha_1(m^2) = 0$ of "one-body" resonances.

When a chain of resonances occur simultaneously for $\alpha_2(s_{12}) = k$, $\alpha_3(s_{123}) = m$, $\alpha_4(s_{1234}) = l$, etc., the corresponding residue is given by the coefficient of $x^k y^m z^l \ldots$ in the product of factors $(1 - x)$, $(1 - xy)$, $(1 - xyz)$, ... in the integrand of Eq. (4), and it is this coefficient which we would like to factorize as

$$\sum \langle n| \cdots |l\rangle \langle l| \Gamma |m\rangle \langle m| \Gamma |k\rangle \langle k| \Gamma |1\rangle. \tag{7}$$

Here the summation is over an as yet unspecified set of states which make up each of the levels $k, m, l \ldots$ Note that we are regarding the process as a chain of transitions $1 \to k$, $k \to m$, $m \to l$, etc., induced by the external particles $2, 3, \ldots n - 1$. Thus we are not treating the lines 1 and n on an equal footing with the rest, but this is a natural consequence of the way Eq. (4) is constructed.

As the next step, consider the factor $(1 - x)^{-2p_2 \cdot p_3 - c_2}$. For a 4-point amplitude this would be the only factor to be dealt with. We rewrite it as

$$(1 - x)^{-2p_2 \cdot p_3 - c_2} = \exp\left[-(2p_2 \cdot p_3 + c_2) \ln(1 - x)\right]$$
$$= \exp\left[(2p_2 \cdot p_3 + c_2) \sum_{r=1}^{\infty} \frac{x^r}{r}\right]. \tag{8}$$

The coefficient of x^n is then to be equated with

$$\sum_{-k} \langle 4| \Gamma(p_3) |k\rangle \langle k| \Gamma(p_2) |1\rangle, \tag{9}$$

where p_2 and p_3 are the respective momentum transfers. Let us now introduce a set of Bose-like creation and annihilation operators $a_\alpha^{+(r)}$ $a_\alpha^{(r)}$, $(\alpha = 1, 2, \ldots, 5)$ $r = 1, 2, \ldots$; with the property

$$[a_\alpha^{(r)}, a_\beta^{+(s)}] = -\delta_{rs} g_{\alpha\beta} \quad (\text{metric } - - - + -)$$
$$[a_\alpha^{(r)}, a_\beta^{(s)}] = [a_\alpha^{+(r)}, a_\beta^{+(s)}] = 0. \tag{10}$$

For each r, $a_\alpha^{+(r)}$ and $a_\alpha^{(r)}$ correspond to a 5-dimensional harmonic oscillator in which the 5th dimension is purely fictitious, and the timelike (4th) excitations are "ghosts" with a negative metric. Observe then that the well known formula

$$e^A e^B = e^B e^A e^{[A,B]}$$

(valid for a c-number commutator) gives

$$\langle 0|\, e^{ik' \cdot a/\eta}\, e^{ik \cdot a^+ \xi}\, |0\rangle = e^{k' \cdot k\xi/\eta}\,, \tag{11}$$

where $k \cdot a^+ = k^\alpha a^+_\alpha$. It is obvious that if a complete set of intermediate states characterized by the occupation numbers $\{n_\alpha\}$ are inserted on the left, the states with $n = \Sigma\, n_\alpha$ contribute to the term $\sim(\xi/\eta)^n$ on the right.

Eq. (11) forms the basis of the factorization program. In fact Eq. (8) can be factorized as

$$\exp\,[(2p_2 \cdot p_3 + c_2) \textstyle\sum x^r/r]$$

$$= \left\langle 0 \left| \exp\left[i\sum\sqrt{\frac{2}{r}}\, a^{(r)} \cdot k'\eta^{-r} \right] \exp\left[i\sum\sqrt{\frac{2}{r}}\, a^{+(r)} \cdot k\xi^r \right] \right| 0 \right\rangle, \tag{12}$$

$$k_\alpha = \left(p_2, i\sqrt{c_2/2}\right), \quad k'_\alpha = \left(p_3, i\sqrt{c_2/2}\right), \quad \xi/\eta = x.$$

The term $\sim x^N$ obviously comes from intermediate states such that

$$\sum_r \sum_\alpha r n_\alpha^{(r)} = -\sum_r r a^{+(r)} \cdot a^{(r)} = N. \tag{13}$$

We may therefore write down the vertex as

$$\langle N|\,\Gamma\,|0\rangle = \left\langle N \left| \exp\left[i\sum \sqrt{\frac{2}{r}}\, a^{+(r)} \cdot k \right| 0 \right\rangle,$$

$$k_\alpha = \left(p_N - p_0, i\sqrt{c_2/2}\right), \quad \langle 0|\,\Gamma\,|N\rangle = \langle N|\,\Gamma\,|0\rangle^+. \tag{14}$$

The level $s = s_N$ then consists of all possible excitations satisfying the condition (13). The 5th dimension was introduced for the purpose of handling the term c_2, assumed to be >0. If $c_2 < 0$ we have to change the metric of the 5th dimension. (The factor i in k_5 is not really necessary; its purpose is to bring Γ to the form (18) below.)

How will this scheme work out for the general n-point function? We can see without much difficulty that if all the c's are equal, we can factorize

Eq. (4) by the ansatz

$$\langle N'|\Gamma|N\rangle = \left\langle N' \left| \exp\left[i\sum\sqrt{\frac{2}{r}}\,a^{+(r)}\cdot k\right]\exp\left[i\sum\sqrt{\frac{2}{r}}\,a\cdot k\right]\right| N\right\rangle,$$

$$k = \left(p_{N'} - p_N,\quad i\sqrt{c_2/2}\right). \tag{15}$$

This vertex operator is Hermitian except for the timelike excitations, and reduces to Eq. (14) when it operates on $|0\rangle$. In order to see how the ansatz works, we form a product of such Γ's according to Eq. (7), and take its vacuum expectation value. By shifting the annihilation operators step by step to the right we will note that the 5-momenta $k_1, k_2, \ldots$ associated with the external lines get contracted with one another, producing an exponential factor for each contraction. To extract a particular set of intermediate states $N_1, N_2, \ldots$, multiply $a^{(r)}$ and $a^{+(r)}$ in each vertex $\langle N_l|\Gamma|N_{l-1}\rangle$ by the power-counting variables $(\xi_{l-1})^{-r}$ and $(\xi_l)^r$ respectively, and pick up the coefficient of $\xi_1^{N_1}(\xi_2/\xi_1)^{N_2}(\xi_3/\xi_2)^{N_3}\ldots$ from the final result. We can verify raedily that the variables $\xi_1, \xi_2/\xi_1, \xi_3/\xi_2, \ldots$ correspond to $x, y, z, \ldots$ in Eq. (4).

In actuality, the constants c_n are not all equal. If all intercepts a_n are equal, i.e. all multiparticle resonances belong to the same trajectories, we have $c_m = 0$ for $m \geq 4$, but c_2 and c_3 are different. So the 5th dimension is necessary only for the nearest neighbor and the next nearest neighbor contractions, and we need to devise a scheme which treats these special cases. Another example is the case where we distinguish between even-n and odd-n resonances so that

$$a_0 = a_2 = \cdots = a_{\text{even}},\quad a_1 = a_3 = \cdots = a_{\text{odd}},\quad c_{\text{even}} = -c_{\text{odd}}$$

for all n. According to Eq. (5), however, this requires $a_{\text{even}} = 1$ and the external lines belong to α_{odd}. The alternating signs of c_n can be realized by a suitable definition of Γ. This example is relevant to the many-pion processes which have, however, $a_{\text{even}} = \frac{1}{2}$ rather than 1. We shall not go into the details of how one would handle these problems more precisely.

We have outlined above the basic procedure for factorizing n-point Veneziano amplitudes in a consistent way. For this we had to pay the price of introducing an infinite set of 5-dimensional harmonic oscillators, although this does not affect the leading trajectory. It is clear that each of them generates a tower of non-unitarity representations of $U(4, 1)$, leading to the unpleasant presence of ghosts. However, this does not necessarily mean that the amplitudes themselves are unphysical since there can be cancellations from various intermediate states.[5] For example, the elastic pion–pion scattering amplitude seems to be free of ghosts.[8] But the general answer to this question is not known yet.

The appearance of harmonic oscillators in our problem is intriguing since the simple bound-state picture of quarks with a harmonic oscillator potential would naturally give rise to linear trajectories and the $U(3, 1)$ level scheme. We can bring out this analogy more clearly in the following way. Let us introduce a Bose field $\phi_\alpha(\xi)$ and its canonical conjugate $\pi_\alpha(\xi)$, which are even and periodic with periodicity 2π in ξ. In analogy to the ordinary field theory, we decompose it into plane waves:

$$\phi_\alpha(\xi) = \sum_{r=1}^{\infty} \frac{1}{\sqrt{2r}} (a_\alpha^{(r)} + a_\alpha^{+(r)}) \cos r\xi$$

$$\pi_\alpha(\xi) = \sum_{r=1}^{\infty} i\sqrt{\frac{r}{2}} (a_\alpha^{(r)} - a_\alpha^{+(t)}) \cos r\xi \tag{16}$$

where we have excluded the constant mode $r = 0$. The a's and a^+'s are the operators we have defined above. In view of Eq. (13), the quantum number N which determines the resonance energy can be written

$$\begin{aligned} N &= -\sum_r r a^{+(r)} \cdot a^{(r)} \\ &= \frac{-1}{\pi} \int_0^{2\pi} :(\partial_\xi \phi(\xi) \cdot \partial_\xi \phi(\xi) + \pi(\xi) \cdot \pi(\xi)): d\xi. \end{aligned} \tag{17}$$

Furthermore, the vertex operator Γ can be written

$$\Gamma = \ :\exp[2ik \cdot \phi(0)]: \tag{18}$$

Eq. (17) suggests that the internal energy of a meson is analogous to that of a quantized string of finite length (or a cavity resonator for that matter) whose displacements are described by the field $\phi_\alpha(\xi)$. A master field Ψ representing the mesons will depend on the space-time coordinates x_μ as well as the internal variables $\phi_\alpha(\xi)$, and satisfy the free wave equation

$$(\Box - N - c)\Psi(x; \phi) = 0. \tag{19}$$

Its interaction is described by a Lagrangian $\sim \Psi^+(x)\Gamma\Psi(x)\varphi(x)$ where φ is the field for the external particles.

The form (18) has a striking resemblance to the form factor in a naive quark model, in which one of the quarks interacts with the external field carrying momentum k. If the $q\bar{q}$ system is described by a wave function $\Psi(x; r)$, where x and r are the center of mass and relative coordinates respectively, the interaction will carry a form factor exp $[ik \cdot r/2]$. An unsatisfactory feature of the result is that the external field φ and the master field Ψ have to be distinguished, whereas clearly the starting program was to include everything in Ψ. To achieve this end one would need to find a trilinear coupling of Ψ's which agrees with the coupling that arises when the n-point Veneziano amplitude is in a configuration in which three resonances meet. The above interpretation of the form factor suggests that one might be able to construct such a coupling geometrically from quark diagrams like

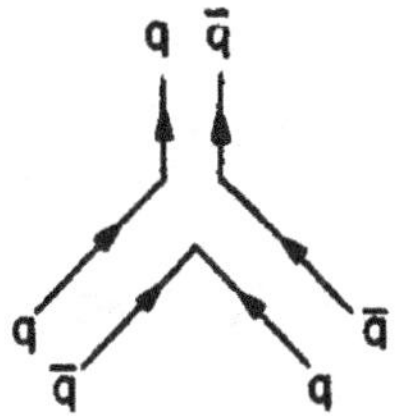

If this can be carried out, we are essentially led to a nonlocal picture of the Veneziano model, though the nonlocality may not manifest itself in a conspicuous way.

Another problem that confronts a field theory of duality is that of the multiple counting of diagrams. With a trilinear coupling of Ψ's one can construct diagrams of both s-channel and t-channel types. Because of duality they will of course give the same contributions, but the amplitude is doubled nevertheless. We can adjust for this by reducing the trilinear coupling parameter, but what about the higher order processes? Even if the factorization may work formally, it is not clear whether or not we should build a field theory in this fashion.[9] That will depend on how one interprets duality, and must eventually be solved through a confrontation with experiment.

Putting aside these problems, we can make some remarks about high energy phenomena based on our results. Since the internal excitations of hadrons resemble those of a cavity resonator, we are able to use statistical concepts like temperature, entropy and the Bose–Einstein distribution in a well defined way. For example, at extreme high energies ($\gg 1$ GeV) where the discreteness of eigenfrequencies becomes unimportant, we find

$$M^2 \sim E = TS + F \sim -F \sim (5\pi^2/6)\,(kT)^2\,, \tag{20}$$

where M is the rest energy of the system, and the other symbols are the standard thermodynamic ones. Thus the density of the levels goes like exp $[S/k]$ $\sim$ exp $[2M\,(5\pi^2/6)^{1/2}]$.

References

1 R. Dolen, D. Horn and C. Schmid, *Phys. Rev.* **166**, 1768 (1968).
2 G. Veneziano, *Nuovo Cimento* **57A**, 190 (1968).
3 H. Harari, *Phys. Rev. Letters* **22**, 562 (1969); J. Rosner, *Phys. Rev. Letters* **22**, 689 (1969); J. Mandula, C. Rebbi, R. Slansky, J. Weyers and G. Zweig, *Phys. Rev. Letters* **22**, 1147 (1969); K. Matsumoto, *Lettere al Nuovo Cimento* **1**, 620 (1969); T. Matsuoka, K. Ninomiya and S. Sawada, Nagoya Univ. preprint.
4 S. Mandelstam, A Relativistic Quark Model Based on the Veneziano Representation, Berkeley preprint; K. Bardakci and M. B. Helpern, A Possible Born Term for the Hadron Bootstrap, Berkeley perprint; K. Bardakci and S. Mandelstam, Analytic Solution of the Linear-Trajectory Bootstrap, Berkeley preprint.
5 S. Fubini and G. Veneziano, Level Structure of Dual Resonance Models, MIT preprint.
6 P. G. O. Freund, *Phys. Rev. Letters* **22**, 565 (1969); K. Kikkawa, B. Sakita and M. A. Virasoro, University of Wisconsin preprint.
7 C. Goebel and B. Sakita, *Phys. Rev. Letters* **22**, 257 (1969); H. M. Chan, *Phys. Letters* **28B**, 425 (1969); H. M. Chan and T. S. Tsou, *Phys. Letters* **28B**, 485 (1969); J. F. L. Hopkins and E. Platt, *Phys. Letters* **28B**, 489 (1969); Z. Koba and H. B. Nielsen, *Nucl. Phys.* **B10**, 633 (1969).
8 Y. Nambu and P. Frampton, University of Chicago preprint EFI 69-56.
9 I thank P. Freund for a discussion on this point.

PMR
(173)

EFI 70-07

COO 264-535

Dual Model of Hadrons*

Yoichiro Nambu
The Enrico Fermi Institute and Department of Physics
The University of Chicago

Talk presented at the
American Physical Society Meeting in Chicago

January, 1970

*Work supported by the U. S. Atomic Energy Commission.

I would like to discuss here some new ideas about the nature of hadrons and their interactions which is based on the recent developments in hadrodynamics. In 1968 Veneziano discovered a remarkable model for hadron scattering amplitudes. It was an elegant mathematical realization of the duality principle developed earlier by Dolen, Horn and Schmid, namely, that the Regge behavior of a scattering amplitude due to the exchange of resonances in the crossed channel can be at the same time a result of the existence of resonance in the direct channel. For this, the Regge trajectories have to be by necessity linearly rising, as actually seems to be the case, and in addition contain an infinite number of parallel daughters, a fact which suggests that there exist a vast number of resonances yet to be discovered. It was also pointed out by Lovelace that the Veneziano amplitude for π-π scattering satisfied the PCAC condition. Subsequently, there has been a great deal of theoretical activity, and a number of groups have contributed to the generalization and analysis of the mathematical and physical content of the model. Much of what I am going to say is not original with me, or otherwise has been conceived by others independently. In order to do justice to these people, I will first give a list of key developments and their main contributors.[1-4]

2

In this talk I cannot go into the very complicated mathematical detail, but I shall be content with presenting a basic physical picture, and then try to build a rather speculative model of hadrons by bringing together the essential elements of the Veneziano model, the three-triplet version of the quark model which Dr. Han and I proposed in 1965,[5] and possibly the magnetic monopole theory of Dirac.

Now first about duality. Essentially it means that one and the same scattering amplitude A(s,t) can be viewed in two different ways simultaneously:

$$A(s,t) = \sum_n \frac{C_n(t)}{s - s_n} = \sum_n \frac{C'_n(s)}{t - t_n} \tag{1}$$

namely as an infinite sum of Breit-Wigner resonance terms in s or t channel, so that it is not possible to tell whether the scattering definitely proceeds through one or the other mechanism as in ordinary perturbation theoretical picture. Turning the argument around, one can further assert that when one sees indeed a series of resonances in one channel, then one must expect also a series of resonances to exist in one of the cross channels. Such a situation can be understood in terms of the quark model as was shown by Harari and Rosner. One draws duality diagrams like

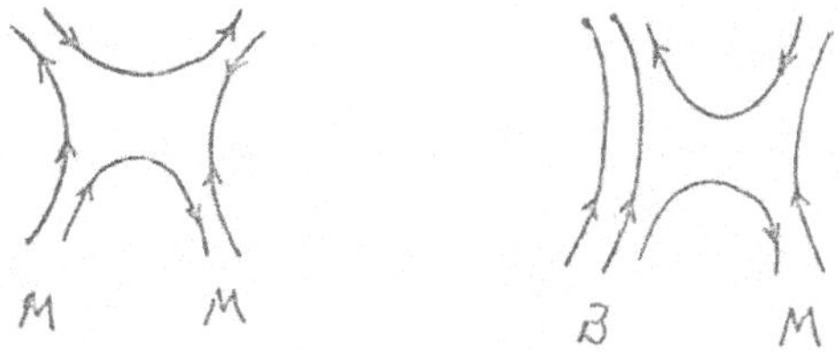

3

for meson-meson and meson-baryon scattering, which exhibits the existence of intermediate hadron states in two channels.

Specific examples of amplitudes satisfying duality have been found by Veneziano. A typical form is

$$A(s,t) = \frac{\Gamma[-\alpha(s)]\ \Gamma[-\alpha(t)]}{\Gamma[-\alpha(s) - \alpha(t)]} \tag{2}$$

where $\alpha(s) = as + \alpha_s(0)$, $\alpha(t) = at + \alpha_t(0)$, implying resonances occurring at $as_n = n - \alpha_s(0)$, $at_n = n - \alpha_t(0)$, $n = 0, 1, 2, \ldots.$ Empirically $a = 1/\text{Gev}^2$. The residues $C_n'(s)$ are polynomials of nth order, so that at the nth level there are states with various spins from 0 to n in general. That is to say, there exist an infinite number of parallel daughter trajectories displaced by integer units. This is one of the immediate consequences of the model which has yet to be tested experimentally. At any rate the Veneziano model suggests a very rich spectrum of levels for hadrons. But it is not quite obvious that the resonance poles in the formula can be intepreted à la Breit and Wigner. For this the residue of the pole coming from a particular intermediate state must be factorizable into a product of coupling parameters of the intermediate state to the incoming and outgoing channels, and for unitarity this product had better be positive if the two channels are the same.

The fact that factorization can indeed be done has been shown by Mandelstam and Bardakci, and by Fubini and Veneziano. To achieve this, however, each level had to be thought of as comprising a great many different states, much more than the

4

number of different spins present. In other words, the hadron spectrum must be even richer and more degenerate than is seen by casual inspection of the Veneziano forms. Of course there is no direct experimental evidence for this yet. But what is interesting is that once factorization is achieved, we can go back to the familiar quantum mechanics and perturbation theory. We know how many different states there are at each level, and what the interaction is that causes the transition. We can then proceed to higher order processes in which a large number of particles are produced. It turns out that the scheme works and one gets indeed the general n-point amplitudes which are multiply dual.

I will now present a simple pictorial interpretation of hadron dynamics which incorporates the above factorization property and the quark diagram selection rules of Harari and Rosner. I would like, however, to deviate from Gell-Mann's fractionally charged quark model, and instead consider the integrally charged 3-triplet model which Dr. Han and I introduced a few years ago. According to this model, there are three different triplets of fundamental spin 1/2 objects, each of which is like the quark triplet except that it is displaced in charge and hypercharge assignments. The situation is illustrated by the diagram

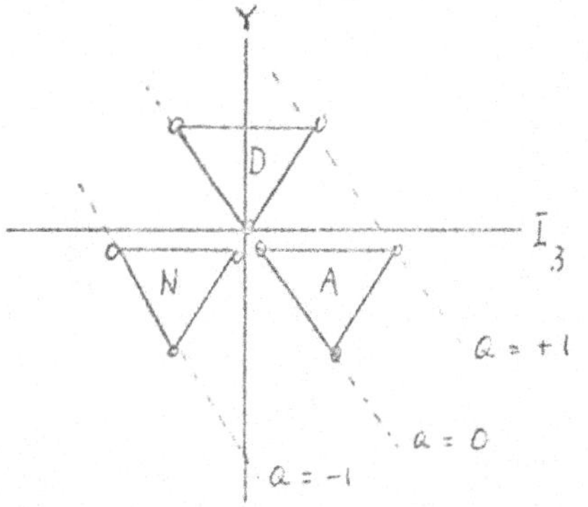

The centers of the three triplets form an inverted triangle, i.e., an antitriplet. Thus if we label the nine members in terms of two separate SU(3) indices, they behave like a representation $(3,3^*)$ of the group $SU(3)' \times SU(3)''$ where the first SU(3)' is the ordinary SU(3) group. To distinguish between the three different triplets, let me call these D, N, and A. I assert then that the known hadrons have the structure

$$\text{Mesons} \sim (D\bar{D} + N\bar{N} + A\bar{A})/\sqrt{3}$$

$$\text{Baryons} \sim DNA$$

Why should this be so? The main reason is that with respect to the second SU(3)'', these states can be in a singlet, and one can argue that only those states are low lying in the scale of superstrong interactions responsible for the binding of these objects, so that one can forget about the second SU(3)'' label. One of the SU(3)'' quantum numbers is what we called charm C, and has the following assignments

	D	N	A	$\bar{D}$	$\bar{N}$	$\bar{A}$
C =	1	-2	1	-1	+2	-1

To within a factor 3, C is then similar to the electric charge in the quark model. Thus the baryons and mesons are all charm-neutral; in general one can show that the so-called triality zero states have always C = 0. It is an attractive possibility then to attribute the occurrence of only triality zero states in hadrons to a dynamical mechanism in the SU(3)'' space. It is not surprising that the zero charm states are the most stable ones if there is a superstrong interaction coupled to charm number. What is this interaction?

6

At this point we go back to the Veneziano model, and postulate that the bond that makes hadrons out of the triplets is a "rubber string." Imagine for example that an elastic string is stretched between D and $\bar{D}$ in a meson.

The SU(3) quantum numbers as well as the quark spin is attached to the end points, but the energy resides in the string, the inertial mass of the triplets being zero or negligible.
In a relativistic treatment, we first designate each element along the string by a parameter ξ, $0 \leq \xi \leq 1$. Each such element sweeps out a world line, so that the whole string sweeps out a two-dimensional sheet. Let us therefore parameterize the sheet by ξ and another timelike parameter η, so the space-time coordinates $x_\mu \equiv \sqrt{2} y_\mu$ of any point on the sheet become a function of ξ and η. The wave equation for the string is then a two-dimensional one

$$\frac{\partial^2 y_\mu}{\partial \eta^2} - \omega^2 \frac{\partial^2 y_\mu}{\partial \xi^2} = 0 \qquad (\mu = 1, 2, 3, 4) \tag{3}$$

with the boundary condition $\partial y_\mu / \partial \xi = 0$ at $\xi = 0, 1$. Thus each coordinate oscillates harmonically with characteristic frequencies ω, 2ω, 3ω, ..., except for the zeroth mode corresponding to the uniform center-of-mass motion. The Hamiltonian of the system (with respect to the internal time parameter η) becomes

$$H_o = \frac{p_{o\mu}{}^2}{2} + \frac{1}{2} \sum_{n=1}^{\infty} (P_{n\mu}^2 + n^2\omega^2 Q_{n\mu}^2) \equiv p_\mu^2 + \omega N \quad \text{(Metric +++-)} \tag{4}$$

where $p_\mu = p_{o\mu}/\sqrt{2}$ may be interpreted as the center-of-mass momentum. When quantized, each harmonic mode may be excited to an arbitrary level, so that N takes eigenvalues

$$N = \sum_{n=1}^{\infty} \sum_{\mu=0}^{3} nN_{n\mu} \tag{5}$$

where $N_{n\mu}$ is the occupation number for the mode (n,μ).
If we now restrict ourselves only to those states having a fixed eigenvalue $H_o = \alpha$, we get the mass spectrum relation[6]

$$-p_\mu^2 = \omega N - \alpha \equiv M^2. \tag{6}$$

So we can set ω equal to the slope of the trajectories, $\simeq$ 1 Gev^2. Each level consists of a large number of states satisfying the condition (5). In fact the degree of degeneracy increases exponentially with M.

Now comes the question: how do these states interact, i.e. how does a hadron produce another hadron? First we will make a qualitative answer: A string may be cut in two by creating a pair of "quarks" at the point where it is cut, and thereby one string becomes two as is shown here.[12]

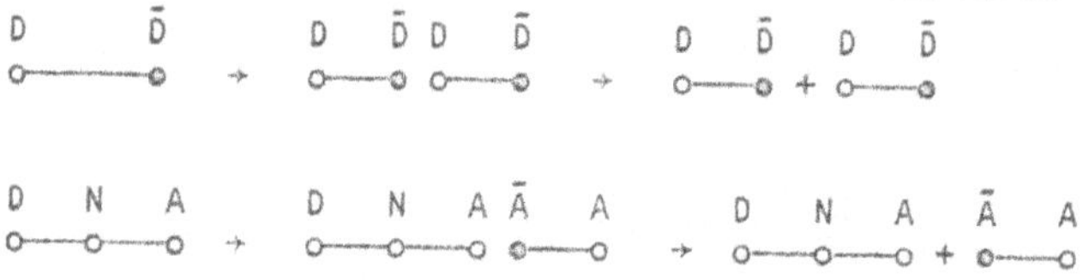

One can see that this picture is equivalent to the duality diagram if we imagine a string stretched between quark and antiquark lines in the diagram.

In order to make the picture quantitative, we must write down an interaction Hamiltonian for this division mechanism.

8

Presumably such an interaction will depend on the degree of overlap of the initial and final configurations, i.e. on the overlap of three strings. As a special case, let us assume that one of the created mesons in the process $M \to M + M$ is always in the ground state, and that the cut occurs at one of the very ends of the string. It is as if a string gets excited when an external force hits it by the end. We can then use perturbation theory to compute the amplitudes for processes like

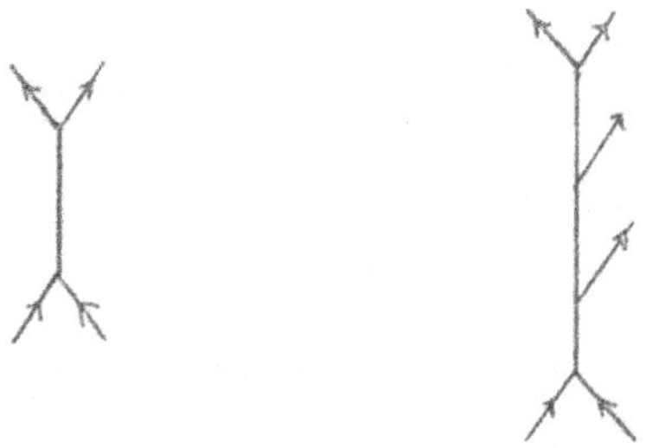

Miraculously, it turns out that we get the dual Veneziano amplitudes!

I will not attempt to describe the detailed mathematics of this problem. Certainly it is not an obvious thing that duality comes out of the rubber string model, at least at first glance. On second thought, however, one might say it is very natural, because one can imagine a two-dimensional world sheet stretched four ways by the four external lines like

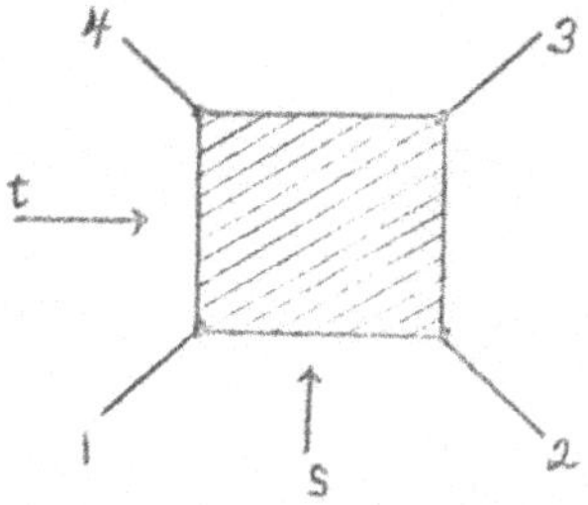

Since two-dimensional wave equation is symmetric between space and time coordinates, there must be symmetry between s and t channels in the above diagram; both must exhibit resonances. This seemingly very simple explanation, however, becomes completely nontrivial when one gets down to the details, and one realizes that things are not that simple at all. Without the Veneziano formula, one would not have found duality this way. It is true, nevertheless, that a two-dimensional wave equation has more symmetry than the four-dimensional one. Since it is related to the two-dimensional Laplace equation, one can perform analytical mapping of the coordinates without affecting the action integral, and manifestly exhibit the duality property of Veneziano amplitudes. An elegant formulation of this kind has been given by Koba and Nielsen.

We have ignored many subtleties that appear in the Veneziano model. We will now add a few remarks on some of them. One is the problem of "ghosts". As we already mentioned, the residues at the resonances must satisfy positivity conditions. This would be true in the factorized form if the Hamiltonian were Hermitian. Actually the four-dimensional harmonic oscillator contains timelike components for which we have to assign an indefinite metric. These are ghost states having negative intrinsic probability. However, it has been shown by Fubini and Veneziano, and more recently by Virasoro, that there is a possibility that some or perhaps all of these ghosts may be effectively cancelled by the non-ghost states just as in quantum electrodynamics ghost photons are killed as a consequence of gauge invariance. In fact because of the large amount of invariance properties inherent in the two-dimensional wave propagation and in the particular form of interaction, the Veneziano model does seem to allow certain gauge transformations which can be utilized for this purpose.[7]

The second point is the problem created by the Regge intercept parameter. In order to reproduce the correct pole structure in the crossed channel by factorization, it is actually necessary to introduce additional modes of excitation, e.g. as oscillations in an extra 5th dimension, which are designed to adjust the intercept parameter properly. But the existence of ghosts depends critically on the intercept values. So far, Virasoro's ghost-killing condition works only at the price of having a new kind of ghost, i.e., a spacelike particle (tachyon) lying on a trajectory with $\alpha(0) = 1$.

Another point is the problem of multiple counting and higher order effects. If we are to build a field theory of duality by means of factorization, we are necessarily forced to take all the higher order effects into account, like self-energy, rescattering, etc. But even in the Born approximation, the amplitude must be the sum over all possible configurations even if they are all equivalent because of duality. Thus the magnitude of an amplitude is multiplied by the number of such dual configurations.[8] Since this is different for different processes, it cannot be absorbed into a renormalization of the basic coupling constants, and therefore is observable.

Finally, let me become even more speculative, and discuss a possible connection between the dual three-triplet model and Dirac's magnetic monopoles. Anybody who is familiar with Dirac's theory of monopoles will recall that it also uses strings as dynamical variables.[9] In Dirac's case, a string is attached to each monopole, but it does not carry energy and momentum, and its motion is arbitrary. In our case the string is a more dynamic object obeying its own equation of motion. At any rate, since we have the string, why not take advantage of it, and add Dirac's electromagnetic Lagrangian to ours! Thus the quarks at the ends of a meson string will be assigned equal and opposite magnetic charges, so there will be a very strong magnetic interaction between them if the magnetic charges are to satisfy Dirac's quantization condition. One realizes easily that the charm number C in our DNA model can be interpreted as the magnetic charge. Then magnetically neutral states will be the most stable ones.

Recently Schwinger[10] proposed a very intriguing model of hadrons in which nine fundamental constituents (called dyons) possess electric as well as magnetic charges, and the magnetic interactions are responsible for the hadron mass spectrum. Biedenhahn and Han[11]

If this can be carried out, we are essentially led to a non-local picture of the Veneziano model, though the nonlocality may not manifest itself in a conspicuous way.

Another problem that confronts a field theory of duality is that of the multiple counting of diagrams. With a trilinear coupling of Ψ's one can construct diagrams of both s-channel and t-channel types. Because of duality they will of course give the same contributions, but the amplitude is doubled nevertheless. We can adjust for this by reducing the trilinear coupling parameter, but what about the higher order processes? Even if the factorization may work formally, it is not clear whether or not we should build a field theory in this fashion.[9] That will depend on how one interprets duality, and must eventually be solved through a confrontation with experiment.

Putting aside these problems, we can make some remarks about high energy phenomena based on our results. Since the internal excitations of hadrons resemble those of a cavity resonator, we are able to use statistical concepts like temperature, entropy and the Bose-Einstein distribution in a well defined way. For example, at extreme high energies (>> 1 Gev) where the discreteness of eigenfrequencies becomes unimportant, we find

$$M^2 \sim E = TS + F \sim -F \sim (5\pi^2/6)\ (kT)^2 \qquad (20)$$

where M is the rest energy of the system, and the other symbols are the standard thermodynamic ones. Thus the density of the levels goes like $\exp\,[S/k] \sim \exp\,[2M(5\pi^2/6)^{1/2}]$.

13

References

1. Generalization of the Veneziano formula to n-particle processes.
   K. Bardakci and H. Ruegg, Phys. Letters 28B, 342 (1968).
   Chan Nong-Mo and T.S. Tsou, Phys. Letters 28B, 485 (1968).
   C. Goebel and B. Sakita, Phys. Rev. Letters 22, 257 (1969).
   Z. Koba and H. B. Nielsen, Nucl. Phys. B10, 633 (1969).
2. Quark model interpretation of duality.
   H. Harari, Phys. Rev. Letters 22, 562 (1969).
   J. Rosner, Phys. Rev. Letters 22, 689 (1969).
   S. Mandelstam, Phys. Rev. 184, 1625 (1969).
3. Factorization property of Veneziano amplitudes and its dynamical interpretation.
   K. Bardakci and S. Mandelstam, Phys. Rev. 184, 1640 (1969).
   S. Fubini and G. Veneziano, Nuovo Cimento, to be published.
   Y. Nambu, Proc. Int. Conf. on Symmetries and Quark Models, Wayne U., 1969.
   S. Fubini, D. Gordon and G. Veneziano, Phys. Letters 29B, 679 (1969).
   L. Susskind, Yeshiva Univ. Preprint.
4. Symmetry properties inherent in the Veneziano amplitudes.
   S. Fubini and G. Veneziano, loc. cit.
   Z. Koba and H. B. Nielsen, loc. cit.
   C. B. Chiu, S. Matsuda and C. Rebbi, Cal Tech Preprint 68-231.
   M. A. Virasoro, U. Wisconsin pr print.
   F. Gliozzi, U. Torino preprint.
   D. Amati, M. LeBellac and D. Olive, CERN preprint Th 1102.
   C. B. Thorn, Berkeley preprint.
5. M. Y. Han and Y. Nambu, Phys. Rev. 139, 1006 (1965).
6. This restriction is compatible with dynamics since the Hamiltonian is a constant of motion. In the presence of interactions we can replace H_o by the total Hamiltonian H as long as the parameter η does not appear explicitly in H. (Note that H is not the energy in the ordinary sense.) The conservation of energy-momentum p_μ can be viewed in this formalism as resulting from the invariance of the Hamiltonian under the displacement of the variables: $y_\mu \to y_\mu + c_\mu$.

14

7. These transformations may be expressed in terms of the energy-momentum tensor $T_{\alpha\beta}$ $(\alpha,\beta = 0,1)$ in the internal space (η,ξ). Take $T_{\alpha\beta}(\xi)$ at a fixed time η, integrate it over ξ with a periodic test function $f_n(\xi)$, and call it $T_{\alpha\beta}[f_n]$. The set of $T_{\alpha\beta}[f_n]$'s generate a Lie algebra when quantized. In particular the combination

$$T_+[f_n^+] + T_-[f_n^-],$$

$$T_\pm = T_{00} \pm T_{01}, \qquad f_n^\pm(\xi) = 1 - e^{\pm in\pi\xi}$$

has the property that it leaves the end point $\xi = 0$ inaffected, and can be used to generate various gauge transformations.

8. The exact number has been obtained by P. Frampton.

9. P. A. M. Dirac, Phys. Rev. 74, 817 (1948).

10. J. Schwinger, Science 165, 757 (1969); 166, 690 (1969).

11. M. Y. Han and L. C. Biedenhahn, Phys. Rev. Letters 24, 118 (1970).

12. According to the drawing, the N-triplet plays a different role from that of D- and A- triplets (because of the charm number assignments). This picture therefore does not give an SU(3)'' singlet. In order to produce correct SU(3)'' symmetry one would have to add other configurations. A triangular structure with three bonds (rather than a linear structure) might look more appealing, but it does not lead to simple production mechanisms.

Duality and Hadrodynamics†

Y.Nambu

Enrico Fermi Institute and

Department of Physics

University of Chicago

Notes prepared for the Copenhagen
High Energy Symposium, Aug. 1970
(Unpublished and undelivered)

(Supported in part by the U.S.Atomic Energy
Commission under the Contract No.AT(11-1)-264)

†Retypeset by M. Okai, Nov. 1993

1 Introduction

We would like to explore here possible dynamical properties of hadrons suggested by the duality principle of Dolen, Horn and Scmid, and in particular by its mathematical realization due to Veneziano. Our primary concern will be not in the very interesting details of the mathematical properties that the Veneziano model seems to possess in abundance, but rather in guessing at the internal structure and dynamics of hadrons which underlie the Veneziano model, recognizing that the latter is in all likelihood only an approximate and imperfect representation of the former.

On crucial step we take in interpreting the Veneziano model is factorization. Of course this is a trivial problem, at least in principle if not in practice, if one is dealing with a given four-point amplitude. But it becomes a very restrictive condition if one demands that the same set of resonances saturate all $n-$point amplitudes. Perhaps such a condition is unwarranted; in discussing as revolutionary a concept as duality one may have to give up all the conventional notions about resonances, e.g. that the intrinsic properties of a resonance is independent of how it is prepared, and a multiple resonance can be analyzed in terms of the individual resonances, etc. Nevertheless, the fact that a set of resonances have been found to saturate all the $n-$point Veneziano amplitudes of the standard variety is very significant. It suggests that our conventional notions about resonances still make some sense, and the know techniques of field theory can be applied to them.

The condition of factorizability immediately rules out an ad hoc addition of satellite terms to a scattering amplitude because it would in general destroy factorization unless more and more resonances are introduced. If satellite terms do arise, they must do so in a well defined and self-consistent way.

As it stands now, the Veneziano model is still beset with many difficulties. For all its mathematical elegance, its practical successes are few. It would therefore be appropriate to list here its basic predictions as well as difficulties without going into the details.

Basic predictions.

1) Linearly rising trajectories. This is in accordance with observation. Whether the trajectories really keep rising indefinitely or not is of course an open question, but it seems to be a valid simplification to assume that they do.

2) Regularly spaced daughter trajectories, implying a highly degenerate level structure. Again the actual degeneracy may be only approximate, but these

daughters must exist if the model makes sense at all. So far there is some evidence for ϵ (daughter of ρ), but none for ρ' and ϵ' (daughter of f).
3) The factorized Veneziano model implies an even higher degree of degeneracy. At each level many distinguishable states having the same spin exist. The number of states increases with energy as $\sim \exp(cE)$. This situation bears a striking similarity to Hagedorn's model of high energy reactions.

Unsolved practical problems.
1) Baryons. We do no yet have a quantitatively satisfactory picture of meson-baryon and baryon-baryon scattering based on the Veneziano model, in spite of the fact that duality was first discovered in meson-baryon scattering.
2) A general satisfactory unification of the quark model (or its $SU(3)$, $SU(6)$ and chiral $SU(3)$ aspects) with the Veneziano model does not exist yet.
3) No convincing theory of form factors exists.
4)We do not know what the Pomeron is within the framework of the Veneziano model.

We must emphasize, however, that there are numerous attempts and speculations regarding all these problems.

More fundamental difficulties.
1) Unitarity. The original Veneziano model is a zero width approximation. The amplitude wildly oscillates with energy, and only after averaging over an interval does it reproduce the smooth Regge behavior at high energy. If a basic Hamiltonian for the model is given, unitarization might be formally carried out by taking into account higher order processes, although the highly singular nature of the Hamiltonian casts doubt about its meaningfulness.
2) A more serious problem, however, is the existence of ghosts. Here we mean by ghosts unphysical particles having a) negative probability, or b) spacelike momenta (tachyons), or both. The levels of the factorized Veneziano model contain those of four-dimensional harmonic oscillators, where the timelike excitations have an intrinsic negative norm. This does not necessarily mean that the Breit-Wigner residue of a partial wave projection of a given amplitude is not positive. Contributions from many degenerate states can add up to a positive value, but there is no guarantee that this will always happen. Fubini and Veneziano have found a Ward-like identity which accomplishes this cancellation to a certain extent, and Virasoro has extended their result. They rely, however, on very special properties of the Hamiltonian, and it is not clear whether and how these can be preserved in general. Besides, Virasoro's scheme still leaves us with a tachyon ghost ($m^2 = -1$). The killing of

tachyons is easy in special cases like $\pi - \pi$ scattering (where the ρ trajectory has a potential tachyon), but a general prescription in a factorized model seems very complicated, if not impossible.

We have emphasized the ghost problem because this is not peculiar to the Veneziano model alone, but it is rather a common disease afflicting all attempts at a description of hadron states as infinite multiplets. The program of current algebra saturation, as well as the use of infinite-component wave equations, have floundered on the same difficulty. To a certain extent, the two kinds of ghosts seem to be complementary: The use of finite-dimensional Lorentz tensors (as in the factorized dual model) involves negative probabilities, whereas infinite-dimensional unitary representations in general lead to tachyons.

If these ghosts are so difficult to eliminate, why not accept them and look for them? Maybe ghosts of one kind or the other do exist, which would make either T.D.Lee or G.Feinberg happy (or both, plus Sudarshan and others). But the trouble is that the extent to which these ghosts appear in a particular problem seems to depend on one's cleverness and ability to avoid them. How many of the ghosts are "real"? This is the most serious question of principle that haunts us, especially us the theorists.

2 Factorized Veneziano model

As has been shown by various people, the $n-$point dual amplitude for scalar external particles can be factorized in terms of a set of harmonic oscillators corresponding to an elastic string of finite intrinsic length. In classical theory, the motion of a free mass point can be derived from the action integral

$$I = -m \int d\tau, \quad d\tau^2 = -dx_\mu dx^\mu \quad (\text{metric } (- + ++)) \tag{1}$$

Alternatively one may take

$$I' = \frac{1}{2} m \int \left(\frac{dx_\mu}{d\tau} \frac{dx^\mu}{d\tau} - 1 \right) d\tau \tag{2}$$

τ being an independent parameter, and $x^\mu(\tau)$ the dynamical variables. (The constant in the integrand is added for convenience.) This form is more suitable for the transition to quantum theory. Because of the translational invariance under $\tau \to \tau + c$ and $x^\mu \to x^\mu + a^\mu$, both the Hamiltonian and the momenta

$$\begin{aligned} H &= (p_\mu p^\mu + m^2)/2m \\ p^\mu &= m dx^\mu/d\tau \end{aligned} \tag{3}$$

are conserved. By imposing the constraint

$$\frac{dx_\mu}{d\tau}\frac{dx^\mu}{d\tau} = -1 \tag{4}$$

we can normalize the parameter τ, which then becomes the proper time of the mass point. This condition (4) amounts to

$$H = 0. \tag{5}$$

In quantum mechanics, one postulates the commutation relations

$$[x^\mu, p^\nu] = ig^{\mu\nu} \tag{6}$$

and the Schrödinger equation

$$i\frac{\partial}{\partial\tau}\Psi = H\Psi \tag{7}$$

Eq.(5) is to be replaced by

$$H\Psi = 0 \tag{8}$$

which is nothing but the Klein-Gordon equation

$$(p^\mu p_\mu + m^2)\Psi = 0. \tag{9}$$

We see thus the usefulness of the proper-time formalism. If we just integrate Eq.(7), we get

$$\Psi(\tau) = \exp[-i(p^2 + m^2)\tau/2m]\Psi(0) \tag{10}$$

or

$$\begin{aligned} (\Psi(x';\tau), \Psi(x;0)) &= \langle x'|\exp[-i\tau(p^2 + m^2)/2m]|x\rangle \\ &= (-im^2/4\pi^2)\exp[imx^2/2\tau - i\tau m/2] \\ &= (-im^2/4\pi^2)\exp[im\tau\{(x/\tau)^2 - 1\}/2] \end{aligned} \tag{11}$$

This is the transition amplitude $x \to x'$ after an elapsed "time" τ. Integrating over τ from 0 to ∞ (we might say it does not matter how much time it has elapsed between the events x and x'), we obtain the Feynman propagator

$$\frac{1}{2m}\int_0^\infty e^{-i\tau(p^2+m^2)/2m}d\tau = -i(p^2+m^2-i\epsilon)^{-1}. \tag{12}$$

It is well known that the above results may be interpreted or derived from the path integration method.

After this digression, let us come to the elastic string. We can imagine it to be the limit of a chain of N mass points as $N \to \infty$. Each mass point will trace out a world line, so that in the limit we are dealing with a two-dimensional world sheet. This sheet may be parametrized by two intrinsic coordinates ξ ($0 \le \xi \le \pi$, let us say) and τ ($-\infty < \tau < \infty$), corresponding to spacelike and timelike coordinates. We assume the action integral

$$I = \frac{1}{4\pi}\iint\left(\frac{\partial x_\mu}{\partial\tau}\frac{\partial x^\mu}{\partial\tau} - \frac{\partial x_\mu}{\partial\xi}\frac{\partial x^\mu}{\partial\xi}\right)d\xi d\tau \tag{13}$$

from which follows the equation

$$(\partial^2/\partial\tau^2 - \partial^2/\partial\xi^2)x^\mu = 0 \tag{14}$$

with the boundary condition

$$\partial x^\mu/\partial\xi = 0 \qquad \text{at } \xi = 0,\ \pi. \tag{15}$$

Duality is essentially a result of the symmetry between ξ and τ though it is not yet a perfect symmetry because of the differences in domain and metric. We have chosen the hyperbolic form because only then can one formulate the Hamiltonian principle. Actually it turns out that in computing the scattering amplitudes a switch to an elliptic form through the change $\tau \to -i\eta$ brings out duality more explicitly.

Eq.(13) is invariant under the translations $\tau \to \tau + c$ and $x^\mu \to x^\mu + a^\mu$, which imply the conservation laws

$$\begin{aligned} H &= \int[\pi p_\mu p^\mu + \frac{1}{4\pi}\left(\frac{\partial x_\mu}{\partial\xi}\right)\left(\frac{\partial x^\mu}{\partial\xi}\right)]d\xi = \text{const.} \\ p^\mu &= (1/2\pi)(\partial x^\mu/\partial\tau), \\ P^\mu &= (1/2\pi)\int(\partial x^\mu/\partial\tau)d\xi = \int p^\mu d\xi = \text{const.} \end{aligned} \tag{16}$$

in direct analogy with Eq.(3). A normal mode decomposition of Eqs.(14) and (15) yields

$$\begin{aligned} x^\mu &= x_0^\mu + 2\sum_{n=1}^{\infty} x_n^\mu \cos n\xi \\ p_n^\mu &= \partial x_n^\mu/\partial\tau \ \ (n \neq 0), \ \ p_0^\mu = \frac{1}{2}\partial x_0^\mu/\partial\tau \\ H &= p_{0\mu}p_0^\mu + \frac{1}{2}\sum_{n=1}^{\infty}(p_{\mu n}p_n^\mu + n^2 x_{\mu n}x_0^\mu \\ P^\mu &= p_0^\mu \end{aligned} \tag{17}$$

We may interpret x_0^μ and p_0^μ as the center-of-mass coordinates and momenta.

When the system is quantized, we get the familiar expression

$$\begin{aligned} &H = P_\mu P^\mu + H_0, \ \ H_0 = \sum_{n=1}^{\infty} n a_{\mu n}^\dagger a_n^\mu (+\text{c number}) \\ &x_n^\mu = (a_n^\mu + a_n^{\mu\dagger})/\sqrt{2n} \\ &p_n^\mu = -i(a_n^\mu - a_n^{\mu\dagger})\sqrt{n/2} \ \ (n \neq 0) \\ &[a_n^\mu, a_n^{\nu\dagger}] = g^{\mu\nu}\delta_{nm} \end{aligned} \tag{18}$$

By imposing the subsidiary condition

$$(H - \alpha)\Psi = 0 \tag{19}$$

we can single out an infinite tower of states with the mass spectrum

$$M^2 = H_0 + \alpha = \sum_{\mu,n} nN_{\mu n} + \alpha \tag{20}$$

In order to construct dual scattering amplitudes we introduce an external scalar field φ and postulate

$$H = P_\mu P^\mu + H_0 + g : \varphi(x(\xi = 0)) : \tag{21}$$

We will not discuss how this leads to n−point dual amplitudes in the multi-peripheral configuration since it is well known.

We now add several remarks.

1) The condition $0 \leq \xi \leq \pi$ fixes a fundamental scale of length. ξ is here

measured in $(GeV/c)^{-1}$ to fit the trajectory slope of ~ 1. Whether the system is to be interpreted as a rubber string or a rubber band depends on the boundary condition to be imposed. The rubber band, as twice as many modes as the string, but the difference does not show up in a case like Eq.(21) because those modes which have nodes at $\xi = 0$ cannot be excited. Clearly there will be differences if one tries to extend the model, and this will be an important point in constructing a general theory of hadrons.

2) For a point particle Eq.(1) has a purely geometric meaning (as the length of a world line), but Eq.(2) does not since it depends on the scale (gauge) of the unphysical parameter τ. In the case of the string, on the other hand, Eq.(13) is invariant under the scaling $\tau, \xi \to \lambda\tau, \lambda\xi$. This is one aspect of the conformal invariance of two-dimensional Laplace equations, a property which has widely been utilized to study the Veneziano model.

Nonetheless, Eq.(13) is not a purely geometrical quantity. For curiosity, then, let us try to construct a geometric action integral as one does in general relativity. Obviously a natural candidate for it is the surface area of the two-dimensional world sheet; another would involve its Riemann curvature. The sheet is imbedded in the Minkowskian 4-space, so one can parametrize its points as $y^\mu(\xi^0, \xi^1)$, $(\xi^0 \sim \tau, \xi^1 \sim \xi)$. The surface element is a $\sigma-$tensor

$$\begin{aligned} d\sigma^{\mu\nu} &= G^{\mu\nu} d^2\xi, \\ G^{\mu\nu} &= \partial(y^\mu, y^\nu)/\partial(\xi^0, \xi^1) \end{aligned} \tag{22}$$

whereas its line element is

$$\begin{aligned} ds^2 &= g_{\alpha\beta} d\xi^\alpha d\xi^\beta \qquad (\alpha, \beta = 0, 1) \\ g_{\alpha\beta} &= (\partial y_\mu/\partial \xi^\alpha)(\partial y^\mu/\partial \xi^\beta) \end{aligned} \tag{23}$$

A possible action integral would be

$$I = \int |d\sigma_{\mu\nu} d\sigma^{\mu\nu}|^{1/2} = \iint |2\det g|^{1/2} d^2\xi \tag{24}$$

to be compared with the old one (13) which can be written $(y \to x)$

$$I = -\frac{1}{4\pi} \iint g_{\alpha\beta} \overset{\circ}{g}{}^{\alpha\beta} \, d^2\xi, \quad \overset{\circ}{g}{}^{\alpha\beta} = \begin{pmatrix} -1 & 0 \\ 0 & 1 \end{pmatrix} \tag{25}$$

It is obvious that Eq.(24) leads to nonlinear equations. More complicated equations involving curvature would be not only nonlinear, but also have higher derivatives.

3) It is sometimes useful to consider the "energy-momentum tensor" in the (ξ, τ) space:

$$T_{\alpha\beta} = \frac{1}{2\pi}(g_{\alpha\beta} - \frac{1}{2}\overset{\circ}{g}_{\alpha\beta}\, g_{\gamma\delta}\, \overset{\circ}{g}^{\gamma\delta}) \tag{26}$$

In particular, let us take its space integral over a test function $f(\xi)$,

$$\bar{T}_{\alpha\beta}[f] = \int_0^\pi T_{\alpha\beta}(\xi) f(\xi) d\xi. \tag{27}$$

By virtue of the canonical commutation relations, they generate a commutator algebra

$$\begin{aligned} &\left[\ (\bar{T}_{00} \pm \bar{T}_{01})[f],\ \ (\bar{T}_{00} \pm \bar{T}_{01})[g]\ \right] = -2i(\bar{T}_{00} \pm \bar{T}_{01})[h], \\ &\left[\ (\bar{T}_{00} \pm \bar{T}_{01})[f],\ \ (\bar{T}_{00} \mp \bar{T}_{01})[g]\ \right] = 0, \\ &h = f'g - fg' \end{aligned} \tag{28}$$

as an integral form of the Schwinger conditions. These relations, when applied to the set $f_n = 1 - e^{-2in\xi}$, amount to

$$\begin{aligned} &\left[\ L_n^\pm,\ L_m^\pm\ \right] = 2(n-m)L_{n+m}^\pm \\ &\left[\ L_n^\pm,\ L_m^\mp\ \right] = 0 \\ &L_n^\pm = (\bar{T}_{00} \pm \bar{T}_{01})[e^{2in\xi}],\ \ L^\pm[f_n] = L_0^\pm - L_{-n}^\pm \end{aligned} \tag{29}$$

These operators $L^\pm[f_n]$ have been found useful in generating the various gauge operations.

4) As we have mentioned already, the most serous defect of the above formulation is the indefinite metric that appears in defining the covariant commutation relations (18). The mass operator (20), however, acquires as a result the nice property of being positive. The transition from a classical to quantum picture of 4-dimensional harmonic ocsillators is a drastic one. A wave function in coordinate space would behave like $\exp[-c^2(\underline{x}^2 - x_0^2)]$, which explodes in the timelike direction. We could actually insist that it should behave instead like $\exp[-c(\underline{x}^2 + x_0^2)]$. This would amount to interchanging the creation and annihilation operators and using positive metric for the time component. But then the mass operator would not be positive, and moreover

each level would become infinitely degenerate. In group theoretical terms, the former corresponds to non-unitary, and the latter to unitary representations of $U(3,1)$. The former have negative probability ghosts while the latter have negative mass squared ghosts. The Veneziano model seems to prefer the former. Such a choice is necessary to ensure a Regge behavior à la Van Hove, but runs into trouble with form factors. This is a general agony of making the choice, not restricted to Mr.Veneziano alone.
5) We have ignored the problems of the extra scalar excitations which are needed to incorporate the proper trajectory intercepts in the dual channel. This is another rather unphysical aspect of factorization. These extra modes may be taken either as a set of harmonic oscillators in a fifth dimension or as a modification of the propagator. Whether these states have positive metric or not depends on the intercept and the external masses. Actually a sixth dimension would be necessary to take care of two-particle trajectories correctly. We have no illuminating interpretations to offer on this subject.

3 Quarks and the dual model

What we propose here is a program of building a general picture of the structure of hadrons on the basis of the factorized Veneziano model. It has been noted by Harari and Rosner that the duality may be interpreted schematically in terms of quark diagrams, which have a strong predictive power, albeit of qualitative nature. These diagrams are indeed very suggestive. First of all they agree with the foregoing picture that the hadrons form two-dimensional sheets in space-time. Furthermore, they imply that quarks and antiquarks form the boundary lines of the sheets. A meson system, for example, is then a $q-\bar{q}$ molecule bound by an elastic string. We could also imagine a rubber band in which q and $\bar{q}$ are attached to diametrically opposite points. To go on further, we have to make a choice.

On the basis of the duality diagram picture, we will adopt the linear molecule rather than the benzene ring. An advantage of the linear picture is that a linear chain can be broken in two linear chains, thereby accounting for the production mechanism

$$A \longrightarrow B + C \tag{30}$$

In fact the duality diagrams can be interpreted exactly in this way.

To make things a bit more sophisticated, we will present a modified version of the quark model. This is the three-triplet model proposed by Dr. Han and me some time ago. It had the advantage of a) having integral charges, b) naturally accounting for the zero triality of known hadrons, as well as c) for the $SU(6)$ classification of the baryons. In this scheme there are nine fundamental fermions grouped into three $SU(3)$ triplets. We may use the notation T_i^n; $i, n = 1, 2, 3$, where n distinguishes between different triplets. T_i^n behaves like a triplet representation in the ordinary $SU(3)$ space (lower index), and like an antitriplet representation in the new $SU(3)$ space (upper index). These $SU(3)$ spaces are denoted as $SU(3)'$ and $SU(3)''$ respectively. Each space has its own isospin and hypercharge, and the electric charge is the sum of two charge operators $\lambda_Q' + \lambda_Q''$. We call $3\lambda_Q''$ the charm number C. We will also make things a bit more exciting by naming the three different triplets D, N, and A. Their quantum number assignments are given in the table.

	C		Q	
D_i	1	1,	0,	0
N_i	-2	0,	-1,	-1
A_i	1	1,	0,	0

In the lowest approximation, $SU(3)'$ and $SU(3)''$ are separately good symmetries. In particular, all low lying hadron states are assumed to belong to $SU(3)''$ singlets, which turn out to correspond to only zero triality states in $SU(3)'$. (The same is accomplished also by assuming zero charm for hadrons.) Baryons and mesons have the usual pattern TTT and $T\bar{T}$. More precisely,

$$\begin{aligned} &B \sim DNA \\ &M \sim D\bar{D} + N\bar{N} + A\bar{A} \end{aligned} \tag{31}$$

where B is completely antisymmetric in the $SU(3)''$ space in order to be a singlet, and this takes care of the Pauli principle.

The preference of zero triality is thus reduced in this model to the $SU(3)''$ symmetry, which one may attribute to a dynamical property of superstrong interactions having a larger scale of masses ($\gtrsim 1GeV$) than for the strong interactions ($\sim 1GeV$).

The combination of duality and the three-triplet model will then produce the following picture. The hadrons are "molecules" bound by superstrong interactions, with the bond structure

$$
\begin{aligned}
&B \sim T - T - T \\
&M \sim T - \bar{T}
\end{aligned}
\tag{32}
$$

The bonds must have a saturation property for zero triality. In the original scheme demanding perfect $SU(3)''$ symmetry, the baryon would have to have either a ring structure

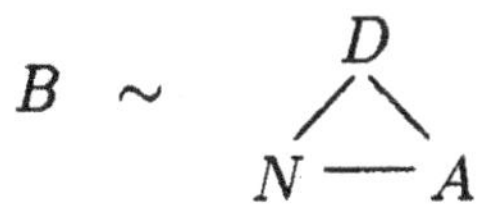

(33)

or a resonating linear structure

$$B \sim D - N - A \quad + \quad \text{permutations.} \tag{34}$$

If $SU(3)''$ symmetry is abandoned and only the neutral charm condition is imposed, we may simply assume

$$B \sim D - N - A \tag{35}$$

In the latter case the charm number is equivalent to valency. But then we lose the distinction between D and A, drifting back to a two-triplet model. Our tentative preference is in Eq.(34) though the other two possibilities should not be ignored. The meson scheme would follow Eq.(31).

What can we do with this model? We have now hadrons endowed with $SU(3)$ and Dirac spin. These are more or less localized at certain points along the string whose function is to carry bulk of the energy and momentum of the system. The triplets (or simply quarks) themselves are massless, or have only small masses. Several remarks are in order.

1) The interaction process (30) is viewed as a creation of pair $T\bar{T}$ at a point where the break occurs. After the cut, each portion subsequently grows into a full grown string, like an earthworm! This would not be possible if the

string is made up of a fixed number of mass points. We must conclude that the number is not only large but also indefinite. Let us examine this situation a little further. Take a string stretched between two fixed points in space with a distance L apart. If the number of mass points is N, the potential energy is

$$V \sim N(L/N)^2 = L^2/N \tag{36}$$

which depends on N. However, if the number of discrete steps in the "time" direction τ also increases with N to sustain duality, the action integral $I \sim NV \sim L^2$ will be independent of N. This is the scale invariance we have discussed. In units of the fictitious time τ, a system lives longer the larger the number N. The actual state of a hadron would be a linear superposition of configurations having different values of N, but their contributions are all proportional to each other.

If this view is accepted, we can define the Hamiltonian (or vertex) responsible for the process (30) as an overlap integral of the three wave functions corresponding to the states A, B, and C. There is a selection rule

$$N_A = N_B + N_C \qquad (N_i > 0) \tag{37}$$

and its two cyclic permutations, either one of which must be satisfied. A more explicit expression satisfying (37) would look like

$$\sum_{0<N'<N} C_{N'N} \int \cdots \int \Psi_B^*(x^{(1)}, \cdots, x^{(N')}) \Psi_C^*(x^{(N'+1)}, \cdots, x^{(N)})$$
$$\Psi_A(x^{(1)}, \cdots, x^{(N)}) F(x^{(1)}, \cdots, x^{(N)}) \prod d^4x^{(i)} \tag{38}$$

where F is some scalar function. This integral can be appropriately rewritten in terms of the wave functions $\Psi[x(\xi)]$ in the limit N, $N' \to \infty$.
In the actual Veneziano model, the interaction is such that the breaking of a string occurs only at one of the ends, which may be interpreted to mean the limit

$$N \to \infty; \quad N'/N \to 0 \quad \text{or} \quad N'/N \to 1$$

In general, however, there is no reason to impose such a condition. We would still get a dual theory of sorts; an amplitude corresponding to one

configuration will have singularities in the crossed channels too, though there may not be a symmetry between dual channels in individual amplitudes.
2) The triplet or the quark fields introduce extra spins to the system in conformity with the $SU(6)$ type theories, so π and ρ mesons belong to the s wave states of the string. But this also brings in the old headaches of relativistic $SU(6)$ theories as well. How can one eliminate half of the Dirac components in order to avoid parity doubling and ghost states? The difficulty is compounded by the fact that we would like to maintain duality too. A possible scheme based on the Carlitz-Kisslinger type cut mechanism has been developed by Freund et al. We will not discuss it here. Instead, we would like to propose a general formalism which tries to accomplish this in a dynamical way. The basic idea is as follows. Instead of regarding the triplets and the strings as separate entities, let us take a unified picture and replace the string with a chain of $T\bar{T}$ pairs, so that it would look like a polarized medium with opposite charges created at its ends. There will be interactions between neighbors which depend on their Dirac, $SU(3)'$, and $SU(3)''$ spins. To be dual, these interactions must occur in the "time" direction too, thus forming a two-dimensional polarizable medium. These interactions would contribute to the action integral in addition to the kinetic term represented by the string Hamiltonian. Roughly speaking, that would produce a spin and $SU(3)$ dependence of various trajectories. It is conceivable that parity doubling and other problems can be reduced in the same way to ones of dynamical stability. An example of $T\bar{T}$ interaction might be

$$I = g \sum_{(n,n')} (\gamma_\mu^{(n)} \gamma^{\mu(n')} + \text{const}) \tag{39}$$

where $\gamma_\mu^{(n)}$ refers to a triplet sitting on a site n of a two-dimensional lattice, and n' refers to one of its neighbors. We must choose the constant g in such a way that a chain $T\bar{T}T\bar{T}\cdots\bar{T}$ will have the lowest energy (which may be adjusted to zero). To be dual, the same patter must be repeated in the time direction too. We thus end up with a pattern

$$\begin{array}{c} T\bar{T}T\bar{T}\ldots\ldots\ldots\ldots\bar{T} \\ \bar{T}T\bar{T}T\ldots\ldots\ldots\ldots T \\ T\bar{T}T\bar{T}\ldots\ldots\ldots\ldots\bar{T} \\ \bar{T}T\bar{T}T\ldots\ldots\ldots\ldots T \end{array}$$

....

....

In other words, it is a two-dimensional antiferromagnet or ionic crystal! External particles should couple to it like a magnetic field couples to the spin:

$$I' = \sum_n \gamma_\mu^{(n)} \phi^{\mu(n)} \tag{40}$$

The scattering amplitude would then be obtained, following the Feynman principle, from an expression like

$$\mathrm{Tr} \quad \exp[I + I'] \tag{41}$$

the trace being taken with respect to the γ matrices. Eq.(41) is nothing but a partition function! It is not scale invariant, but rather an extensive quantity proportional to the number of constituents (unless $I = 0$). Thus the probability of creating exotic states having many disordered atoms (Large $|I|$) would be severely cut down.

The simplest mathematical model of the above type is the well known Ising model with its glorious Onsager solution. Perhaps we can adopt the Ising model here as a prototype. The transition from a Lagrangian to a Hamiltonian formalism is accomplished by means of the so-called transfer matrix.

3) The world sheet formalism presented here may accommodate Dirac's monopoles, because the monopoles can be described, according to Dirac, in terms of the same kind of world sheets swept out by strings attached to them. If this is the case, the triplets can have magnetic charges. Since all hadrons are magnetically neutral, these charges must add up to zero, which reminds us of the fact that the charm number is also zero for them. Thus we are tempted to identify the charm with the magnetic charge, which would cause strong binding between opposite charges. The two spaces $SU(3)'$ and $SU(3)''$ may be called electric and magnetic $SU(3)$ respectively, corresponding to the 3×3 ways of assigning electric and magnetic quantum numbers.

Such a formalism has been independently proposed by Schwinger from a different motivation. He calls these nine objects dyons. The direct parallelism between dyons and the triplets has been pointed out by Biedenhahn and Han.

Of course there are all sorts of problems associated with monopoles. The most serious and perhaps most intriguing is the large P and T violation one must expect off hand. This could give a natural explanation for CP violation, but the problem is how to suppress it to a degree $\lesssim 10^{-10}$ which is required by neutron electric dipole moment.

At any rate, the close mathematical connection between the dual model and the monopoles lies in the fact that the Maxwell field due to a pair of monopoles is given by

$$F_{\mu\nu}(x) = g\epsilon_{\mu\nu\lambda\rho} \iint \delta^4(x-y)G^{\lambda\rho}(y)d^2\xi \tag{42}$$

where $G_{\mu\nu}d^2\xi = d\sigma_{\mu\nu}$ is the surface element, Eq.(22), of a world sheet spread between the world lines of the pair. Eq.(42) is independent of the choice of the sheet; one gets the correct equations if Eq.(42) is substituted in the Maxwell Lagrangian and the y's are varied, including the end points.

4 Statistical approximation

We will briefly discuss here a high energy approximation to dual amplitudes which was originally based on an intuitive argument, but can also be justified more rigorously. The point is that in a high energy process a very large number of states are available according to the factorized dual theory. In fact the number of states $\rho(s)$ increases like $\exp[c\sqrt{s}]$, which one can easily derive from the Stephan-Boltzmann law for a one-dimensional black body radiation. The only difference is that $s =$ (center-of mass energy)^2 takes the place of the ordinary energy. The constant c is numerically

$$c = 2\pi\sqrt{n/6} \tag{43}$$

where n is the dimensionality (4, or 5, or more) of the oscillator vectors. As has been pointed out by Fubini et al, this happens to give the same number ($c \approx 1/160 MeV^{-1}$ for $n = 6$) as the Hagedorn constant.

The absorptive part of a scattering amplitude consists of a sum over intermediate states with fixed s, i.e. over a microcanonical ensemble:

$$A = \sum_{s_n = s} \langle p'|j(-q')|n\rangle\langle n|j(q)|p\rangle \tag{44}$$

If s is large, one is tempted to replace it with a sum over a canonical ensemble. More precisely

$$\begin{aligned} A &\sim c(\beta)\sum_n \langle p'|j(-q')|n\rangle e^{-\beta s_n}\langle n|j(q)|p\rangle \\ &\equiv c(\beta)\sum_n F_n e^{-\beta s_n} \end{aligned} \tag{45}$$

Here the s_n's are the eigenvalues of the mass operator (20), not the actual $s = (p+q)^2$. This amounts to relaxing the subsidiary condition (19). $c(\beta)$ is a normalization factor. If the sum (45) had a sharp peak around $s_n = s$, it would be a good approximation to Eq.(44) as in the usual statistical mechanics. The parameter β should then be the inverse temperature ($\sim 1/\sqrt{s}$) of the string. Actually, things do not work out that way because F_n does not grow like $\rho(s_n)$ but much more slowly like s^α, which is the Regge behavior. If $\alpha > 0$, still there will be a peak around $s_n = s$ where

$$\alpha/s \approx \beta \tag{46}$$

So we can use this as the definition of β.

In the operator formalism, Eq.(45) can be explicitly evaluated from

$$A = c(\beta)\langle 0|\Gamma' e^{-\beta H_0}\Gamma|0\rangle \tag{47}$$

where Γ and Γ' are appropriate interaction vertices. We find

$$A \sim (1-e^{-\beta})^{-1-\alpha(t)} \sim \beta^{-1-\alpha(t)} \tag{48}$$

which give the correct Regge behavior in view of Eq.(46), if $c(\beta) \sim \beta$. The above idea makes physical sense, perhaps better than the Veneziano model itself, because we apply it to the absorptive parti and smear the resonance peaks as in the discussion of finite energy sum rules. But the real part, when smeared, should also show a similar behavior for reasons of analyticity. Justification of the method depends on the assumption that the absorptive part grows with s ($\alpha > 0$). The formula may still be valid for $\alpha \leq 0$, but that does not follow from the above argument. Assuming the general validity of the procedure, we can also handle the high energy behavior of many particle processes. The main point is that any high energy ("hot") propagator $\sim 1/(s - H_0)$ is replaced by a Boltzmann factor $\sim \beta\exp(-\beta H_0)$

where $\beta \sim 1/s$, which would give a correct answer as far as the s−dependence is concerned.

At this point let us indulge in some speculations. The problems are unitarity and the Pomeron, both of which are lacking in the Veneziano model. It is generally assumed that the Pomeron (in the t−channel) is equivalent to non-resonant background in the s−channel. But the background is after all made up of hadrons, so a unitarized dual theory should naturally contain the Pomeron. Now the unitarization means taking account of dissociation and recombination of resonances among themselves. Wouldn't it be reasonable, then, to consider a grand canonical ensemble of resonances? The main problem is of course how to define the vertex operator in a scattering problem. We would like to suggest the ansatz that instead of Eq.(46), β^{-1} should be the thermodynamic temperature, or

$$\beta \approx c/\sqrt{s} \tag{49}$$

because the distribution of M^2 would be decided by the interaction among the strings in the ensemble, and not by the coupling of these states to the external channels. We would then get the result

$$A \sim s^{\alpha(t)/2} \tag{50}$$

suggesting that the Pomeron trajectory has half the universal slope of resonances. Whether this is a pole or a cut, and what the intercept is, cannot be decided in such a crude picture. It is interesting that the same behavior as Eq.(50) has been obtained by explicitly computing certain higher order diagrams of the dual model.

There is still another possibility for the Pomeron which will be discussed later.

5 Electromagnetic interactions and inelastic $e-p$ scattering

We would like to tackle the problem of inelastic $e-p$ scattering on the basis of our model. Before doing this, we have to make some remarks about the electromagnetic interactions in general. If we just have an elastic spring with a charge distribution along the string but without the quark spin, the problem of setting up the gauge principle is very simple. In the Hamiltonian (16) one makes the replacement

$$p^\mu(\xi) \longrightarrow p^\mu(\xi) - \rho(\xi)A^\mu(x(\xi)) \tag{51}$$

where $\rho(\xi)$ is an arbitrary distribution function. But this is not the most general form. Instead of $x(\xi)$ and $p(\xi)$, we should choose the set $\{x_n, p_n\}$, or any other orthogonal basis, and apply the gauge principle. Because $A(x)$ is nonlinear in x, we get inequivalent results.

The trouble with this method is that the form factors are all Gaussian, an undesirable characteristic of harmonic oscillators. For a pointlike charge distribution the Gaussian peak is infinitely sharp (in momentum space) since the string has an infinite zero-point length. If we renormalize away the Gaussian factor, on the other hand, what is left is in general a polynomial which is not good either.

Another popular approach to form factors is the spurion method which allows one to obtain pole dominance form factors. A difficulty here is the gauge invariance, or the conservation law, which cannot be automatically guaranteed. At any rate there is a large amount of arbitrariness in either method.

Another serious problem is dealing with external fields in a factorized dual model. It arises from the fact that one needs fifth-dimension oscillators for factorization of a general dual amplitude, but the parameters of the fifth dimension depends on the individual external masses in a non-factorizable way. Thus one cannot maintain duality and factorizability for arbitrary external fields. Full duality must then be abandoned in our model when dealing with electromagnetic interaction. For example, a virtual Compton amplitude should not be dual;

One might argue that in the pole dominance model the lines p and p', instead of the photons, may be treated as external:

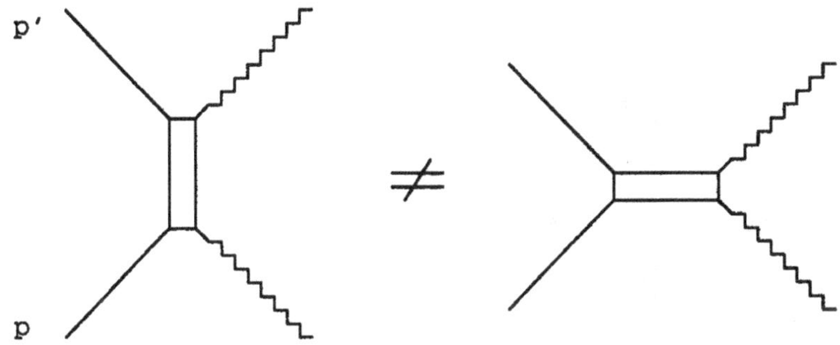

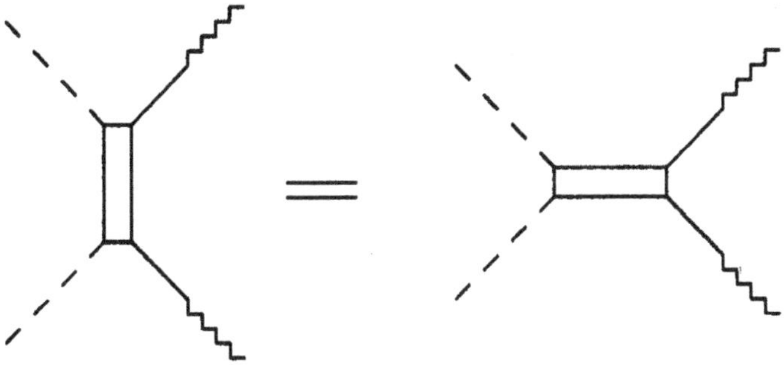

but this does not work for many photon cases.

In spite of the unpleasant Gaussian nature of form factors in the first method described above, let us see what will come out if duality is not demanded. The crossed channel singularities in the usual case arises from the singular nature of the vertex operator

$$\Gamma(q) \sim \exp[iq \cdot x(0)] = \exp\left[iq \cdot x_0 + 2iq \cdot \sum_{n=1}^{\infty} x_n\right] \tag{52}$$

In order to blunt the singularity, we assume that the charge is located at a coordinate

$$\bar{x} = \int x(\xi) f(\xi) d\xi = x_0 + 2\sum_{n=1}^{\infty} x_n f_n \tag{53}$$

such that

$$\sum f_n^2 < \infty \tag{54}$$

According to the gauge principle, the electromagnetic interaction is obtained by the substitution $p_0^\mu \to p_0^\mu - eA^\mu(\bar{x})$, $p_n^\mu \to p_n^\mu - 2ef_n A^\mu(\bar{x})$ in the Hamiltonian. The current operator is then

$$j^\mu = e\left\{ p_0^\mu + \sum_{n=1}^{\infty} p_n^\mu f_n \ , \ e^{iq\cdot\bar{x}} \right\}_+ \tag{55}$$

For the ground state (spin zero), this gives a vertex

$$\Gamma_\mu = e(p+p')_\mu \exp\left[-q^2 \sum_{n=1}^{\infty} f_n^2/n\right] \tag{56}$$

Now let us discuss off-diagonal elements $\langle n|j_\mu|0\rangle$ corresponding to inelastic processes. We would like to compare them with the $e-p$ scattering data ignoring the effect of spin. The familiar structure functions W_1 and W_2 should be obtained from

$$\sum_{s_n=n} \langle 0|j_\nu(-q)|n\rangle\langle n|j_\mu(q)|0\rangle \tag{57}$$

where the statistical method could be applied for large s.

There still remains a problem. In the SLAC data, W_1 and W_2 seem to have an s dependence ($W_1 \sim s$, $W_2 \sim 1/s$) which is consistent with the Pomeron picture. On the other hand, Eq.(55) will not give any Regge (or power) behavior. We propose to fix this by including the fifth dimension:

$$\exp[iq\cdot\bar{x}] \longrightarrow \exp[iq\cdot\bar{x} + iq_5 x_5(0)] \tag{58}$$

where q_5 is an appropriate constant. Since $x_5(0)$ is not smeared out, it gives rise to a fixed power behavior $\sim s^\alpha$ where α is constant, corresponding to a flat trajectory. We do not know what all this means, but the Pomeron might be of this nature. At any rate we can evaluate Eq.(57) with the new ansatz, and obtain the result

$$W_1 \sim (s-m^2)^{\alpha_1} \exp\left[-2q^2 \sum_{n=1}^{\infty} \frac{1-e^{-n\beta}}{n} f_n^2\right] \tag{59}$$

and similarly for W_2. Here m^2 is the initial Hadron mass. Actually we have lost gauge invariance from a) addition of the fifth dimension and b) the statistical approximation, so we cannot get the ratio W_1/W_2 exactly. But the main feature of Eq.(59) is that for large s we have

$$\beta \sim c/(s-m^2) \tag{60}$$

so that

$$\begin{aligned} W_i &\sim (s-m^2)^{\alpha_i} \exp[-2q^2\beta_i \sum f_n^2] \\ &\sim (s-m^2)^{\alpha_i} \exp[-\lambda_i q^2/(s-m^2)] \\ &= (s-m^2)^{\alpha_i} \exp[-\lambda_i/(\omega-1)], \quad i=1,2 \\ \lambda_i &= 2c_i \sum_{n=1}^{\infty} f_n^2, \quad \omega = 1+(s-m^2)/q^2 \ (= 2m\nu/q^2) \end{aligned} \tag{61}$$

Which $\alpha_1 = 1$, $\alpha_2 = -1$, we get the scaling law (for large s)

$$\begin{aligned} (s-m^2)W_2 &\sim \exp[-\lambda_2/(\omega-1)] \\ \text{or} \qquad 2m\nu W_2 &\sim (\omega/\omega-1)\exp[-\lambda_2/(\omega-1)], \\ (s-m^2)^{-1}W_1 &\sim \exp[-\lambda_1/(\omega-1)] \end{aligned} \tag{62}$$

The general behavior of the exponential factor in Eq.(62) is:

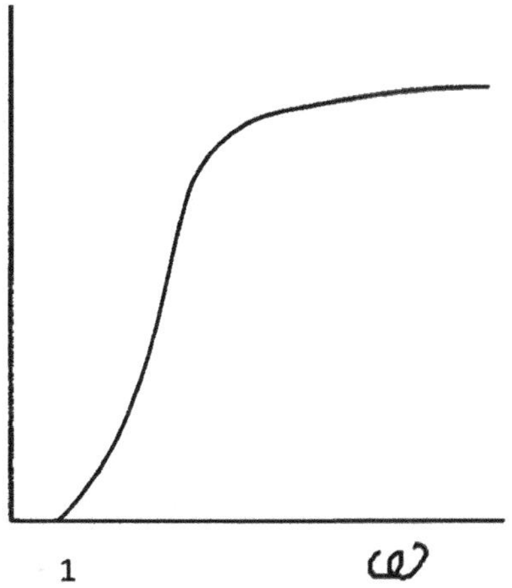

www.ingramcontent.com/pod-product-compliance
Lightning Source LLC
Chambersburg PA
CBHW081535190326
41458CB00015B/5559